Springer Geography

The Springer Geography series seeks to publish a broad portfolio of scientific books, aiming at researchers, students, and everyone interested in geographical research. The series includes peer-reviewed monographs, edited volumes, textbooks, and conference proceedings. It covers the entire research area of geography including, but not limited to, Economic Geography, Physical Geography, Quantitative Geography, and Regional/Urban Planning.

More information about this series at http://www.springer.com/series/10180

Yukio Himiyama
Editor

Exploring Sustainable Land Use in Monsoon Asia

Springer

Editor
Yukio Himiyama
Hokkaido University of Education
Asahikawa, Hokkaido
Japan

ISSN 2194-315X ISSN 2194-3168 (electronic)
Springer Geography
ISBN 978-981-10-5926-1 ISBN 978-981-10-5927-8 (eBook)
https://doi.org/10.1007/978-981-10-5927-8

Library of Congress Control Number: 2017947703

Printed on acid-free paper

This Springer imprint is published by Springer Nature
The registered company is Springer Nature Singapore Pte Ltd.
The registered company address is: 152 Beach Road, #21-01/04 Gateway East, Singapore 189721, Singapore

Preface

Future Earth is a new international research platform launched in June 2012 by the International Council for Science (ICSU), the International Social Science Council (ISSC), and other international academic or funding organizations for a sustainable world. The Global Land Project (GLP) has been chosen as one of the first such international projects it has endorsed that originated from the International Geosphere–Biosphere Programme (IGBP) and/or the International Human Dimensions Programme (IHDP). The choice is a clear indication of the importance of the issues related to land, its use, and change. Monsoon Asia, with its huge and still increasing population and rapid socioeconomic changes, is regarded as a major hot spot of global change in general and of land use change in particular. This book is the outcome of a project entitled "Towards Sustainable Land Use in Asia (SLUAS)," which was the pilot undertaking for development in a series of projects on land use carried out between May 2009 and March 2014. The study areas included China, India, the Russian Far East, Indonesia, the Philippines, Thailand, and Japan.

The major issues of the SLUAS project included sustainability, urbanization, rural development, land-related problems such as food problems, and land-related disasters. It was, therefore, natural for the project to incorporate the Great East Japan Disaster which was triggered by the M9.0 earthquake of 11 March 2011. SLUAS carried out a number of field surveys in East Japan, observing the extremely uneasy situation of the tsunami-stricken areas and the hopelessly battered land and people's lives in the areas contaminated by the radioactive substances emitted from the Tokyo Electric Company Fukushima Daiichi Nuclear Power Plant.

Future Earth, which is intended to restructure the framework of global change research internationally, has been another great concern of SLUAS, and the project has been deeply involved in the process of framing Future Earth in Asia. This book reflects these concerns and activities.

Land use change is an essential driving force of environmental change, a result of socioeconomic and environmental changes, and is a major environmental change itself. Because of this complex and multifaceted nature and the difficulties in obtaining relevant data with historical depth, this phenomenon has not been studied fully in the context of global change or sustainability. It is hoped that this book is of use to those who are concerned about the present and future land use in the world as well as in their locality.

March 2017

Yukio Himiyama
SLUAS Project Leader

Contents

Part I
Introduction

Chapter 1
Towards Sustainable Land Use in Asia

Yukio Himiyama

Abstract Land use and its change have been a great concern of humanity for decades, and they have been studied widely mainly by geographers. An international project on land use change in Monsoon Asia called SLUAS (Towards Sustainable Land Use in Asia) was conducted by a team of geographers in Asia during 2009.5–2014.3 with focus on sustainability. The project had its background in the traditional geographic land use studies, but at the same time it was closely associated with more recent global change research programs, namely LUCC, GLP, and Future Earth. Historical and recent developments of the studies on land use change are reviewed, and SLUAS project is introduced. It is argued that both traditional and modern approaches are necessary, and that with the joint effort of them can land use studies make substantial contribution to global sustainability programs such as Future Earth, as partly demonstrated by SLUAS.

Keywords Land use · LUCC · GLP · Sustainability

1.1 LUCC (Land Use/Cover Change Programme)

Land use became a major issue of geography in the early twentieth century when the world population and economic activities grew fast, and large-scale conflicts devastated the land and the society in many parts of the world. Dudley Stamp, the world pioneer of land use study, conducted the first land use survey of Britain in the 1930s, i.e., soon after the Great Depression of 1929. Its primary aim was to make a map of Britain in order to secure domestic food production under an isolated situation which was feared to occur due to the war. In his masterpiece "the Land of Britain - its use and misuse" (Stamp 1948), he discussed not only the present but also the historical change of land use. Since then, land use had been a major issue of geography until the 1970s, as reflected on the continuous presence of the IGU

Y. Himiyama (✉)
Tokyo, Japan
e-mail: himiyamay@kkd.biglobe.ne.jp

Y. Himiyama (ed.), *Exploring Sustainable Land Use in Monsoon Asia*,
Springer Geography, https://doi.org/10.1007/978-981-10-5927-8_1

commission on land use, and the Stamp's work had far-reaching effect worldwide. In Britain, Alice Coleman conducted the Second Land Use Survey in the 1960s with the support of Stamp, and identified the decline of agricultural land from the 1930s to the 1960s (Coleman 1977).

In China, Wu Chuanjung took leadership in producing "1:1,000,000 Land Use Map of China" (Wu et al. 1990) covering the whole country, and a few years later in publishing "Land Use in China" (Wu and Guo 1994), a grand book of land use in China published in association with the 1:1,000,000 land use map. The description of land use and its social and physical background in each region are detailed and comprehensive, and are useful for land use planning. Despite these and other achievements, however, the traditional land use studies by geographers gradually lost momentum in the 1980s, and the study group on land use, then called the IGU Study Group on the Dynamics of Land Use Systems, was terminated in 1988.

In 1994 an epoch making book entitled "Changes in Land Use and Land Cover - a global perspective" (Meyer and Turner II 1994) was published, and it was soon followed by the establishment of LUCC (Land Use/Cover Change Programme) by IGBP (International Geosphere-Biosphere Programme) and IHDP (International Human Dimensions Programme of Global Environmental Change). LUCC had the following three foci of research (Turner II et al. 1995):

Focus 1: Land use dynamics—comparative case study analysis
Focus 2: Land cover changes—direct observations and diagnostic models
Focus 3: Regional and global models—integrative assessment

LUCC did not have its root in the traditional geographic land use studies, but IGU decided to establish IGU-LUCC (IGU Commission on Land Use/Cover Change) in order to back it up because of its affiliation to ICSU and ISSC which sponsored LUCC. IGU-LUCC had the following objectives:

- To promote geographical research on land use/cover changes, at scales ranging from the local to the global.
- To stimulate the production and the use of land use information bases of both the present and the past.
- To coordinate the comparative study and the model study of land use/cover changes and their driving forces in different regions.

In order to put them into practice, IGU-LUCC organized a number of international meetings and publications, as well as assisted initiation of the research projects related with LUCC at regional to national levels. The notable publications include "Land-use Changes in Comparative Perspective" (Himiyama et al. 2002) and "Understanding Land-Use and Land-Cover Change in Global and Regional Context" (Milanova et al. 2005). Himiyama (2004) reviewed research on land use change in China, which was then a great concern of the world LUCC community. "Our Earth's Changing Land—an encyclopedia of land-use and land-cover change"

(Geist 2006), which the editor describes as being "a reference tool for those who want to gain a better understanding of land change as a forcing function in global environmental change" is a comprehensive and collaborative product of LUCC, to which IGU-LUCC contributed with other affiliated projects.

1.2 GLP (Global Land Project)

LUCC was merged with GCTE (Global Change and Terrestrial Ecosystems) and was restructured in 2005 as GLP (Global Land Project). Partly because LUCC was the latest starter among the IGBP core projects, and partly because it was a literally interdisciplinary project which natural scientists alone could not manage, considerable part of the mission of LUCC had not been completed and had to be succeeded in GLP or elsewhere (Lambin and Geist 2006). In fact the Science Plan of GLP (2005) stated that "GLP will build on the research of more than a decade within IGBP and IHDP core projects, especially GCTE and LUCC, along with other projects sponsored by the international global change programs". IGU-LUCC played a unique role in filling the gap caused by this transformation of core project.

GLP started with the following three broad themes:

Theme 1: Dynamics of land systems
Theme 2: Consequences of land system change
Theme 3: Integrated analysis and modeling for land sustainability.

In order to promote research related with these themes, GLP set up four nodal offices with different research foci, i.e., Sapporo nodal office for vulnerability, resilience, and sustainability of land systems, Beijing office for land use and ecosystem interactions, Taipei office for nature of social-ecological linkages and their implications for the resilience of land systems at various scales, and Europe office for land management, land use planning and land use policies. Each office organized academic conferences and educational seminars, as well as functioned as regional hub of GLP research community (Watanabe et al. 2014). It is noted that three out of the four offices are in East Asia, indicating strong interest in land issues in the region. "Environmental Change and Social Response in the Amur River Basin" (Haruyama and Shiraiwa 2015) represents high-level achievement of a multi-national research project in East Asia sponsored by GLP Sapporo nodal office.

"Land Use Science," a new journal starting in 2006, made a good progress side by side with GLP, and it played an important role in promoting land use science, particularly in mapping and data compilation at national to multi-national level, as seen in "Land Use Information for Integrated Natural Resources Management—A coordinated national mapping program for Australia" (Lesslie et al. 2006), "A Comprehensive Global 5 min Resolution Land-use Data Set for the Year 2000 Consistent with National Census Data" (Erb et al. 2007), or "Mapping Contemporary Global Cropland and

Grassland Distributions on a 5 × 5 min Resolution" (Goldewijk et al. 2007). However, it was centered in Europe, and flow of papers to it from Asia was limited.

CORINE (coordination of information on the environment), which was initiated by European Union in 1985, i.e., a decade earlier than LUCC, represents another remarkable achievement of multi-national integrative work for grasping land cover changes. CORINE database is available for most areas of Europe, and has been used widely in Europe, e.g., in "European Landscape Dynamics—CORINE Land Cover Data" (Gabriel and Jaffrain 2016). CORINE shows what a regional union of nations can do, but in East Asian context it is still unrealistic to realize, partly because of lack of unity of the nations there, and partly because of diversity and complexity of land use/cover in the region.

1.3 Disasters and Land Use

Disasters caused by natural hazards often affect land and its use seriously and extensively, and in turn they can be a result of the state of land and/or its use. United Nations Environment Programme (UNEP) issues a comprehensive report on the world environmental situation every 5 years, and its 2002 report entitled "GEO 3—Past, present and future perspectives" (UNEP 2002) warned the very rapid increase of economic costs of great natural disasters during 1950–2000 based on the information provided by Munich Re (2001). It indicated that global warming, destruction of the natural environment because of logging or inappropriate land uses for short-term economic gain, migration of population to urban and coastal areas, etc. can be the reasons for the increase. "GEO 4—Environment for development" (UNEP 2007) just confirmed what were said previously (Munich Re 2006). However, unlike IGU-LUCC, both LUCC and GLP put relatively low priority on disasters.

In Japan where there are many disasters affecting, or being caused by, land use, the importance of incorporating disaster into the study of land use/cover has been recognized for years, and SLUAS (Towards Sustainable Land Use in Asia) Project put high priority on it when it started in 2009. It was justified in a painful but convincing way in Japan on March 11, 2011, when the East Japan Disaster triggered by the M 9.0 earthquake and the huge tsunami that followed attacked the vast Pacific coastal areas in Eastern Japan, and it not only devastated the over 1,000 km coastal area, but also damaged and shocked the whole country. "The Impact of the Great East Japan Tsunami of 2011 on Land Use" (Himiyama et al. 2014) is one of the articles written on the East Japan Disaster from the standpoint of land use.

1.4 SLUAS (Towards Sustainable Land Use in Asia) Project

SLUAS is a 5-year project sponsored by the JSPS (Japan Society for Promotion of Science) starting in April 2009. In Asia sustainable land use is closely linked with sustainable society, and it is essential for its survival. Science Council of Japan issued a proposal entitled "Towards Sustainable Co-existence of Nature and Human on the Land and the Coastal Sea" in July 2008 (Science Council of Japan 2008). SLUAS is part of the effort to implement the proposal with focus on the sustainability issues related with land use in Monsoon Asia. It is a project of holistic nature, emphasizing consolidation of global information base, promotion of integrative research and improvement of related policy and education dealing with Asia, with particular focus on China, India, Far East Russia, Japan, and the other countries in Monsoon Asia.

Extensive field surveys were carried out in order to grasp land use changes, their mechanisms, and the related problems in various places in Monsoon Asia accurately. The project also encouraged construction of geographic land use information bases based on both new and old maps and statistics of various kinds and the information collected through field surveys, and analyze them. It was thought important to grasp long-term changes, i.e., about a century or even longer period, where information is available, in order to cover not only modernizing or developing period, but also the period preceding them, because otherwise the regions at different development levels, i.e., those with time lags of development, cannot be compared in a meaningful way. The study was carried out under the framework of regional comparison, so that better understanding of such large area as Monsoon Asia can be achieved, and the environmental problems, large-scale hazards and food problems related with land use in Asia can be handled more properly, and better proposals can be made to solve those problems.

SLUAS carried out three or four major joint surveys and numerous smaller surveys a year. In every survey thousands of geo-referenced photos were taken, field notes were taken, and observations and interviews were recorded each day, and were used for ground-truth, documentation, mapping, identification of changes in land use/cover and their causes and consequences, etc. at detailed scales. In 2010 academic year, for instance, major joint surveys were carried out in Miyazaki Prefecture, Japan (June), in eastern Northeast China (August), and in Russian Far East (September). In Miyazaki we observed devastation of one of the main livestock regions in Japan due to foot–mouth disease. In Sanjiang Plain in eastern Northeast China we saw massive conversion of wetland to paddy field, which meant elimination of a major wildlife haven in the world. In the Vladivostok-Hanka Lake region in Far East Russia we witnessed "land grab", or land acquisition by foreigners, which was changing the agricultural landscape of the region in a large scale, to mention only a few.

This book is a product of all those research activities of SLUAS, which covers substantially larger areas and longer periods with more detailed ground observations than most other projects of land use/cover changes in Asia. However, the bulk of the research results was published already in the form of journal articles, and cannot be included in this book. The readers of this book are therefore requested to refer to those papers absent in this book in order to have an integrated holistic view of what has been achieved by SLUAS. The following list of the shortened titles of already published articles resulted from SLUAS would be useful in supplementing this book and gaining an overall view of the project and the current setting of the studies on sustainable land use in Asia.:

I. Land Use Change in China

Himiyama et al. (2010): land use change in southern North China Plain
Himiyama et al. (2011): land use in southeastern North China Plain
Himiyama et al. (2012): land use change in western Heilongjiang Province China
Himiyama et al. (2015): land use change in the plain areas in eastern China

II. Land Use Change in India

Kimoto (2011): land use in southern India
Himiyama and Hasegawa (2013): land use/cover change in northwestern India
Himiyama and Igarashi (2015): land use change in Rajasthan India

III. Land Use Change in Japan

Himiyama (2011): land use change and disaster in Japan
Doi (2011): land use change in Miyazaki Prefecture Japan
Himiyama and Takei (2012): land use change and vulnerability in Osaka Japan
Himiyama (2013): Great East Japan Disaster and land use
Himiyama and Takamatsu (2013): land use in Chiba Prefecture Japan
Himiyama and Takase (2014): survey of land use trend of Japan
Himiyama et al. (2014): Great East Japan Tsunami and land use

IV. Land Use Change in Other Countries/Regions in Asia

Himiyama and Ieiri (2011): land use change in Hanka Region, Far East Russia
Himiyama and Maeda (2012): land use/cover change in West Java Province Indonesia
Himiyama and Fujima (2013): land use/cover change and flood disasters in Central Thailand
Himiyama and Abe (2014): land use/cover change in Central Luzon Plain Philippines

1.5 Concluding Remarks

SLUAS was a medium-sized project intended to be a pilot project which would pave the way for a major framework of projects on sustainable land use in Asia. The importance of such projects may be recognized if one looks at the foci of the eight global sustainability challenges of Future Earth, as shown below (Future Earth 2014):

1. Delivering water, energy, and food for all.
2. Decoupling carbon emission from economic growth.
3. Safeguarding land, freshwater, and marine natural assets.
4. Building healthy, resilient, and productive cities.
5. Promoting sustainable rural futures.
6. Improving human health by incorporating global change concerns.
7. Encouraging sustainable consumption and production patterns.
8. Improving governance and early warning systems to respond to complex future threats.

It is obvious without saying that land use is related with all the eight challenge foci of Future Earth. Therefore, specialists of land use should be ready to work together with other specialists with different backgrounds and research interests in each research focus. Furthermore, they have to brush up their skills of observation of land use/cover changes, their causes and effects, and spatiotemporal analyses, by using direct observation on the ground and by interviews, remote sensing, printed and digital maps, statistics, historical documents, etc. There are a few additional suggestions from SLUAS based on its research experience, as follows:

1. Global change programs, such as Future Earth, tend to be problem oriented, so one should be careful not to fragment research on land use in the absence of basic understanding of it.
2. Land use/cover changes are closely related with disasters, and their relations should be studied more thoroughly in global change programs.
3. Land use maps, both in printed and digital forms, are important, and they are desired to be made at different scales or resolutions at certain time intervals. The past records of land use, physical features, and disasters at each place are extremely important when security and sustainability are considered.
4. Efforts should be made to reconstruct land use of the past, and to identify changes of land use by comparing the maps of different periods.
5. Comparisons of the land use changes in the countries of different development stages suggest that they follow quite similar sequences of change with certain time lag in between. This is important when one makes prediction of future land use at each place.

It is hoped that this book offers useful information and viewpoints on land use and the related matters and deep insights in them in Monsoon Asia to those who are interested in the issue of sustainable land use in Japan, Asia or elsewhere.

References

Coleman A (1977) Land use planning—success or failure? Archit J 3(165):91–134
Doi H (2011) Land use change in Miyazaki Prefecture since the Meiji Period. Chiri 56(9):74–83 in Japanese
Erb KH, Gaube V, Krausmann F, Plutzar C, Bondeau A, Haberl H (2007) A comprehensive global 5 min resolution land-use data set for the year 2000 consistent with national census data. J Land Use Sci 2(1-4):191–224
Future Earth (2014) Future Earth strategic research agenda 2014. International Science Council (ICSU), Paris
Gabriel JTG, Jaffrain FSH (eds) (2016) European landscape dynamics—CORINE land cover data. CRC Press, London
Geist H (ed) (2006) Our Earth's changing land—an encyclopedia of land-use and land-cover change. Greenwood Press, Westport
GLP (2005) Science plan and implementation strategy. IGBP Report No. 53/IHDP Report No. 19, IGBP Secretariat, Stockholm, Sweden
Goldewijk KK, Van Drecht G, Bouwman AF (2007) Mapping contemporary global cropland and grassland distributions on a 5 × 5 minute resolution. J Land Use Sci 2(1–4):167–190
Haruyama S, Shiraiwa T (eds) (2015) Environmental change and social response in the Amur River Basin. Springer, Tokyo, p 262
Himiyama Y (2004) Achievements and tasks of research on land-use change in China. Geogr Rev Jpn 77(12):783–799
Himiyama Y (2011) Land use change in Japan. J Jpn Soc Eros Control Eng 63(5):62–72 (in Japanese)
Himiyama Y (2013) The Great East Japan earthquake and tsunami of 2011 and land use. Chikyu Kankyo 18(1):35–44 (in Japanese)
Himiyama Y, Abe K (2014) Land use/cover change in the Central Luzon Plain and Its Surroundings in the Philippines since the 1950s. Reports of the Taisetsuzan Institute of Science, No. 48, pp 9–26
Himiyama Y, Fujima M (2013) Land use/cover change and flood disasters in Central Thailand since the 1950s. Reports of the Taisetsuzan Institute of Science, No. 47, pp 55–72
Himiyama Y, Hasegawa T (2013) Land use/cover change in North-western India from ca. 1985 to ca. 2010. In: Himiyama Y, Bicik I (eds) Land use/cover changes in selected regions in the world volume VIII, IGU-LUCC. IGU-LUCC, pp 39–45
Himiyama Y, Ieiri M (2011) Land use change in the Vladivostok-Khanka Lake Area since ca. 1980. Reports of the Taisetsuzan Institute of Science, No. 45, pp 55–70
Himiyama Y, Igarashi Y (2015) Land use change in Rajasthan, India, from ca.1985 to ca.2010. Science Reports of the Taisetsuzan Institute of Science, No. 49, pp 7–20
Himiyama Y, Maeda R (2012) Land use/cover change in West Java Province in Indonesia since the 1990s. Reports of the Taisetsuzan Institute of Science, No. 46, pp 17–32
Himiyama Y, Takamatsu K (2013) Land use in Chiba Prefecture, its change and vulnerability to natural hazards viewed from the topographic maps. Reports of the Taisetsuzan Institute of Science, No. 47, pp 13–32
Himiyama Y, Takase K (2014) The use and prospects of the survey of land use trend of Japan, Reports of the Taisetsuzan Institute of Science, No. 48, pp 49–62
Himiyama Y, Takei T (2012) Land use in Osaka Prefecture, its change and vulnerability to natural hazards viewed from the topographic maps. Reports of the Taisetsuzan Institute of Science, No. 46, pp 53–73 (in Japanese with English abstract)
Himiyama Y, Hwang M, Ichinose T (eds) (2002) Land-use changes in comparative perspective. Science Publishers, Enfield
Himiyama Y, Doki T, Hayashi N (2010) Land use change in the Southern Northeast China Plain since ca.1930. Reports of the Taisetsuzan Institute of Science, No. 44, pp 73–87 (in Japanese with English abstract)

Himiyama Y, Shiratori Y, Higuchi S (2011) Land use in Southeast Heilongjiang in the 1980s and ca. 2010. Reports of the Taisetsuzan Institute of Science, No. 45, pp 37–54 (in Japanese with English abstract)
Himiyama Y, Kasai M, Yabe H (2012) Land use in Western Heilongjiang in the 1980s and ca.2010, Reports of the Taisetsuzan Institute of Science, No. 46, pp 33–51 (in Japanese with English abstract)
Himiyama Y, Watanabe T, Hara K (2014) The impact of the Great East Japan tsunami of 2011 on land use. Glob Environ Res 18(2):191–206
Himiyama Y, Ono T, Nakai Y (2015) Land use change in the plain areas in Eastern China since the 1980s. Reports of the Taisetsuzan Institute of Science, No. 49, pp 21–48
Kimoto K (2011) Land use in Southern India—government and governance. Chiri 56(9):57–63 (in Japanese)
Lambin EF, Geist HJ (eds) (2006) Land-use and land-cover change: local processes, global impacts. Springer, Berlin
Lesslie R, Barson M, Smith J (2006) Land use information for integrated natural resources management—a coordinated national mapping program for Australia. J Land Use Sci 1(1):45–62
Meyer WB, Turner BL II (eds) (1994) Changes in land use and land cover—a global perspective. University Press, Cambridge
Milanova E, Himiyama Y, Bicik I (eds) (2005) Understanding land-use and land-cover change in global and regional context. Science Publishers, Enfield
Munich Re (2006) Topics geo annual review: natural catastrophes 2005. Munich Re Group, Munich
Munich Re (2001) Topics 2000: natural catastrophes—the current position. Special Millennium Issue. Munich Re Group, Munich
Science Council of Japan (2008) Proposal—towards sustainable co-existence of nature and humanity on the land and in the coastal sea. Science Council of Japan, Tokyo, Japan (in Japanese)
Stamp LD (1948) The land of Britain—its use and misuse. Longmans & Geographical Publications, London
Turner BL II, Skole D, Sanderson S, Fisher G, Fresco L, Leemans R (1995) Land-use and land-cover change science/research plan. IGBP Report 35/HDP Report 7. Prepared by IGBP, ICSU, HDP, ISSC, Stockholm and Geneva
UNEP (2002) GEO 3—past, present and future perspectives. Progress Press Ltd., Valletta, p 446
UNEP (2007) GEO 4—Environment for development. Progress Press Ltd., Valletta, p 540
Watanabe T, Shibata H, Ishihara M, Yoshimura N, Kohyama T (2014) Global Land Project Sapporo Nodal Office addressing land system studies on resilience, sustainability and vulnerability in Asia. Glob Environ Res 18(2):105–112
Wu C et al (eds) (1990) 1:1,000,000 land use map of China. Science Press, Beijing
Wu C, Guo H (eds) (1994) Land use in China. Science Press, Beijing

Part II
Land Use Change in China

Chapter 2
Urbanization in Jiangsu Province and Zhejiang Province in China Since ca. 1930

Yukio Himiyama, Miho Ikeshita and Tetsuya Shinde

Abstract This chapter reports on the compilation and preliminary analyses of the land use datasets of the fourteen cities in Jiangsu Province and Zhejiang Province conducted as part of the series of study, and on the comparison of urbanization and its background. 1:50,000 topographic maps of China produced by Japan ca. 1930, *Atlas of Cities of China* which shows land use at ca. 1990 and Landsat ETM+ satellite images of ca. 2000 have been used as basic information sources. The study has shown that the urban expansion in the 1990s was about fourfold in 10 years, i.e., at the speed twice as fast as that during the rapid economic growth in Japan in the 1960s, that expansion was faster in medium to smaller cities than in big cities, and that the agricultural land around cities gave way to urban expansion rapidly. The datasets have been proved to be extremely useful in showing the urban area of each city in ca. 1930, ca. 1990, and ca. 2000 and the trend of change in a graph, and in comparing the fourteen cities to find out their general tendencies and the specific characteristics of each city far more accurately and easily than before. The study also revealed that the statistical areal units in China, namely built-up area, city, and urban district, do not represent "urban area" as accurately as DID, or Densely Inhabited District, used in Japan.

Keywords East China · Land use change · Urbanization · Urban area Japanese topographic map

2.1 Introduction

China has been urbanizing very fast since the 1980s along with its modernization and economic development. The rapid urbanization, particularly the expansion of urban areas, has been causing various land use and other kinds of problems. However, our understanding in urbanization in China has not been sufficient due to

Y. Himiyama (✉) · M. Ikeshita · T. Shinde
Hokkaido University of Education, Asahikawa Campus, Asahikawa 070-8621, Japan
e-mail: himiyamay@kkd.biglobe.ne.jp

Y. Himiyama (ed.), *Exploring Sustainable Land Use in Monsoon Asia*, Springer Geography, https://doi.org/10.1007/978-981-10-5927-8_2

the lack of needed reliable data and the delay in related study. Bulk of research on urbanization in China relies on satellite images for spatial and quantitative information on urban expansion, and it inevitably restricts study period to the 1980s onward. Although the land use changes in China during the last two decades have been great and their environmental impacts severe, more attention should be directed to long-term changes as well as the recent ones, as without knowing the former the present situation and future prospect cannot be accurately grasped (Himiyama 2004a). For those reasons, Institute of Geography, Asahikawa Campus, Hokkaido University of Education, started a project on urbanization in China in 2003, in which 1:50,000 topographic maps of China produced by Japan ca. 1930, which are generally called *Gaiho-zu* in Japanese, *Atlas of Cities of China* (Zhou et al. 1994), and Landsat ETM+ satellite images were used as basic information sources. The present paper first reviews urbanization study on China, and then reports on the compilation and preliminary analyses of the land use datasets of the fourteen cities in Jiangsu Province and Zhejiang Province as part of the series of study, and on the comparison of urbanization and its background. An objective of the present paper is to show the value of the 1:50,000 topographic maps for LUCC study in general, and for the study of urbanization in particular.

2.2 Urbanization Study on China

Urbanization is not only a major type of land use change itself, but also a major driver of the change of other types of land use. It is often a threat to agriculture, and at the same time it affects the people's lives, society, economic activities, and the environment outside as well as inside cities. From the standpoint of LUCC, expansion of urban areas is particularly important among the land use changes related to urbanization, because of its strong effect on other land uses and the environment (Himiyama 2004b).

Ji and Torii (2000) studied the expansion of built-up area of Suzhou City in Jiangsu Province since 1980. They made land cover grid maps of 1980, 1985, 1990, 1995, and 1999 based on Landsat images, and identified major changes of land cover. They then carried out field surveys on the processes and mechanisms of urbanization, as well as analyses of available maps and statistics. They concluded that the major driving forces of urban expansion of Suzhou City west to Shanghai were industrialization backed by foreign capitals and mitigation of congestion within the built-up areas, and that the change was strongly policy-driven. Similar satellite-based studies were conducted elsewhere, such as in Su–Xi–Chang City during the 1990s (Liu et al. 2004), or Kunshan City since 1987 (Long and Li 2001).

The study of socioeconomic and political aspects is an essential part of LUCC urbanization research. The series of research on the expansion of medium to smaller cities in the Changjiang Delta area conducted by Ji offers detailed accounts of the processes and mechanisms of the conversions of rural land uses to urban land uses, not only qualitatively, but also quantitatively, fine maps of the conversions and photographs to show what are actually taking place at each site (Ji 2003, 2004a, b). Although these studies focus only on the period since 1980, the extension of them to the other parts of the country would be an extremely meaningful challenge for LUCC.

Land Use and Land Cover Changes in Beijing 1903–1999 (Lu and Wu 2002) offers comprehensive account of the long-term land use/cover changes of Beijing Municipality. The article presents 1:650,000 land use/cover maps of ca. 1903, 1950, 1980, and 1999, land use/cover change maps between these periods, land use/cover structure tables and structure change tables, together with outline descriptions and explanations of the state, change, and background of the land use/cover during 1903–1999.

2.3 Study Areas and Method

From the discussion above, it is obvious that the study of urbanization in China should now put high priority on comparison of many cities in the country, and an extension of the time range to longer past in order to grasp much of the changes taken place during the country's modernization in the twentieth century. The present study deals with Jiangsu Province and Zhejiang Province. As mentioned above, it uses the 1:50,000 topographic maps of China produced by Japan ca. 1930, *Atlas of Cities of China* and Landsat ETM+. The cities to be chosen for the study have to have all these.

Considering these conditions concerning data availability, as well as size, location, and other features of the cities in Jiangsu Province and Zhejiang Province, the following fourteen cities have been selected as study sites (Fig. 2.1):

- Jiangsu Province: Nanjing, Changzhou, Xuzhou, Yixing, Dongtai, Nantong, and Wuxi;
- Zhejiang Province: Hangzhou, Ningbo, Huzhou, Jianxing, Shaoxing, Jinhua, and Linhai.

Basic information of Jiangsu Province and Zhejiang Province is outlined below according to *Atlas of Cities of China.* Jiangsu Province is 100,000 km^2 large, with much of its territory located in the fertile Chanjiang Delta. There are freshwater lakes such as Taihu Lake, and some low mountains and hills in the southwest and north. The climate is subtropical to temperate wet monsoon type. Population is about 73,549,000. Zhejiang Province is 101,800 km^2 large, with 70.4% of its territory in mountains and hills, 23.2% in plains and basins, and 6.4% in water surface. The climate is subtropical monsoon type, and precipitation is around 1000–2000 mm. Population is about 46,800,000. This province is not rich in natural resources or cultivable land, but its proximity to Shanghai and its coastal location attract many industries not only domestically, but also from abroad.

Fig. 2.1 Study areas

The following three types of land use map have been produced for each study site, as illustrated in Fig. 2.2:

(1) land use map ca. 1930 based on the 1:50,000 topographic map;
(2) land use map ca. 1990 based on the general city map in *Atlas of Cities of China;*
(3) land use map ca. 2000 based on Landsat ETM+ image taken through Google Earth.

The area covered by the map in the Atlas is just large enough to contain the urban area of ca. 1990, but the urban area of ca. 2000 surpasses it in most cities, as urban expansion during the 1990s was extreme.

As is shown in Fig. 2.2a, the 1:50,000 topographic map is colored by color pencils according to land use identified by the symbols shown on the map. Thirteen types of land use, namely "settlement", "road", "railway", "paddy field", "dry field", "orchard", "mulberry", "tea", "forest", "bamboo", "rough land", "water", and "other" were identified. The colored map is then scanned and processed on the computer by using Adobe Photoshop, a popular image processing software. Figure 2.2b shows the resultant land use map of ca. 1930.

Atlas of Cities of China is the first large-scale comprehensive city atlas of China published in 1994 by the Construction Ministry of China. The atlas includes information dated from the birth of the People's Republic of China in 1949–1989. It includes information on 450 Chinese cities, including the cities' natural environment and human geographical features, their history, and present conditions of urban development, and presents their overall city plans. It is intended not only to

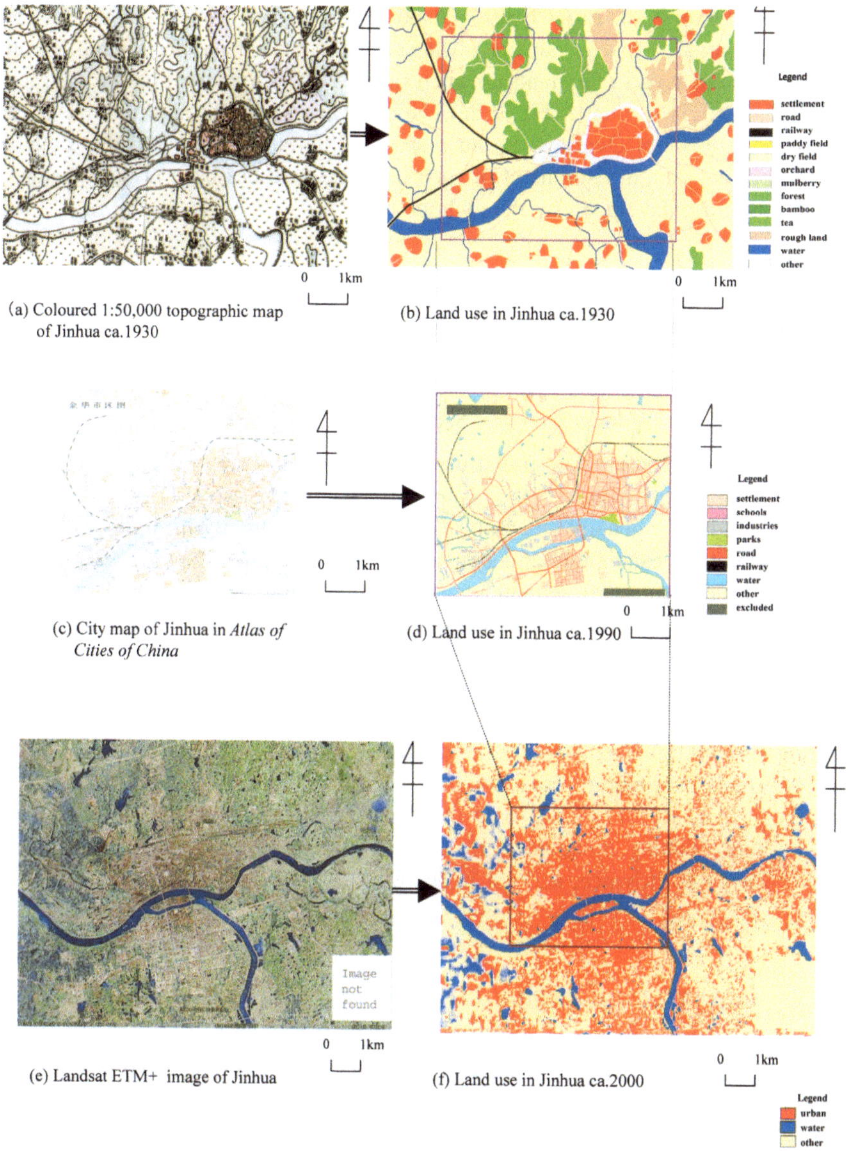

Fig. 2.2 Provision of land use maps of Jinhua representing ca. 1930, ca. 1990, and ca. 2000

provide solid fundamental materials for city planning, construction, and administration, but also to offer comprehensive information of cities to the government departments and to the general public.

The general city map in *Atlas of Cities of China* shown in Fig. 2.2c is similarly used as the information source for the digital land use map representing ca. 1990, as shown in Fig. 2.2d. Eight types of land use, namely “urban”, “schools”, “industries”, “parks”, “road”, “railway”, “water”, and “other” have been identified and put on different layers. The land use map of ca. 2000 shown in Fig. 2.2f is made from Landsat ETM+ image (Fig. 2.2e), which is accessible through Google Earth, by using Photoshop. Overlays of the maps of different periods are also done in Photoshop, but area measurement of each land use type has been done by using a BASIC program developed by the authors.

2.4 Urbanization in the Fourteen Study Areas

2.4.1 Land Use ca. 1930

Tables 2.1 and 2.2 show land use structure of each study area in Jiangsu Province and Zhejiang Province ca. 1930, respectively. Each study area is set equal to the area covered by the general city map of the corresponding city in *Atlas of Cities of China*. Therefore, each study area contains the old settlement that later grew up to major city, adjacent rural settlements, agricultural, and marginal land uses. The cities that had larger urban land use among the fourteen cities include Hangzhou (57 km^2) and Nanjing (37 km^2), but the average area of urban land use in the fourteen study areas was only 12.6 km^2 at that time. Agricultural land use occupied 52% in average of a study area, and most of it is now in urban use, as seen later. Among agricultural land uses, dry field was by far the largest, occupying 45% of the study area in average. Hangzhou, a historical power center is unique in many ways, e.g., it is the only study area that had tea garden, and it also had orchard together with Nanjing, an another city of historical importance. Bamboo was found only in Zhejiang Province, and not in Jiangsu.

2.4.2 Expansion of Urban Areas During Ca. 1930–Ca. 2000

Figure 2.3 shows the expansion of urban areas of the fourteen study areas in Jiangsu and Zhejiang during ca. 1930–ca. 2000. The urban areas of ca. 1930 and ca. 1990 are those within the areas set by the general city maps in *Atlas of Cities of China*, while those of ca. 2000 include the contiguous urban areas beyond these boundaries. As the urban areas of ca. 2000 are quite large in most study areas and they are engulfing the adjacent settlements, precise delineation of contiguous urban areas is an important but a difficult task, and is left as a future theme of study. The urban area data for ca. 2000, which have been calculated from the land use map produced from Landsat ETM+ image, are not fully objective, and are presented here only for a reference purpose. The data of the area of ‘built-up area’, which are presented in *Urban Statistical Yearbook of China* (Statistical Bureau of China ed. 2000),

Table 2.1 Land use ca. 1930 in each study area in Jiangsu set by *Atlas of Cities of China*

Land use	km^2	%
(a) Nanjing		
Urban	37.23	16.73
Settlement	17.90	8.05
Road	19.33	8.69
Railway	1.38	0.62
Agricultural	62.37	28.04
Paddy field	0.02	0.01
Dry field	62.02	27.88
Orchard	0.19	0.09
Mulberry	0.14	0.06
Tea	0.00	0.00
Forest	9.30	4.18
Bamboo	0.00	0.00
Rough land	0.67	0.30
Water	25.19	11.32
Other	88.64	39.43
Total	222.49	100.00
Land use	km^2	%
(b) Changzho		
Urban	11.09	13.84
Settlement	6.43	8.03
Road	4.66	5.82
Railway	0.00	0.00
Agricultural	59.36	74.10
Paddy field	0.00	0.00
Dry field	59.36	74.10
Orchard	0.00	0.00
Mulberry	0.00	0.00
Tea	0.00	0.00
Forest	0.92	1.15
Bamboo	0.00	0.00
Rough land	0.00	0.00
Water	5.66	7.07
Other	3.08	3.84
Total	80.11	100.00
Land use	km^2	%
(c) Xuzhou		
Urban	8.94	11.60
Settlement	3.62	4.70
Road	4.10	5.32
Railway	1.22	1.59
Agricultural	34.27	44.48

(continued)

Table 2.1 (continued)

Land use	km^2	%
Paddy field	0.00	0.00
Dry field	34.27	44.48
Orchard	0.00	0.00
Mulberry	0.00	0.00
Tea	0.00	0.00
Forest	0.62	0.81
Bamboo	0.00	0.00
Rough land	0.00	0.00
Water	2.27	2.95
Other	30.94	40.16
Total	77.04	100.00
Land use	km^2	%
(d) Yixing		
Urban	1.36	13.42
Settlement	0.73	7.19
Road	0.63	6.24
Railway	0.00	0.00
Agricultural	5.66	55.75
Paddy field	0.00	0.00
Dry field	5.63	55.46
Orchard	0.00	0.00
Mulberry	0.03	0.29
Tea	0.00	0.00
Forest	0.00	0.00
Bamboo	0.00	0.00
Rough land	0.00	0.00
Water	2.57	25.35
Other	0.56	5.48
Total	10.15	100.00
Land use	km^2	%
(e) Dongtai		
Urban	3.47	13.67
Settlement	1.71	6.74
Road	1.76	6.93
Railway	0.00	0.00
Agricultural	14.45	56.93
Paddy field	0.00	0.00
Dry field	14.45	56.93
Orchard	0.00	0.00
Mulberry	0.00	0.00
Tea	0.00	0.00
Forest	0.00	0.00

(continued)

Table 2.1 (continued)

Land use	km^2	%
Bamboo	0.00	0.00
Rough land	0.00	0.00
Water	1.33	5.24
Other	6.13	24.16
Total	25.38	100.00
Land use	km^2	%
(f) Nantong		
Urban	7.32	10.84
Settlement	1.87	2.77
Road	5.45	8.08
Railway	0.00	0.00
Agricultural	51.23	75.90
Paddy field	0.00	0.00
Dry field	51.23	75.90
Orchard	0.00	0.00
Mulberry	0.00	0.00
Tea	0.00	0.00
Forest	0.09	0.13
Bamboo	0.00	0.00
Rough land	0.00	0.00
Water	7.70	11.41
Other	1.16	1.72
Total	67.50	100.00
Land use	km^2	%
(g) Wuxi		
Urban	16.09	12.23
Settlement	4.35	3.31
Road	11.73	8.92
Railway	0.00	0.00
Agricultural	83.09	63.16
Paddy field	0.00	0.00
Dry field	81.70	62.10
Orchard	0.00	0.00
Mulberry	1.39	1.06
Tea	0.00	0.00
Forest	0.08	0.06
Bamboo	0.00	0.00
Rough land	1.21	0.92
Water	18.22	13.85
Other	12.80	9.73
Total	131.56	100.00

Table 2.2 Land use ca. 1930 in each study area in Zhejiang set by *Atlas of Cities of China*

Land use	km^2	%
(a) Hangzhou		
Urban	57.03	21.39
Settlement	18.77	7.04
Road	36.13	13.54
Railway	2.13	0.81
Agricultural	89.97	33.73
Paddy field	0.00	0.00
Dry field	54.76	20.53
Orchard	1.47	0.56
Mulberry	33.38	12.51
Tea	0.36	0.13
Forest	16.69	6.26
Bamboo	5.30	1.99
Rough land	0.00	0.00
Water	45.57	17.07
Other	52.23	19.56
Total	266.79	100.00
Land use	km^2	%
(b) Ningbo		
Urban	10.86	28.19
Settlement	6.48	16.81
Road	4.11	10.68
Railway	0.27	0.70
Agricultural	19.85	51.53
Paddy field	0.00	0.00
Dry field	19.85	51.53
Orchard	0.00	0.00
Mulberry	0.00	0.00
Tea	0.00	0.00
Forest	0.04	0.12
Bamboo	0.00	0.00
Rough land	0.04	0.10
Water	5.82	15.10
Other	1.92	4.96
Total	38.53	100.00
Land use	km^2	%
(c) Huzhou		
Urban	4.82	21.77
Settlement	3.04	13.74
Road	1.78	8.03
Railway	0.00	0.00
Agricultural	12.59	56.84

(continued)

Table 2.2 (continued)

Land use	km²	%
Paddy field	0.00	0.00
Dry field	11.82	53.35
Orchard	0.00	0.00
Mulberry	0.77	3.49
Tea	0.00	0.00
Forest	0.00	0.00
Bamboo	0.00	0.00
Rough land	0.00	0.00
Water	2.64	11.91
Other	2.10	9.48
Total	22.15	100.00
Land use	km²	%
(d) Jianxing		
Urban	6.00	18.36
Settlement	2.98	9.12
Road	2.76	8.44
Railway	0.26	0.80
Agricultural	17.82	54.55
Paddy field	0.00	0.00
Dry field	14.01	42.88
Orchard	0.00	0.00
Mulberry	3.81	11.67
Tea	0.00	0.00
Forest	0.27	0.83
Bamboo	0.65	1.99
Rough land	0.00	0.00
Water	3.84	11.74
Other	4.09	12.53
Total	32.67	100.00
Land use	km²	%
(e) Shaoxing		
Urban	6.06	20.71
Settlement	3.55	12.14
Road	2.51	8.57
Railway	0.00	0.00
Agricultural	12.44	42.55
Paddy field	0.00	0.00
Dry field	12.44	42.55
Orchard	0.00	0.00
Mulberry	0.00	0.00
Tea	0.00	0.00
Forest	0.09	0.30

(continued)

Table 2.2 (continued)

Land use	km²	%
Bamboo	0.00	0.00
Rough land	0.00	0.00
Water	5.71	19.52
Other	4.95	16.92
Total	29.25	100.00
Land use	km²	%
(f) Jinhua		
Urban	3.75	15.24
Settlement	2.37	9.61
Road	1.21	4.90
Railway	0.18	0.73
Agricultural	13.00	52.81
Paddy field	0.00	0.00
Dry field	13.00	52.81
Orchard	0.00	0.00
Mulberry	0.00	0.00
Tea	0.00	0.00
Forest	3.75	15.23
Bamboo	0.00	0.00
Rough land	0.79	3.22
Water	2.38	9.66
Other	0.95	3.84
Total	24.63	100.00
Land use	km²	%
(g) Linhai		
Urban	1.88	16.18
Settlement	0.90	7.76
Road	0.98	8.42
Railway	0.00	0.00
Agricultural	4.79	41.28
Paddy field	0.00	0.00
Dry field	4.29	36.98
Orchard	0.00	0.00
Mulberry	0.50	4.31
Tea	0.00	0.00
Forest	0.23	2.00
Bamboo	0.00	0.00
Rough land	0.14	1.22
Water	1.31	11.32
Other	3.25	27.99
Total	11.60	100.00

are also presented in Fig. 2.3 for comparison. Urban growth rates for the periods ca. 1930–ca. 1990, ca. 1990–ca. 2000, and ca. 1930–ca. 2000 are also given for each study area. Figure 2.4 shows the change rates of "urban", "water", and "other" uses

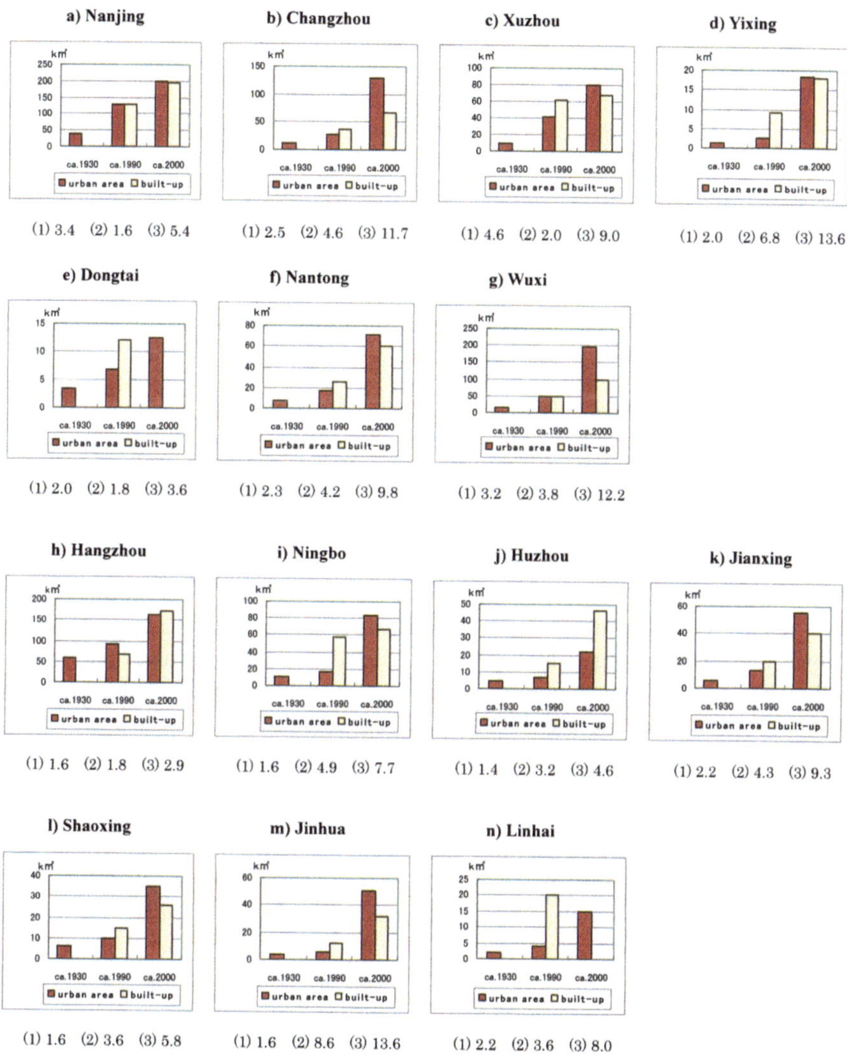

(1) urban growth rate, ca.1930 – ca.1990
(2) urban growth rate, ca.1990 – ca.2000
(3) urban growth rate, ca.1930 – ca.2000

Fig. 2.3 Urban growth in the fourteen study areas in Jiangsu and Zhejiang, ca. 1930–ca. 1990–ca. 2000

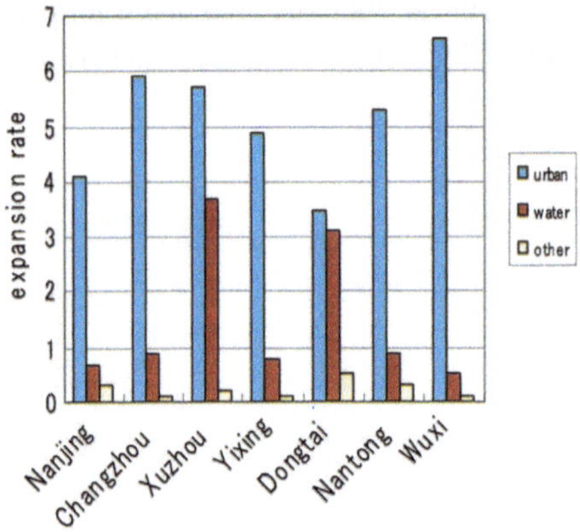

(a) Jiangsu

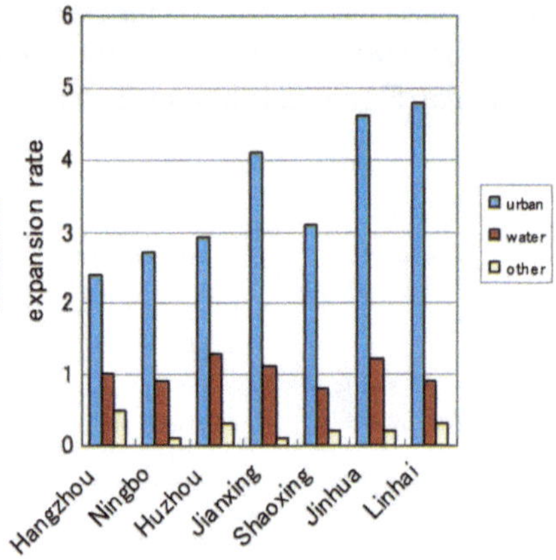

(b) Zhejiang

Fig. 2.4 Land use change in the fourteen study areas in Jiangsu and Zhejiang, ca. 1930–ca. 2000

in the fourteen study areas during ca. 1930–ca. 2000. It is noted that the comparison of the two periods is made within the area set by Atlas of Cities of China.

The average growth rate for the period ca. 1930–ca. 1990 for all the fourteen study areas is 2.30, while it is 2.86 for Jiangsu and 1.74 for Zhejiang. It indicates that the cities in Jiangsu Province, which occupies the Changjiang Delta and is closer to Shanghai, grew more than those in Zhejiang before ca. 1990. On the other hand, the average growth rate for the period ca. 1990–ca. 2000 for all the fourteen study areas is 3.91, while it is 3.54 for Jiangsu and 4.29 for Zhejiang, i.e., the growth rate for Zhejiang is higher than that for Jiangsu. The growth rate of 3.91 within ten years is considered to be extreme, as even Japan under the historic rapid economic growth during the 1960s–1970s experienced the growth rate of DID (Densely Inhabited District) per ten years of only 1.87, i.e., less than half of the Chinese average. It is worth noting, however, that much of the urban expansion in China was driven by local governments, and that there appear to be more dormant 'developed' areas in China than in Japan.

For the period ca. 1930–ca. 2000, the average growth rate for all the fourteen study areas is 8.37, while it is 9.33 for Jiangsu and 7.53 for Zhejiang. Yixing and Jinhua show the highest growth rate of 13.6 for the period of ca. 1930–ca. 2000. Dongtai on the other hand shows the lowest growth rate of 3.6 for ca. 1930–ca. 2000.

Figure 2.3 suggests that the data for ca. 1930 are extremely important when one intends to grasp the general long-term trend of urban growth of each city and to foresee its future trend.

2.4.3 Loss of Agricultural Land

The expansion of urban areas during ca. 1930–ca. 2000 was in no doubt a serious threat to the surrounding agricultural land, if not to the suburban rural community. Recent news reports from China indicate that there are many incidents of conflict between local governments and suburban rural communities. Figure 2.5 shows an example of urban expansion, which is eating up the surrounding agricultural land. According to Table 2.2, all the agricultural land of Ningbo ca. 1930 within the area set by *Atlas of Cities of China* was dry field, and it occupied 51% of the area. It was almost completely eliminated by urban expansion. As shown in Fig. 2.3, the urban growth rate of Ningbo during ca. 1930–ca. 2000 is 7.7, which is close to the average and is not particularly high. It means that the loss of agricultural land by urban expansion is substantial in most cities, if not all. It is feared that the effect of this fast expansion of urban areas is far-reaching and complicated, and much more than what mere reduction of area of cultivated land might indicate (Himiyama 2005).

2.4.4 Physical Barriers to Urban Expansion

Urbanization is constrained directly or indirectly by physical barriers, such as rivers or mountains. Figure 2.6 shows the case of Huzhou, which exemplifies how urbanization was contained within the area surrounded by rivers and canals till ca. 1990, and the sweeping advance of urbanization beyond these barriers during the

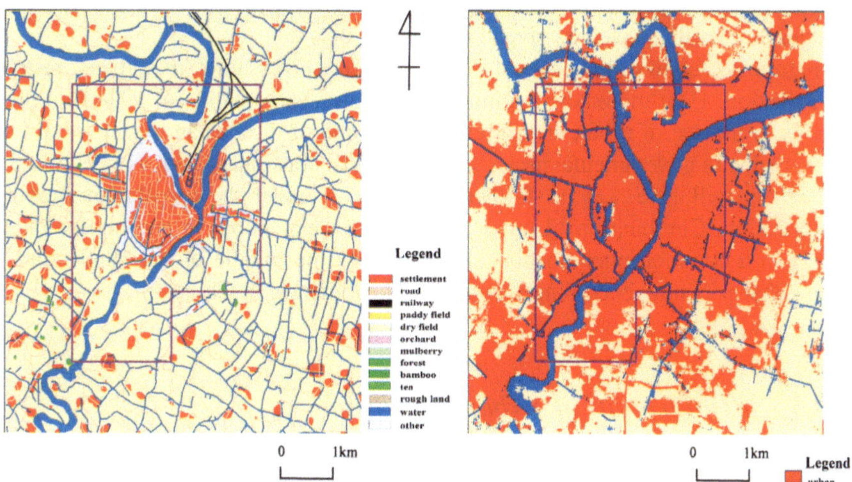

(a) Land use in Ningbo ca.1930 (b) Land use in Ningbo ca.2000

Fig. 2.5 Farmland loss by urban expansion in Ningbo ca. 2000

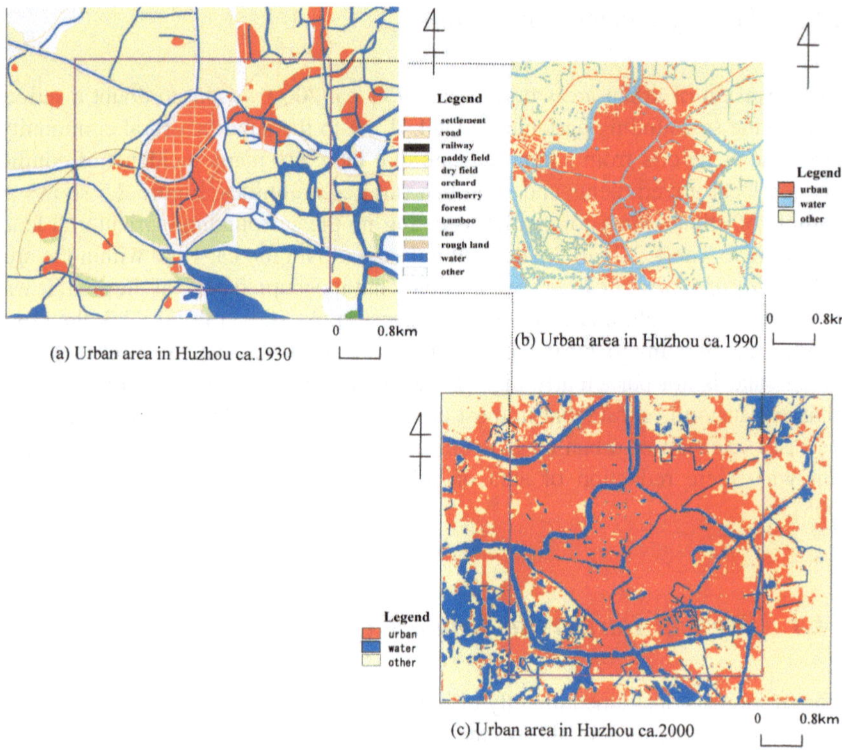

Fig. 2.6 Urbanization in Huzhou and the rivers blocking it

1990s. This advance is obviously backed by construction and improvement of bridges and roads as well as spread of cars.

Figure 2.7 shows the case of Wuxi, which shows how mountains restrict urbanization. Here mountains within and around urban area in ca. 1990 look like islands in the sea of urban area in ca. 2000. Some of these islands will remain as they are due to governmental protection or physical constraints such as steep slopes, but others will give way to further urbanization under the rapidly progressing motorization. Climbing up of urban area on less steep slopes in and around cities has been a common phenomenon in Japan since the middle of the 1960s, and it seems that it is now becoming a common phenomenon in China.

2.4.5 Identification of Urban Expansion

Delineation of an urban area is not an easy task. It is particularly so where there is no clear visible boundary in between urban area and the surrounding countryside. It is generally believed by specialists that 'city area' does not represent urban area

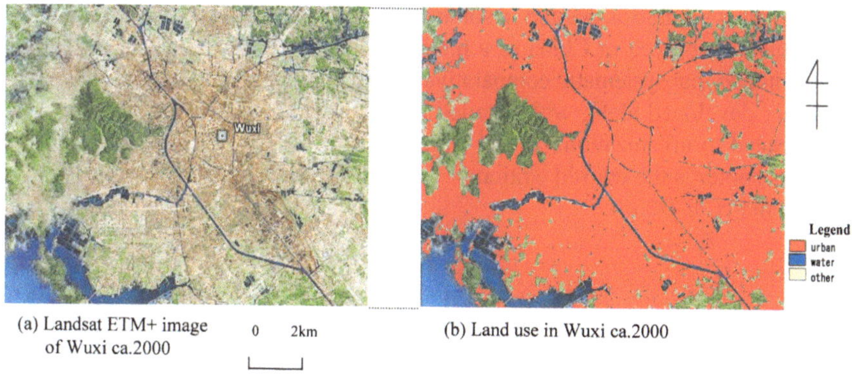

(a) Landsat ETM+ image of Wuxi ca.2000

(b) Land use in Wuxi ca.2000

Fig. 2.7 Urbanization in Wuxi ca. 2000

well. An alternative to it may be 'built-up area', but again specialists are often reluctant to use it, partly because they are not convinced with its reliability, and partly because it has only a couple of decades of history. For these and other reasons, Ji (2004b) and many other specialists used satellite images, particularly night-light images, in order to identify urban areas and their changes.

As mentioned above, Fig. 2.3 shows built-up area together with urban area defined in the present study. It is shown only for ca. 1990 and ca. 2000, as it did not exist ca. 1930. Built-up area and urban area are almost identical in some cases, e.g., Nanjing ca. 1990 and ca. 2000, Yixing ca. 2000, Xuzhou ca. 2000, and Nantong ca. 2000, but large differences are also seen, e.g., Linhai ca. 1990, Changzhou ca. 2000, Wuxi ca. 2000, and Huzhou ca. 2000. Both urban area and built-up area appear to need further improvements.

In Japan, DID (Densely Inhabited District) has been defined and mapped since the 1960s in order to delineate urban area as accurately as possible based on the population census data. As Himiyama and Morishita (2003) demonstrated, it is an extremely useful tool for urban planning and the study of urbanization, although it has certain limitations. It is hoped that DID or other means equivalent to or better than DID is defined and used in China, so that urbanization is grasped more accurately and necessary planning measures are taken based on the facts.

2.5 Conclusions

Urban expansions of fourteen cities in Jiangsu Province and Zhejiang Province of China since ca. 1930 have been grasped spatially and quantitatively by producing and comparing the digital land use maps of ca. 1930, ca. 1990, and ca. 2000. The study has shown the long-term trend of urban expansion, the extreme speed of it during the 1990s, especially that of smaller cities, and the resultant farmland loss. It was argued that the existing areal unit for urban area, such as city area or built-up area, is not as reliable as desired in some cities, and that the DID used in Japan may be useful if employed in China as a means to identify urban area.

It is hoped that systematic comparative study of long-term urbanization in China is further developed along with detailed regional studies, and helps avoiding conflicts related to urbanization.

This chapter is the revised edition of CJLUC Project Report Vol. 5 pp. 191–202 (Himiyama et al. 2006).

References

Himiyama Y (2004a) Achievements and tasks of research on land use change in China. Geogr Rev Jpn 77(12):783–799

Himiyama Y (2004b) The trend, mechanism and impact of urban expansion in Japan. Reports of the Taisetsuzan Institute of Science, no. 38, pp 53–60

Himiyama Y (2005) Land use and its historical change of Japan—since the Meiji Era (1868–2012). In: Nakamura K, Arai T, Yonekura N (eds) Geography of Japan—physical overview. Asakura Shoten, Tokyo, pp 293–308

Himiyama Y, Morishita Y (2003) Urbanization in Japan since 1960 viewed from the DID Statistics. Reports of the Taisetsuzan Institute of Science, no. 37, pp 37–52

Himiyama Y, Ikeshita M, Shinde T (2006) Urbanization in Jiangsu Province and Zhejiang Province since ca. 1930. In: Himiyama Y (ed) China–Japan comparative study of land use/cover changes (V). CJLUC, Asahikawa, pp 191–202

Ji Z, Torii K (2000) The expansion of the built-up areas of regional capital cities in the Yangtze River Delta area—a case study of Suzhou City in Jiangsu Province, China. In: Otsubo K (ed) LUGEC project report VI, CGER, pp 64–82

Ji Z (2003) Rearrangement of land use in the suburbs of Wuxi City, China. In: Himiyama Y (ed) China–Japan comparative study of land use/cover changes (II). CJLUC, Asahikawa, pp 51–72

Ji Z (2004a) Development of industrial estates and the response of the surrounding rural communities in Shanghai Metropolitan Area: the case of Kunshan City, Jiangsu Province. In: Himiyama Y (ed) China–Japan comparative study of land use/cover changes (III). CJLUC, Asahikawa, pp 25–42

Ji Z (2004b) Changes of cities and villages in China. Ashi Shobo, Tokyo

Liu J, Zhao T, Deng X, Zhan J (2004) Study in Su–Xi–Chang city zone. In: Himiyama Y (ed) China–Japan comparative study of land use/cover changes (III). CJLUC, Asahikawa, pp 67–74

Long H, Li X (2001) Land use changes under the pressure of rapid industrialization and urbanization: the case of Kunshan, the Yangtse River Delta Economic Area. In: Shi P (ed) Proceedings of international conference on land use/cover change dynamics. Organizing Committee of LUCCD2001, Beijing, pp 211–222

Lu Q, Wu Peilin (2002) Land use and land cover changes in Beijing 1903–1999. In: Himiyama Y (ed) Land use/cover changes in selected regions in the world (II). IGU-LUCC, Asahikawa, pp 45–62

Statistical Bureau of China ed. (2000) Urban statistical yearbook of China. China Statistical Press, Beijing

Zhou G, Chu C et al (eds) (1994) Atlas of cities of China. China Map Publishers, Beijing

Chapter 3
Reorganization of Suburban Areas in Terms of Real Estate Utilization and Transformation of Evicted Farmers

Zengmin JI

Abstract In the suburban regions of the metropolitan cities in emerging countries of Asia, there has been a drastic large-scale change arising for a short period of time with economic growth, the establishment of development zones, and population inflow since the end of the twentieth century. Suburban regions, where the mixture of land utilization and the blending of residents are defined as their characteristics, are not only academic research targets as mere urban studies but also the real focusing point of social and regional issues on a global scale. This thesis attempted to link the social research method, the elucidation of the process, and the life history of life recovery of indigenous farmers and the geographical method focusing on the fluctuation of housing space where its trajectory and results are projected, that is, a collaboration method of social research and geography.

Keywords Reorganization · Suburban areas · Real estate utilization
Spatial migration trajectory · Yangtze River Delta · China

3.1 Introduction

3.1.1 Role and Characteristics of This Research

As is widely known, since the beginning of the twenty-first century, the Yangtze River Delta area centering on Shanghai has rapidly changed from the world's pre-eminent grain belt to an area of the most remarkable urbanization in the world. One of the noticeable characteristics of urbanization seen here is that the suburban region has drastically expanded the existing urban area at such a speed and on such a scale, which has not been seen in the past. The result of such a dramatic change in suburban space has deeply reflected and affected the land utilization and landscape, the regional structure, and social lives.

Z. JI (✉)
School of Culture-Information Studies, Sugiyama Jogakuen University, Nagoya, Japan
e-mail: zengmin@sugiyama-u.ac.jp

Y. Himiyama (ed.), *Exploring Sustainable Land Use in Monsoon Asia*,
Springer Geography, https://doi.org/10.1007/978-981-10-5927-8_3

The writer has aimed at deepening research on the suburbs in the metropolitan regions of China and constructing academic theories as part of understanding the actual spatial–temporal conditions of the regional transfiguration in the Yangtze River Delta and has researched these for more than ten years and publicized the research results in Table 3.1. This thesis clarifies the ownership status of housing of the indigenous farmers in the suburban regions and the process of its utilization change with the change of social and economic conditions after 2000 and analyzes its background. The reason to have focused on the ownership of housing and utilization status is closely discussed by JI (2013)[1] and thus, refer to it with this thesis.

In the geography field, research that focuses on changes in suburban areas has become more and more common. As examples of research similar to that of this author, there are (1) a study by Li analyzing the so-called "Chengzhongcun" land market (2008); (2) a thesis on land development and inequality in villages surrounded by cities by Wu (2009); (3) a case study by Wang et al. (2009) of an urban village in Shenzhen; and (4) Lin and Wang (2011) research on the transition of the suburban village in the course of rapid urbanization, based on a case study of Shuangqiao Village, Ningbo.

The present thesis adopts a micro-approach and analyzes the changes in the ownership and utilization of housing in order to clearly elucidate the process by which farmers have reconfigurated their lives in suburban areas.

More specific characteristics of this thesis are as follows.

(1) The process of farm retirement of the evicted farmers due to urbanization has been categorized into three time periods: commercialization of farm products, commercialization of the labor force, and commercialization of farmland and housing sites. Accordingly, the process of utilization change of the housing sites can be divided into three stages: (1) utilization of land and housing sites during the rural community period before 2000, (2) housing site utilization of concentrated dwellings at the middle-rise farmers' housing park due to eviction after 2001 till around 2012, and (3) housing utilization after transferring to condominiums around 2012.

    Based on the above, the writer has attempted to understand the details of part of the transfiguration of suburban space and landscape and regional social change in the metropolitan area by connecting the processes of eviction and housing site utilization change through the ties of life recovery after eviction and clarifying the dynamic corresponding relationships which are seen in both processes.

(2) The above-mentioned change process and trajectories will be visualized (diagrammatic representation). That is, the following will be illustrated as diagrams: the change from a two-dimensional utilization in the rural community

[1]Zengmin (2013).

Table 3.1 Structure flow of development of academic theory on study of suburbs in China

Order	Year	Study point	Target area	Study achievement	Method
1	2010	Raised concept of the 3rd space	Overall suburban areas	Background to raise the 3rd space and definition were mentioned and basic study viewpoints and characteristics of the 3rd space were presented	Presentation and thesis[a]
2	2011	Selection method of target region	Suburbs, towns, and sub districts	Study scheme, method to extract the 3rd space, and target region were selected and examination method through combination of study on multi scale and survey of intentions	Presentation and thesis[b]
3	2012	Transition of housing possession and utilization	Farmers' housing complex (community) Housing towers (former village)	Clarification of actual conditions of change occurred in housing utilization and ownership condition and analysis background of its change by relating to process of life recovery of evicted farmers.	Presentation and thesis[c]
4	2013	Integrated analysis of residents and space	Housing tower (settlement), by floor, by housing type	Establishment of comprehensive study method to integrally relate people and region, clarification of all the process of moving out to condominium apartments through the process of farmers' housing from rural settlement age and its background	Presentation and thesis[d]
5	2014	Future forecast by combination of survey on intention	Indigenous farmers, migrant workers, citizens	Utilize the result of survey of intentions responding to native farmers, migrant workers, and citizens as residents for clarification of spatial structure and social composition of the third space and future forecast	Presentation and thesis[e]

[a]Zengmin (2010), (2011)
[b]Zengmin (2012)
[c]Zengmin (2013)

era to a three-dimensional one and furthermore, the location and distance of the condominiums which the farmers moved to.

In order to do the above, based on the detailed information of hearing investigations from a former village mayor, village elders, people in charge of statistics, the land utilization of the rural community era before 2000 was restored. Next, the location of each farm household in the target communities was specified and they were diagramed according to the ownership and utilization statuses of the housing buildings after moving to the farmers' housing park after 2001.

In addition, hearing investigations on the change of ownership and utilization statuses of the housing sites were implemented to the community members. By visualizing its change, steps of life recovery will be divided based on ownership and utilization statuses where the results of life rehabilitation as well as the process of life recovery will be reflected.

(3) In the past, as a matter fact, regional reorganization in suburban regions has been examined separately in terms of sociology and geography. Since "residents' act" and "its spatial trajectory" cannot be considered separately, the establishment of research methods where the fields of sociology and geography are collaborated and combined is considered to be a significant issue for more in-depth research.

This thesis classified the change of ownership and utilization statuses of 13 farm households who transferred from settlement 5 of former Lihuangcun belonging to the Qunyi Community to Zhonghuadongcun tower 25 by farm household statistical data and hearing investigations. Based on the results, they were categorized into five patterns of life recovery and the corresponding relation with the spatial transfer was examined.

(4) The status of ownership and utilization of housing is an epitome representing the change of economic conditions and social status in the life stage of each farm household. Each former farmer who transferred to the farmers' housing park has their own original drama such as life in the former community era, transfer time, career change, and life recovery after transfer. The writer will reveal the actual conditions of the life recovery of the indigenous farmers, which are difficult to see from the outside, through a number of hearing investigations with the case study farm households. While the economy decides the housing utilization and ownership status of indigenous farmers on a priority basis, its opposite direction, that is, how it impacts on the customs of rural villages where the elderly people are cared for and family ties are emphasized, will also be examined.

In order to search the background of housing ownership and utilization condition changes, with regard to the 13 households, a hearing investigation on family composition, occupation, income, and disposition of housing was implemented and the results were presented as the background of utilization change of housing sites. Moreover, as of 2012, as for the eight farm households who moved out to condominiums from the farmers' housing parks, the transfer time, location, and reasons were investigated.

(5) The commercialization (purchase and sale) of housing is a breakthrough event; it means the separation from rural society, agriculture, and farmland and the "transformation" from farmers to citizens for the evicted farmers. It is a starting point as a "new citizen" on the life stage as well as meaning the end of a life recovery cycle.

Evicted farmers ultimately aim for a detached house through the stages of the farmers' housing park, a high-rise condominium, and a gated community. In essence, the acquisition of a detached house is the end goal to turn into a citizen in both material and spiritual terms. In this thesis, the precedent of Mr. H who completed the cycle from a farmer to a true citizen rather earlier than the others was observed.

By analyzing the ownership of housing and utilization status in three dimensions, through the mixture of the five viewpoints above, the process and goal of life recovery of evicted farmers will be presented.

3.1.2 Target Region

Figure 3.1 describes the distribution of the old settlement, which belongs to the Qunyi Community. Until 1949, in Qunyi, which is a riverside district, 26 old

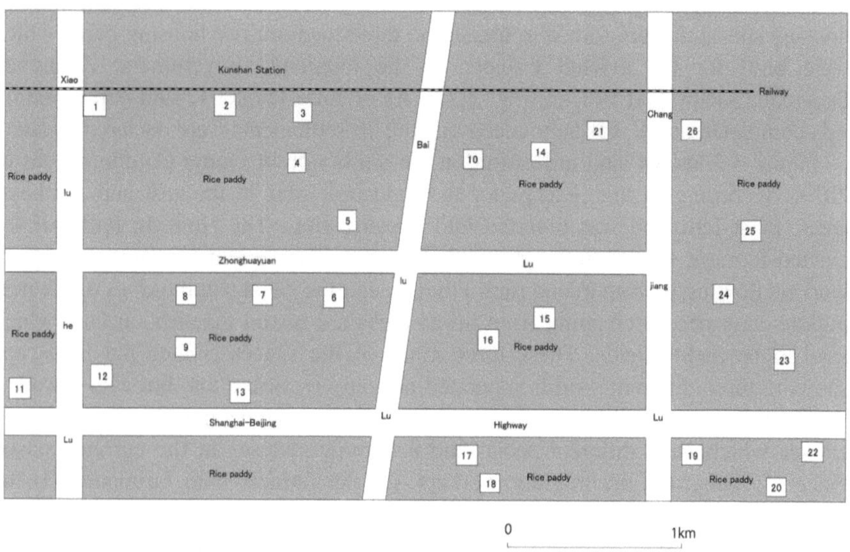

Fig. 3.1 Land utilization and distribution of settlement in Qunyi Community before eviction (as of 2000). *Note* number in □ corresponds to the number of the settlement in Table 3.2. *Source* based on materials of Qunyi Community

settlements were established at the point where the waterways crossed or on hilly land (Fig. 3.2a). Over the long history of hundreds of years, societies of rural communities based on kinship relations and territorial bonding were built-up. Due to the equal field system (fields were equally distributed), any downsizing of owned farmland area is conspicuous. There were no big differences in the natural and social economic conditions and any extreme differences depending on the settlement could not be observed.

After 1949, due to the planned economy system, food production was directed such as rice as the main crop and wheat as an off-season crop, and cultivation of other marketable cropping was not permitted. Nonagricultural side jobs were basically limited to one per farm household and in most cases, the householder or the eldest son was selected.

Since 1978, through the lease of collective ownership of farmland with a limited term, farmland was diverted into plant sites or roads, and township and village enterprises and village companies utilizing the surplus labor within a community were established (Fig. 3.2b). However, in 1994 due to the designation of municipal authorities, 3 ha of farmland was leased to a foundry of *T*, a Japanese manufacturing company for a long time, and farmers in the village were preferentially employed there. Farmland was continuously diverted to bonded areas and foreign capitalized companies, and the Qunyi Community was incorporated in an urban region and furthermore, a global social economic structure.

Table 3.2 describes the land utilization, composition of households, population and destinations of transfer in each settlement belonging to the Qunyi Community before eviction. According to this by 2000, 1228 households and 5100 farmers with land were completely evicted from approximately 748 ha of farmland and 70 ha of housing sites and were forced to transfer to three medium-rise housing parks which were built for the evicted farmers by the municipal government: Zhonghua Dongcun, Xicun, and Beicun (Fig. 3.2c). As of June 15, 2001, land expropriation was completed and it was announced that the 26 settlements were declared vacated.

Figure 3.3 shows land utilization on the north side of Qunyi Community as of 2013. According to this, except for the industrial sites in the east and northeast areas, most farmland was diverted into housing sites. The Housing Parks of the evicted farmers, the exclusive dormitories and collective rental houses of migrant workers flowing in from inland rural villages, and the gated communities of the new middle-class people escaping from urban areas are mixed together on land which used to be paddy fields. These three kinds of life spaces created the landscape showing three different worlds separated not only by a fine line but also a wall.

The mixture of such land utilization is the epitome reflecting the four kinds of classes which have different social and economic statuses in the current coastal cities of China: (1) the migrant workers, (2) the farmers with farmland, (3) the relatively rich middle citizens and higher class farmers with farmland, and (4) the new rich people such as business managers. Moreover, it is providing a condition where the evicted farmers who do not wish to leave their hometown where they were born and raised can improve their housing (lifestyle) condition while remaining to live there.

(a) Settlement spreading to the river side
(b) Corporate park in former Lihuangcun
(c) Evicted farmers' housing park where residents of former Lihuangcun were moved to
(d) Current condition of Lixiang River incorporated in Kunshan bonded ward
(e) Farmer's housing before eviction
(f) Certificate of Title specifying details of the site
(g) High-rise condominium
(h) Luxury condominium

Fig. 3.2 **a** Settlement spreading to the river side, **b** corporate park in former Lihuangcun, **c** evicted farmers' housing park where residents of former Lihuangcun were moved to, **d** current condition of Lixiang River incorporated in Kunshan bonded ward, **e** farmer's housing before eviction, **f** certificate of title specifying details of the site, **g** high-rise condominium, and **h** luxury condominium

Table 3.2 Composition of land utilization, households, and population of Qunyi Community by settlement and destination of transfer before eviction (as of 2000)

Former village name	No. of settlement	Name of settlement	Cultivated land size (mu)	Community land size (m^2)	Expropriated land (mu)	No. of households	Population		Name of destination (farmer's housing park name)
							Sub total	Male	
Chezhancun	1	Majiabang	753	72,000	753	80	366	189	Zhonghuabeicun
	2	Lujiajiao	215	19,200	215	33	147	71	Zhonghuabeicun
	3	Shijiaqiao	239	13,200	239	32	131	58	Zhonghuabeicun
	4	Xihe	251	18,750	251	30	122	65	Zhonghuabeicun
	5	Nanhengtang	225	14,000	225	27	129	58	Zhonghuabeicun/xicun
	6	Beihengtang	357	20,800	357	30	164	82	Zhonghuaxicun
	7	Xujialou	387	15,400	387	44	163	80	Zhonghuabeicun
	8	Hengtang	247	8000	247	22	95	45	Zhonghuabeicun
	9	Yangjing	201	12,600	201	19	73	39	Zhonghuabeicun
	10	Beilian	377	30,000	377	46	226	108	Zhonghuabeicun
Qiongzhuangcun	11	Zhicun	474	9500	474	45	191	101	Baolingxincun
	12	Xiongzhuang	611	18,480	611	84	303	154	Zhonghuaxicun
	13	Wanjiawan	298	20,000	298	27	108	54	Zhonghuaxicun
Qunyicun	14	Dacaixiang	307	15,600	307	32	134	60	Zhonghuadongcun
	15	Sitan	350	17,800	350	38	171	81	Zhonghuadongcun
	16	Sixiang	561	20,500	561	56	272	54	Zhonghuadongcun
	17	Yaoxiang	703	25,000	703	87	314	145	Zhonghuadongcun
	18	Dunxia	877	25,000	877	78	298	152	Zhonghuadongcun/xicun
	19	Shengjing	242	19,000	242	38	160	83	Zhonghuadongcun
	20	hanpo	235	22,000	235	23	108	51	Zhonghuadongcun
	21	Xiaocaixiang	422	26,000	422	46	187	84	Zhonghuadongcun

(continued)

Table 3.2 (continued)

Former village name	No. of settlement	Name of settlement	Cultivated land size (mu)	Community land size (m^2)	Expropriated land (mu)	No. of households	Population		Name of destination (farmer's housing park name)
							Sub total	Male	
Lihuangcun	22	Dongqiao	608	12,000	608	58	227	120	Zhonghuadongcun
	23	Xiaxu	311	30,000	311	33	144	66	Zhonghuadongcun
	24	Baigoucun	248	56,000	248	25	106	50	Zhonghuadongcun
	25	Huangjiadai	878	82,000	878	94	387	186	Zhonghuadongcun
	26	Lihuang	846	84,000	846	101	374	172	Zhonghuadongcun
	Total		11,223	706,830	11,223	1228	5100	2408	

Note 1 mu is approximately 1/15 hectare (approximately 6.67 are). Number of settlement corresponds to the number of Fig. 3.1. *Source* based on materials of Qunyi Community

Fig. 3.3 Land utilization on the north side of Qunyi Community as of 2013. *Source* based on field study

3.2 Land Utilization Change in Settlement 5 of Former Lihuangcun

3.2.1 Land Utilization of the Communities

Figure 3.4 represents an outline of the land utilization in and around settlement 5 in the rural community period before the eviction in 2000. According to this, settlement 5 was an agglomerated settlement, located on the east side of Lixiang River. It faced settlement 1 and 2 from across the river, and in the south it faced settlement 6 from across the village road running from east to west. It was composed of 28 households. The Lixiang River played the main role as the water traffic route until transportation on land such as a highway had been constructed. Rowboats and power sail boats frequently went by and heading north leads to Yushan town, which is the center of Kunshan city while heading south leads to the metropolitan city, Shanghai (Fig. 3.2d). Village people usually took an hour to walk to Yushan town

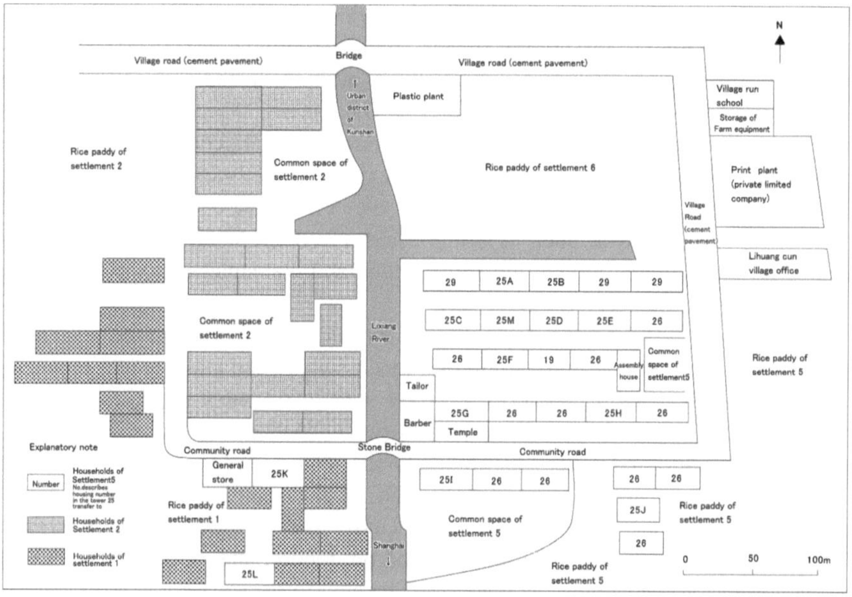

Fig. 3.4 Land utilization in former Lihuangcun and housing distribution of settlement 5 (as of 2000) A–M in the figure are numbers of households of settlement 5, moved to housing tower 25 and correspond to the number of farm households in Fig. 3.5 and Table 3.3. *Source* based on hearing investigations in Qunyi Community in March, May, and September, 2012

via the village road. In case of emergencies such as childbirth or going to hospital, they took a powerboat and it took around 20 min.

Since the settlement was surrounded by swamp, the farmers' housings were located in a low highland area, which was built with 1 m of fill on a natural bank and had been gathered as an irregular clustered village. Of the 28 households, only 2 households, 25K and 25L, were located on the west side of Lihuang River and the rest of 26 households were gathered in an area of approximately 6 square ha land surrounded by village roads (south to north 300 m × east to west 200 m). The two households, 25K and 25L, were moved to the current places when they became branch families since the housing site which the initial owners owned was allocated to them.

Settlement 5 was located in the center of former Lihuangcun (as of 2000, the number of households: 311 and the population was 1238) and across from the village road on the east side and along the road, there were public facilities such as a village office of former Lihuangcun, village primary school, and storage for agricultural machinery and chemicals of the village and on the west side of the village road, there was an assembly hall of the former Lihuangcun.

Near the stone bridge crossing the Lihuang River, commercial facilities whose commercial sphere was the former Lihuangcun or its surrounding villages were gathered. On the west side of the stone bridge, there was a general shop which the "Kunshan General Merchandize Supply and Retail Store" directly operated and besides daily necessities such as everyday sundries, seasoning such as salt and soya sauce, oil for lights, agricultural-related goods including chemicals, fertilizers, seeds, and plastic materials were also sold. In addition to the former Lihuangcun village, people of its surrounding villages of Baigoucun, Dongqiaocun, Xiaxucun, and Huangjialucun also used the shop and its commercial sphere expanded to the range of a 30-min walk. At the northeast foot of the stone bridge crossing Lihuang River, there was a reliquary shrine that was the cornerstone for the village people besides service industries such as a tailor and a barber. At night in the summer, village people gathered around the stone bridge and had small talk while enjoying the cool evening breeze.

On the north sides of the village office of former Lihuangcun and the settlement, there used to be a village run agricultural machinery manufacturing and repairing, and spinning factories. However, they were sold to private companies and were changed to a printing shop and plastic factories.

As settlements 1 and 2 were distributed on the west side of the community, in the surrounding areas of each community, farmland concentrates and the double cropping of rice and wheat were used. Under the planned economy system, farmland utilization was limited to only food production and cultivation to cash in on such things as vegetables and cotton was banned. A 40-square-meter private plot was allocated to each farm household and vegetables for personal consumption were cultivated in fields on 1 m upland. Due to historical reasons, half of the paddy fields and private plots that farmers of settlement 5 owned were far from their settlement and they had to take a 15 or 20 min walk for farming passing through settlement 10, which was located on the south side. Near each settlement, there was a common open space (original word is gongchang) for agricultural land utilization. The common open space of settlement 5 was located in two areas: the east side of the village assembly hall and the south side of the village road. During the harvest seasons in summer and autumn, in the settlement base, grain threshing, rice hulling, and wheat drying, and the distribution of farm produce including straw and sweet potatoes were implemented in the common open space. Usually, it was used as storage for construction materials or agricultural machinery of the local people in the settlement. Just as the village road, it had been paved with mud. In 1980 it was changed to one of the cements. Due to the high population density, any land that could be reclaimed had all been made into farmland and paddy fields were located even just under the roof of farm households.

3.2.2 Utilization of Housing Sites by Farmers

In the rural community period, each farm household was allowed to have 0.3 mu (approximately 200 m^2) of housing site. The most popular housing style was the so-called "three up three down" (three rooms upstairs and three rooms downstairs) (Fig. 3.2e). That is, the first floor was composed of three rooms: a kitchen and dining room in the east, a living and work room in the middle, and a storage room (for agricultural machinery, grains, chemicals, and fertilizers) in the west, and in the second floor there were three rooms for the householder and their children. Each room was 36 m^2 whose width was 4 m and length was 9 m. On the south side of both the first and the second floors, a corridor with a roof whose width was between 1 and 2 m was built. The average floor space including rooms and corridors was 265 m^2.[2]

In addition to the basic style of "three up three down", according to their financial condition, there were some farm households who remained to stay in one story thatched house where there were three rooms and whose floor space was 50 m^2 (e.g., 25H). A somewhat richer farm household would build a "two up three down", that is, two rooms on the second floor: one in the west side and the other in the middle (floor space was 116 m^2) and after securing financing, they added one more room for their child on a kitchen and dining room on the east side of the first floor (e.g., 25M). Moreover, much wealthier people would have added a 36-square-meter building for a kitchen and storage for fuel on the west side of the main house and the room on the east side of the first floor was used as a bedroom for their parents.

The housing size during the rural community period had been the criteria to calculate the compensation space by the local government in the eviction of 2000 (Picture 6). Housing size directly affected the size of housing for compensation and life recovery after farm retirement. Stated differently, life recovery after eviction was an extension of the life conditions of the rural village period and it has taken over the level of life and accumulation during the rural village period.

A house and its front yard besides daily living were used mostly for agriculture. The open space in front of the house was used for threshing grain, drying farm produce such as rice hull and beans, and preserving straw as fuel. In addition, they were used as a breeding place for chickens, sheep, and cows. The living room located in the middle of the first floor has various functions such as a place for the family to spend time together for fun, a social place for guests, a ceremonial place including funerals and weddings, bike parking, storage for agricultural-related equipment: machinery, aerosol sprayers, chemical fertilizer, and agrichemicals, storage for farm products such as unpolished rice, soy beans, and broad beans, and a workplace for carpentry and bamboo work and therefore, it was a wide size of about 30 m^2.

As mentioned above, the farmers' housing during the rural community period was limited for personal use only and agricultural-related use.

---

[2]Regarding housing layout, utilization status, and pictures of "three up three down" (three rooms upstairs and three rooms downstairs), please refer to the thesis below. JI Zengmin (2008): Changes in the lives of farmers brought about by the renting of living quarters to migrant workers in the suburbs of Chinese cities: the case of villages adjacent to the development zone in the city of Kunshan, Jiangsu Province, "Sugiyama Jogakuen University culture Information Department Annals" volume 8 pp. 10-17.

3.2.3 *Relationships of Relatives*

Figure 3.5 shows the relationships of relatives based on the householder of 25M. The house on the right 25B was the family that the father of householder 25M had married into. He was a dressmaker specialist working at a tailor shop at the foot of the stone bridge in Fig. 3.4 and had moved from another village to marry. Therefore, he had no relatives in settlement 5 and its surrounding areas. In settlement 5, there were three households that were the householders' wife's uncles, 7 households of the parent in laws' relatives, and 1 household of the mother's brother making a total of 11. Furthermore, in settlement 2 that was located across the river, there were four households who were the mother's cousins. According to the customs of the settlement, all the relatives of these 15 households voluntarily gathered at a funeral and helped with the cooking and chores. Meanwhile, as for participation in weddings, basically, only a total of eight households attended: the three households of the householders' wife's uncles, the one household of the mother's brother, and the four households of the mother's cousins. Apart from these 15 households, needless to say the entire settlement 5, most residents in settlements 1, 2, 6 that were adjacent to settlement 5 can be said to have been distant relatives.

During the rural community period, residents who had kinship and territorial bonding relationships went to each other's homes for mutual cooperation and daily conversation. After being transferred to the middle-rise farmers' housing parks,

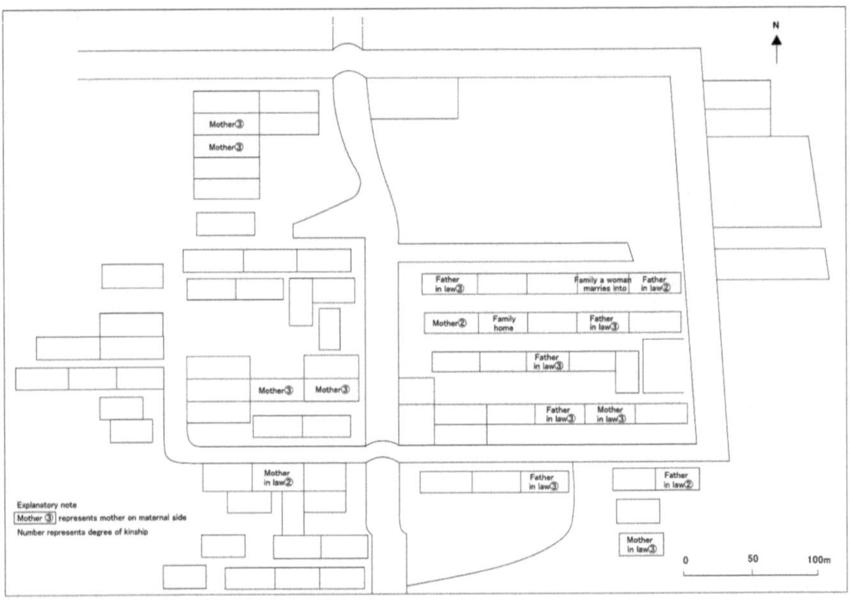

Fig. 3.5 Relationships of relatives of farm household 25M in settlement 5, Lihuang Cun, Kunshan (as of 2000). *Source* based on hearing investigation from farm household 25M in September 2012

they would gather on the first floor to chat about their health and exchange information since it was difficult for seniors to go up and down stairs, after enjoying the cool evening breeze or shopping.

3.3 Housing Utilization and Life Recovery After Eviction

3.3.1 Overall Condition of Residential Members in Settlement 5

In 2000, the eviction started and as a result of a drawing by a representative of each farm household, 28 farm households were transferred to Zhonghuadongcun: 13 households to Tower 25, 12 households to Tower 26, the remaining 3 households to Tower 29. As of June 15, 2001, all of the eviction was completed and the declaration of a vacated village was made. Figure 3.6 represents the ownership and utilization status of farm households who moved to Zhonghua Dongcun Tower 25.

Based on the statistical data of Qunyi Community, the overall conditions of the residential members in settlement 5 as of the end of February 2012 will be mentioned as below. The number of the households increased to 31, the population is

E Room601+401 Move in	Room 602+204+402	Room 603+403	Room 604	A Room 605+505	L Room 606+406
△6 people ◎★ Over 2 years residence Rent 900 yuan	●Rental room/Rental bed	△8 people☆◆◇ 8 months residence Rent 1,000 yuan	全3 people	全3 people	△6 people◎☆◆◇ 1.5 years residence Rent 1,000 yuan
M Room 501 Move in	Room 502	B Room 503	Room 504	A Room 505+605	J Room 506+306 Move in
○4 people	○3 people	○4 people	全3 people	○3 people	△5 people□☆◆◇ Over 2 years residence Rent 1,200 yuan
E Room 401+601 Move in	Room 402+602+204	Room 403+603	F Room 404+405 Move in	F Room 405+404 Move in	L Room 406+606
○4 people	○6 people	△6 people◎☆◆◇ Over 2 years residence Rent 1,000 yuan	△Rental room/Rental bed Planned to be sold by 2013	○3 people	△3 people□☆◆◇ Over 1 year residence Rent 1,200 yuan
D Room 301+201	G Room 302+202 Move in	H Room 303+203 Move in	C Room 304 Move in	K Room 305+205 Move in	J Room 306+506 Move in
○2 people	○2 people	△6 people◎☆◆◇ 9 months residence Rent 1,100 yuan	※ ○2 people △4 people Over 2 years residence Rent 600 yuan	△6 people◎☆◆◇ Over 1 year residence Rent 900 yuan	※○3 people △4 people Rent 600 yuan
D Room 201+301	G Room 202+302 Move in	H Room 203+303 Move in	Room 204+602+402	K Room 205+305 Move in	Room 206
※ ○2 people △6 people☆ Over 1 year residence Rent 900 yuan	△8 people◎★◆◇ 6 months residence Rent 1,200 yuan	△6 people□☆◆◇ Over 2 years residence Rent 1,000 yuan	△Rental room/Rental bed	△8 people◎☆◆◇ 10 months residence Rent 800 yuan	○2 people
Room 101	Room 102	Room 103	Room 104	Room 105	Room 106
Over 5 years residence Rent 1,000 yuan	Over 3 years residence Rent 1,000 yuan	Over 3 years residence Rent 1,000 yuan	Over 3 years residence Rent 1,000 yuan	Over 5 years residence Rent 800 yuan	Over 2 years residence Rent 900 yuan
# Home appliance repair shop △3 people□☆	# Hair salon △3 people◎☆	# General store △2 people□☆, Mahjong parlor	#Hair salon △3 people◎☆	#General store△3 people□☆	#Real estate agency △2 people□☆

Fig. 3.6 Ownership and utilization status of farm households of Lihuang 5, who moved in Zhonghua Dongcun Tower 25. *Explanatory note* ○, owner's personal use; △, rental apartment; ◎, tenant is a colleague; □, tenant is family; ●, new permanent resident; ※, shared living of owner and tenant; 全, sold apartment; ◆, owner left housing complex; ◇, owner resides in Kunshan; #, for shop; ★, good relationship; ☆, normal relationship. *Note 1* Highlighted room number + number means one owner. For example, room 301 + 201 means the same person owns room 301 and 201. *Note 2*: A–M correspond to the number of farm households in Fig. 3.4. *Source* based on a hearing investigation held in March, May, and September, 2012

136, and the number of the labor force is 70 people. Out of these 70 people, 34 are middle-aged (women over 40 years old and men over 50 years old). The number of people with jobs is 67 and the breakdown of the rest of the people without jobs is as follows: one is ailing and two are taking care of their children. In terms of occupations, 25 are engaged in the secondary industry, which accounts for 38% of the total labor force. The number of people who are engaged in the tertiary industry (mainly rental house or shop owner) reached 42, which accounts for 62% of the total. In terms of the average income composition, income from wages is the most and accounts for 56% of the total, followed by income from a rental house, which is 27%. Both account for 83% of the total. Other forms of income are as follows: welfare-related income such as pension or compensation, accounting for 13%, income from business operation 2%, income from investment 1%, and other 1%.

Based on the statistics as of the end of November 2012, all the households of settlement 5 operate a rental house business and the total rental space reached 2385 m^2 and the average size per households reached 77 m^2. As for the annual income from a rental house, three households make less than 50K Yuan while 28 households make over 50K Yuan. The average income from rental houses of all the households in settlement 5 as of 2012 reached 695K Yuan and the average income per household increased to 22,419 Yuan.

The average annual income per person is 23,263 Yuan and the highest is 36,667 Yuan. The lowest is 9600 Yuan. In terms of classification by household, the number of households whose average income per person is below 20K Yuan is 12, 1 household is between 20K to 21.5K Yuan, 5 households are between 21.5K and 25K Yuan, and 13 households are over 25K Yuan.

According to the statistics as of the end of February 2012, 25 households with 86 members reside in the farmers' housing parks and 12 households with 48 members left the farmers' housing parks and transferred to condominiums. In terms of participation in a pension plan, of all the 86 people who remained in the farmers' housing parks, 11 have rural resident basic medical insurance, 57 have urban resident basic medical insurance, and the remaining 7 residents are under 18 years old. Meanwhile, of all the 48 people who moved out from the farmers' housing parks, 2 have rural resident basic medical insurance, 33 have urban resident basic medical insurance, and the remaining 6 are under 18 years old. While the ratio of the people who have rural and urban resident basic medical insurance in the case of people staying in farmers' housing parks is 16 versus 84% that in the case of transferring is 6 versus 94%. The ratio of people with urban insurance is 10% higher, which describes urbanization of former farmers who transferred proceeds more. The reason why the above numbers are different from that of the people who participate in insurance is that university and junior college students are excluded. In case of university or college students, a respite of participation in insurance is granted.

3.3.2 *Pattern of Housing Utilization and Life Recovery*

1. The patterns of housing utilization and life recovery of former farmers can be divided as follows.

    (1) Regarding farm managers or public servants such as farm households 7, 8, and 13, their annual income is by far the highest and they receive 10 times as much income as that of ordinary shop floor workers. As of 2012, the annual income of government office staff is 100K Yuan at division chief level, between 150K and 180K Yuan at deputy township chief level, and 200K Yuan at township chief level. The average annual income per person of these three households is equal to double of the average of settlement 5 and they have already moved from the farmers' housing park out to condominiums or detached houses..

    (2) Household 3 and 5 have succeeded in running a construction or moneylending business after farm retirement. As for these two households, their average annual income per person is more than double the average of settlement 5 and they have managed to move up to the wealthy class from ordinary farmers. These two households also moved out from the farmers' housing park to condominiums (Fig. 3.2g).

    (3) Most farm households belong to this type. Households 2, 4, 6, 10, and 11 are applied to this. Since farm retirement, they have steadily promoted life recovery by combining salary and the income from rent. The average annual income per person of these farm households is close to the average of the settlement 5.

    (4) Since farm household 1 sold a whole house due to a private issue in 2004, they depend only on their salary income for a living and the average annual income per person remains at 20K Yuan.

    (5) Since farm households 9 and 12 have special matters such as the wife's running away from home or the householder's illness, their income is relatively low and life recovery is severe.

2. As clarified in the above types, farm managers and business managers utilize their status and talent after they left the settlement. They succeeded in life recovery the fastest and were able to change from farmers to citizens. In here, the case of a married couple of H (the farm manager and the business manager) who have transferred to a detached house, which is the achieving point of life recovery, will be specifically studied.

Mrs. H was a director of the former Xiongzhuangcun women's association and was transferred to a community office with the merging to the Qunyi Community in 2001, and is currently in charge of business related to women as a deputy director. In 2004 as compensation for the eviction, three 85-square-meter apartments in Zhonghuaxicun Apartment were allocated. After residing there for half a year, she sold one to make a down payment and purchased a condominium of "Sijihuacheng High Quality Apartment" in 2005 and moved in. Her mother-in-law, who is old, uses one of the other apartments and the other one is rented out to migrant workers for 1500 Yuan.

The apartments in the "Sijihuacheng High Quality Apartment" are 125 m^2 and there is a small and a big living room, two toilets, and three other rooms: living conditions have been drastically improved compared to that of the farmers' housing park. At the entrance of Sijihuacheng High Quality Apartment, there are security guards deployed 24 h and the coming and going of people without residential permission are strictly limited. Compared to the farmers' housing park, Sijihuacheng High Quality Apartment is blessed with a quiet environment. There is no situation where people speak in a loud voice and people can feel calm and have a tranquil life. There is secured bicycle parking space for the residents and since security guards go on a patrol, people have no concerns about theft or damage. Besides the original local people in Kunshan, many new comers centering on management level workers live there and its ratio is 6 versus 4. Of all the ten households of the housing tower where Mrs. H resides in, the number of local resident households is 3 and the rest of 7 are new comers. In most cases, three generations live together. The purchasing price in 2005 was 4000 Yuan per square meters and this increased to double, 8000 Yuan in 2012.

In addition, she obtained a store along the main arterial road, Changjianglu, in 2009 for 3 million Yuan at market price and at present it is rented out to a Pingan insurance company for 150K Yuan annually. She purchased 250 m^2 of a detached house near a sports center in the west side of the city center for 3 million Yuan in 2010 (Fig. 3.2h). The housing park, "Zitaibieshu" where many wealthy class people live, is 5 km away from the Sijihuacheng High Quality Apartment and it takes about 20 min by car.

As of 2012, the husband who used to be a tiler works for a communications company as a communication technology manager and his annual income reaches 500K Yuan. Their daughter with university degree, working in a bank, receives 200K Yuan of annual income and Mrs. H's annual income reaches 800K Yuan. The total annual income of the family is nearly 1 million Yuan. Their daughter got married in October 2011 and the married couple moved to "Zitaibieshu" in October 2012. Both the daughter and her husband are an only child and thus, they spend time with husband's family from Monday to Thursday and with daughter's family from Friday to Sunday. That is, they lead a life called "migrant bird type". Mrs. H is planning to transfer to "Zitaibieshu" in 2014 when she reaches mandatory retirement.

The first reason why she lives in the Sijihuacheng High Quality Apartment is the convenience to commute to the community office since it takes only 5 min by car and the second is that she needs to take care of her old mother-in-law. She was born in 1936 and is old and therefore, currently Mr. and Mrs. H sometimes see her after dinner in addition to seeing her four times a week, since the mother lives near them. Moving to Ziyibieshu is a symbolic event for Mrs. H who has changed from a farmer into a rich citizen and they represent one of the success stories of the evicted farmers. On top of the meaning of simplistic, in terms of life restructuring, transferring to Zitaibieshu represents a vast improvement in the quality of life of H's family (e.g., the increase in residential size and the drastic improvement of the living environment) and by living in the same apartment as the daughter's family, Mrs. H can secure support from them. After transferring to Zitaibieshu in 2014, she will utilize the Sijihuacheng High Quality Apartment as a rental apartment.

3. Ten years have passed since the transfer to the farmers' housing park for evicted farmers in settlement 5 as of 2012 and 67% of them, which is 8 out of 12 households, have moved out of the farmers' housing park to condominiums or elsewhere. Table 3.3 shows the time of transfer, destination, and the reasons for moving out for these eight farm households.

The time when they relocated concentrates on two periods due to the fluctuation of the real estate price: around 5 years and 10 years after moving into farmers' housing park. With regards to the destination, apart from two households to second-hand housing, six households relocated to new luxury housing. The housing size is drastically larger than that of farmers' housing park, 85 m^2 and the one 25C obtained is equivalent to 3 times that, 250 m^2. The reason for moving is to improve their living standards triggered by the marriage of successors. As previously mentioned, most transferees are farm managers and business managers.

Meanwhile, in terms of space, 6 households out of 8 relocated to a place, which is within 3 km away from the farmers' housing park and the remaining 2 relocated to a place, which is between 3 and 5 km away. Former farmers have almost no relationships with their neighbors in the condominiums into which they moved. The household of 25M usually returns to the farmers' housing park from noon to five in the evening on weekends. There are three mah-jongg tables in 15 m^2 of bicycle parking of 103 of Tower 25 and 12 old friends enjoy mah-jongg while exchanging information of the residents of settlement 5. The owner of the mah-jongg parlor is also a former resident in settlement 5 and players pay 10 Yuan for use in each time. The person, who wins mah-jongg on the day, pays for the air conditioning fee. Although their residence has been changed from the farmers' housing park to a condominium, the social relations of the transferees and the places where they spend their free time remain the same, which is the farmers' housing park, where the former residents of settlement 5 live. Householder 25M mentions that he feels comfortable as soon as he arrives at the farmers' housing park.

3.4 Observation and Conclusion

(1) Figure 3.7[3] shows a conceptual diagram describing the residential space transition in the suburbs with the social and economic change. Evicted farmers

[3]Figure 3.7 was already included on page. 11 of the above-mentioned thesis. However, due to the essential figure introducing the conclusion of this thesis, it has been included again.

Table 3.3 A list of the destinations of farm households of Lihuang 5, who moved from Farmer's Housing Park

No. of farm household	Time of moving out	Place moving to	Reason	Distance from Farmer's housing park	Size/m^2	Housing type	Purchasing price/10K yuan	Owner	Education of household	Insurance type
25C	2006/10	Luzhong haibieshu	Grand child's marriage/Better living conditions	3 km	250	Newly built luxury housing	105	First son	Junior high graduate	Social insurance
25E	2012/8	Lideguoji	Better living conditions	500 m	160	Existing housing	85	Individual	Junior high graduate	Farm village insurance
25F	2013/3	Chengbei condominium	Son's marriage/Better living conditions	5 km	170	Existing housing	155	Individual	Junior high graduate	Social insurance
25G	2005/8	Xingangwan housing	Son's marriage/Better living conditions	4 km	170	Newly built luxury housing	56	Individual	Junior college graduate	Public worker insurance
25H	2000/8	Yongjingwan housing	Daughter's marriage/Better living conditions	2 km	140	Newly built luxury housing	37	First son	Junior college graduate	Public worker insurance
25 J	2006/10	Xiuyixincun	Better living conditions	3 km	120	Newly built luxury housing	60	Individual	Junior high graduate	Social insurance
25 K	2005/8	Sijihuacheng	Better living conditions	2 km	120	Newly built luxury housing	60	First daughter	High school graduate	Social insurance

(continued)

Table 3.3 (continued)

No. of farm household	Time of moving out	Place moving to	Reason	Distance from Farmer's housing park	Size/m^2	Housing type	Purchasing price/10K yuan	Owner	Education of household	Insurance type
25 M	2013/6	Lidouguangchang	Son's marriage/Better living conditions	3 km	117	Newly built luxury housing	108	Individual/First son	High school graduate	Social insurance

Note The number of farm household corresponds to the ones shown in Figs. 3.4 and 3.6. *Source* based on a hearing investigation held in September 2012

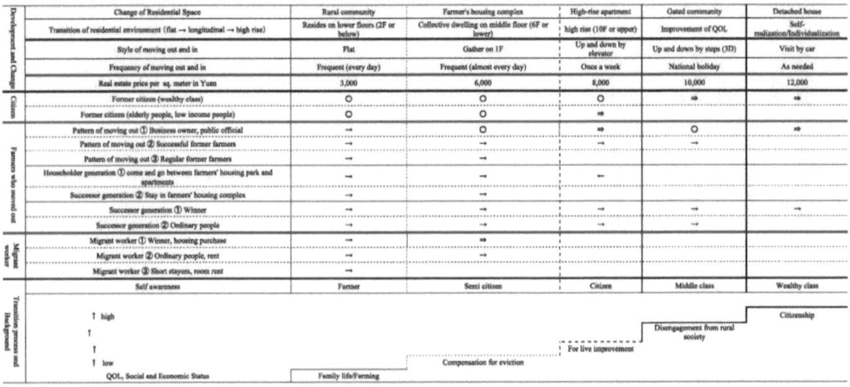

Fig. 3.7 Conceptual diagram of residential space transition of residents in the third space with change of social and economic status. *Explanatory notes* →, move out; ○, no resident; ⇒, Purchase; →←, moving in and out. *Note* Real estate price is based on 2012 prices

entered the third period where two-thirds of farmers reside in high-rise condominium in three dimensions in 2012 through the periods from the rural settlement scattering period (until 2000) to the farmers' housing park dwelling period (from 2001). That is, residing in a condominium means former farmers who were evicted truly turned out to be "citizens" after being "sub citizens" in the farmers' housing park and stepped up to the upward mobility stage. Furthermore, some successors like public servants and business owners were able to enter the new middle class or the wealthy class by starting to reside in detached houses.

(2) In terms of spatial transfer, evicted farmers aim for the fifth stage, the "completion to become citizens", through the fourth stage shown in Fig. 3.6 based on income and life design. It means when they go on to the next stage each time, the quality of life reaches a new stage. Among original citizens and "successors" of indigenous farmers and migrant workers who enjoy the benefit from development and whose social status has surged, the phenomena where some of them go up two stages by jumping over one stage can be seen. On the other hand, the following cases can occasionally occur as well: people who have not been able to move out of the farmers' housing park due to the restriction of the social economy, and failures who had once moved out but had to return to the farmers' housing park due to some problems.

(3) Farmers who have improved their life in the order of clothing, food, and housing have been impacted by the real estate bubble and concerns for living lately, and they not only consider housing as a function of "living" but also use it for moneymaking, improvement of social status, donation of property to successors or generation of grandchildren, and as an investment. The ownership of real estate is now more an indicator describing social status and moneymaking, more so than savings and gold ingots. In the case of their

children's marriage, and as collateral for a loan, the ownership of housing is one of the important requirements.

Ownership and the utilization status of housing is positioned as one of the key indicators to understand the life recovery of the four social classes in suburban regions: (1) the migrant workers, (2) the indigenous farmers, (3) the relatively rich citizen middle class and elites of indigenous farmers, and (4) the new wealthy class such as business managers.

(4) As shown in the lower line of Fig. 3.7, even during the rural community period, housing was often used for living and farm operation. With the eviction, after transferring to the farmers' housing park, housing is partly used as rental housing to earn extra income from the rent for themselves. The purposes of usage have been changed considerably. The income from rent applies to utilities including water and gas bills and the purchase of vegetables: they used to be self-sufficient during the rural community period. It also covers insufficient income from salary.

From the viewpoint of land utilization, until the farmers who depended on land as the cornerstone of life planning (e.g., life after retirement, medication, and accidental insurance), started to reside in housing allocated as compensation after the eviction from the settlements, they were tied to the land including farmland and housing site from generation to generation. Moving out from the farmers' housing park means being independent as citizens by leaving a life style based on land forever and transferring their base of life to purchased housing. During the rural community period, it was common sense that housing must be built by themselves with the support from relatives and friends. However, the farmers' awareness has largely been changed to "housing is sold and bought just like an ordinary product" at present where the market economy is being developed.

(5) This thesis attempted to link the social research method, the elucidation of the process, and the life history of life recovery of indigenous farmers, and the geographical method focusing on the fluctuation of housing space where its trajectory and results are projected, that is, a collaboration method of social research and geography.

If one applies the results of the questionnaires and hearing investigations in addition to the above methods, real opinions on awareness of the current life situation, the choices of location for a future life, and the life design in each of the four classes, which are difficult to confirm through only statistical data, can be understood. Accurately presenting the regional restructure and development of sustainable suburban regions from the standpoint of residents based on these is a future issue.

In particular, based on the examination focusing on the change of community members with "change of ownership", urban development needs to be prepared by always observing the movement of both the "community based population," composed of senior members of indigenous farmers and children, and the "company man community" centering on successors and new comers.

(6) It is required to simultaneously apply a micro-viewpoint based on regions and a macro-viewpoint to see a wider range and to mutually understand both viewpoints. In the future although focusing on the Yangtze River Delta in terms of a region, it is important to apply and verify this research method to the whole of China and moreover, emerging countries such as Indonesia, India, and Thai. Although the situation in each country is different, it is possible to implement comparative research focusing on projection (change) of suburban regions by urbanization expansion. Likewise, comparative case studies of these regions under large-scale development and China will be promoted and points of difference and commonality will be elicited.
It is required to adapt a perspective of regional and social restructuring and comprehensively examine the positioning of the regional structure and the clarification of actual conditions to the presentation of development policy for suburban regions where the actual conditions and issues of population, society, economics, culture and religions which emerging countries have are projected. In addition, clarification of the actual conditions of suburban regions enables us to understand the characteristics of the social, economic, regional, and cultural transformation.

Acknowledgements This research was funded by a grant from the Japan Society for the Promotion of Science (JSPS), Science Fund Basic Research (S) "Towards Sustainable Land Use in Asia" (Topic # 21222003), Research Group Representative Yukio Himiyama (Institute of Geography, Hokkaido University of Education).

References

Li T (2008) The Chengzhongcun land market in China: boon or bane?—A perspective on property rights. Int J Urban Reg Res 2008(2):282–304
Lin X, Wang M (2011) The research on the transition of the suburban village in the process of the rapid urbanization—a case study of Shuangqiao Village, Ningbo. E J China Urban Stud 6 (1):31–36
Wang Y, Yanglin W, Wu J (2009) Urbanization and informal development in China: urban village in Shenzhen. Int J Urban Reg Res 3(4):957–973
Wu F (2009) Land development, inequality and urban villages in China. Int J Urban Reg Res 33 (4):885–889
Zengmin JI (2010) Regional Reorganization of outlying rural village in China-through the case of adjacent region of Kunshan development zone in Jiangsu, Ashi shobo, 355 pp
Zengmin JI (2011) (The 3rd Space) formed surrounding urban areas in China. Perspective on East Asia, vol 22, December issuance, The International Centre for the Study of East Asian Development, pp 6–17
Zengmin JI (2012) Perspectives and methods for researching Chinese urban fringe areas. SLUAS Science Report 2012, pp 19–36
Zengmin JI (2013) Housing utilization and life recovery of evicted farmers at the urban fringe. In: Himiyama Y (ed) Towards sustainable land use in Asia Science Report 2013, pp 5–24

Chapter 4
Economic Development and Land Use Changes in an Inland Area of China: A Case Study of Gansu Province

Haruhiro Doi and Yanwei Chai

Abstract The purpose of this report is to clarify land use changes and driving factors in Gansu Province, which is located in an inland area of modern China. The study is based primarily on field observations. Gansu Province has lagged behind the coastal area of China in economic development due to numerous constraints, such as severe natural conditions and a remote location. Moreover, the scarce water resources of the province are being utilized near capacity, with the amount of water consumed at 50–60% of the total water resources available in the province. However, economic development is now progressing more rapidly from urbanization, industrialization, and new economic activities such as renewable energy production. This rapid development, driven by the National Western Development policy, is affecting land use in the region. Cropping patterns have become more diversified, and croplands are advancing into the limited area of arable land. Higher productivity land use expanded following cultivation of fruit trees in the Loess Plateau. Thus, the province has a varied land use for economic gain based on local conditions.

Keywords Inland area · Semiarid condition · Regional development · Land use

4.1 Introduction

Asia has been experiencing rapid population increase and economic development, surpassing Europe and North America in global economic progress. South Korea and Taiwan in East Asia, along with the countries in East Asia, gained economic power in the 1990s, following the prosperity of Japan between the 1970s and

H. Doi (✉)
Faculty of Education, Oita University, Oita, Japan
e-mail: hdoi@oita-u.ac.jp

Y. Chai
Institute of Urban and Environmental Sciences, Peking University, Beijing, China
e-mail: chyw@pku.edu.cn

Y. Himiyama (ed.), *Exploring Sustainable Land Use in Monsoon Asia*,
Springer Geography, https://doi.org/10.1007/978-981-10-5927-8_4

1980s. These countries all experienced expansion of the residential sector corresponding to population increases, and of the urban sector for maintenance of secondary and tertiary industry, leading to a reduction in farmland and forestland. China has been experiencing rapid economic growth since the introduction of the Reform and Opening Up policy in 1978, which introduced a unique capitalist economic system fused with the socialist political system. China, which has the largest population in the world, is experiencing rapid industrialization and urbanization, and consequently urban to rural transformation (Long 2014).

In consideration of Chinese land use changes, the economic development of inland areas influenced by a regional development policy of the central government is essential, although the economically developed coastal area attracts most of the attention. Spatial expansion of urban land use in the coastal zone is affecting the inland areas, and inland urbanization is progressing not only in major cities, such as the provincial capitals, but also in minor cities (Doi 2014). The national government of China has been encouraging economic growth in the inland area in recent years, though the arid and semiarid climate and remote location are disadvantages for inland regional development. Gansu Province is in a central location, and is typical of the western inland area of China (Fig. 4.1).

The authors intend to understand land use changes in Gansu Province and the driving factors affecting these changes using on-site observations as well as statistics, maps, and land cover data from Global Map.

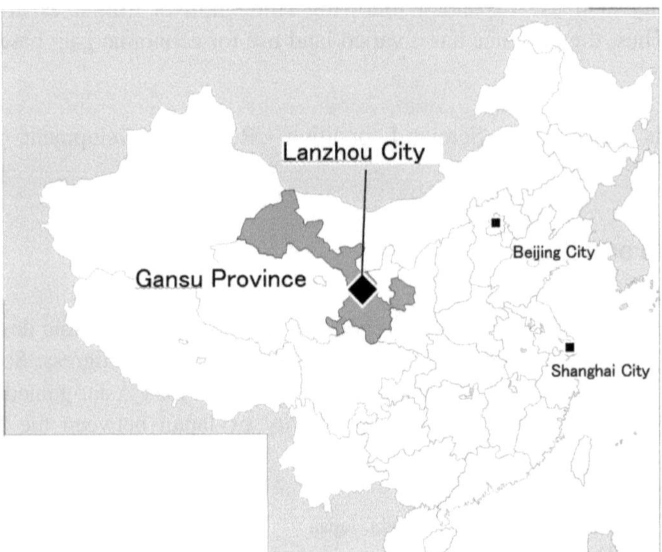

Fig. 4.1 Location of Gansu Province

4.2 Land Use Trends in China

A long history of human activity in China has encouraged altering natural conditions to enable human habitation, while regional trends in agricultural and forestry land use correspond with the physical conditions, such as climatic environment (Wu and Guo 1994). Research using historical satellite imagery combined with field surveys has clarified land use changes and driving factors in China as a whole in recent years (Wang et al. 2012b; Zhang et al. 2014). There have been significant land use changes between forests, farmlands, and urban areas since the 1980s in China. Land use policies of a given period that focused on economic development were important in developing the land use pattern (Wang et al. 2012b).

Lin and Ho (2003) pointed out that not only building lot expansion but also conversion of grassland to farmland occurred in China during the second half of the twentieth century because of urban economic development. Much of the newly developed farmland is located in remote regions of the country, which have high environmental vulnerability, and cropland development was not able to compensate for loss of fertile farmland converted to urban areas. Urban development of quality arable land serves as a wonder for the food supply in China, at the same time it caused unused land and a low quality of arable land (Wang et al. 2012b).

China has been introducing a capitalist economic system on the preexisting socialist political system with the Reform and Opening Up policy in 1978, and has experienced remarkable economic progress. Chinese land use has been subject to the influence of various management techniques, including a ground lease system, land use rights, and land taxation, since 1978 (Long 2014). National development strategy has also been crucial, for instance policies for restoring agricultural land to forest and the Western Development Project. Wang et al. (2012a) also reasoned that Chinese land use is the result of an entangling of national and regional industrial development policies, urbanization policy, and land conservation plans.

Chen et al. (2014) pointed out that China presupposed a land intensification stage equivalent to a transition from premodern to modern society as defined by high rural to urban population mobility and the resultant land use changes, as is seen in many other developing countries. China will continue to experience a demographic shift from rural to urban areas and high population pressure. Hence, the most highlighted issue of current Chinese land use is how to balance the land area requirements for economic development, food security, and conservation (Chen et al. 2014).

4.3 Gansu Province Geography and Land Use

4.3.1 Gansu Province Physical and Political Geography

Gansu Province occupies 455,000 km^2 in central China. The capital of the province is Lanzhou City, located in the eastern part of the region about 1200 km from Beijing. Part of Gansu Province occupies the Tibet Plateau. The province is divided by mountain ranges of about 4000–5000 m elevations (Fig. 4.2). Wushaolin Pass physically separates the region at an altitude of about 3000 m (Fig. 4.3). The western part of the province contains the Huxi Corridor—the western area of the Huanghu (Yellow) River—and the eastern part consists mainly of the Loess Plateau. The land of the province is at mostly high elevations and includes mountainous area. Elevations between 1000 and 2000 m account for about 50% of the province, and elevations above 2000 m account for about 40% of the province. Climatic conditions are arid and semiarid, with moderate temperatures in the western part. The annual precipitation in Dunhuang, located between the Gobi and Taklamakan Deserts, is only 3.3 mm. The eastern area is also dry even though its yearly rainfall is large. For instance, the annual precipitation in Lanzhou is about 317.0 mm.

The total population of the province increased after the establishment of New China. Urban population has been growing rapidly since 1980, but rural populations have decreased steadily since peaking in 1997 (Fig. 4.4). The population has been concentrating increasingly in the cities. Traffic conditions in the province have recently improved, with an expressway now connecting local cities to Lanzhou City.

The regional development policy of the central government has had important effects on remote areas. The petrochemical and other heavy industries have been established as important industries for the city in the early stages of the new China policy. Many factories have moved from coastal cities to inland cities as a result of the Three Line Construction national policy. Therefore, Lanzhou has developed a reputation as an immigrant city. The central government is currently encouraging economic growth in the western inland area, including Gansu Province, under the Western Area Development Program. The goal is for the inland area to catch up to the coastal area in economic status.

4.3.2 Land Use Distribution in Gansu

Figure 4.5 shows the land cover of Gansu Province created by using data from Global Map.[1] Forest, which occupies 5.4% of the province, is limited to the

[1]The Global Map data are from version 2 using satellite imagery from 2008. One point of land cover as shown in Fig. 4.5 is 15 s, corresponding to approximately 450 m around.

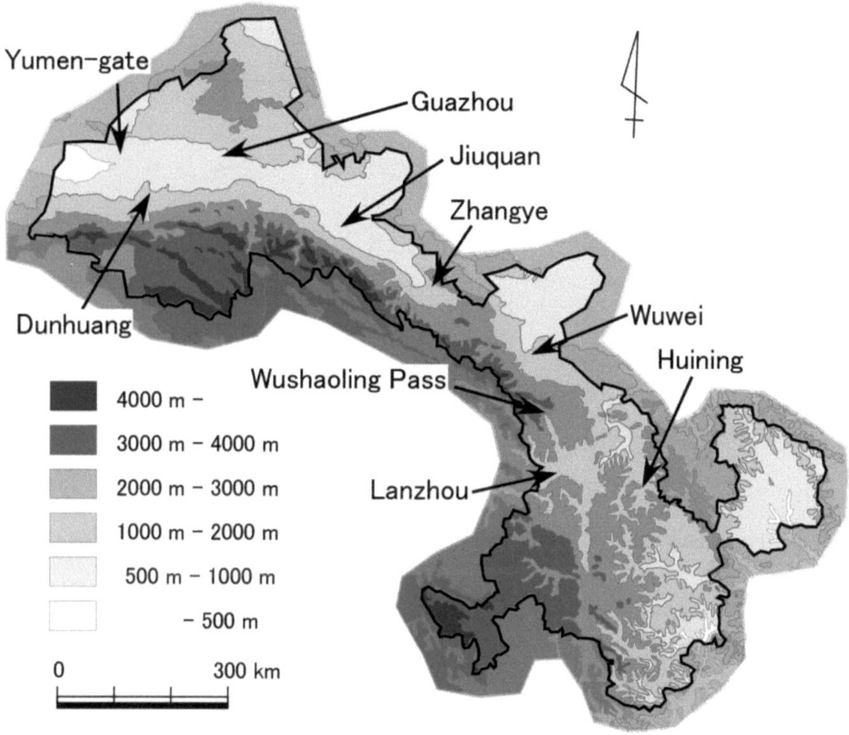

Fig. 4.2 Elevation of Gansu Province. *Source* Editorial Board of Geographic Atlas of China (2009)

Fig. 4.3 Wushaoling dividing mountains (September 29, 2012)

southeastern part of Gansu that neighbors Sichuan Province. Forests in this area supply timber, and much of the arable land is occupied by extensive dry fields according to Wu and Guo (1994), who considered the geographical features of land

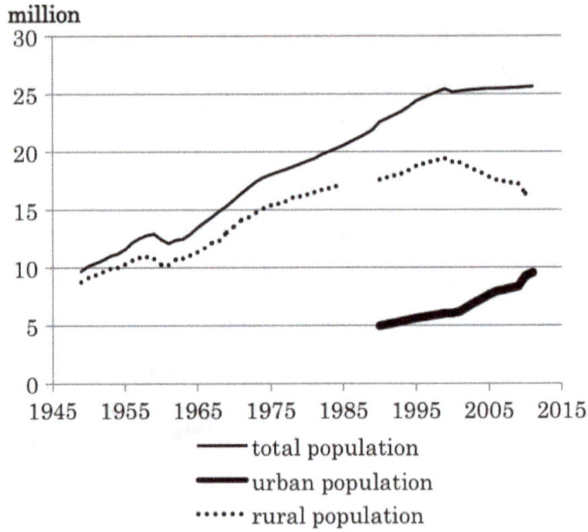

Fig. 4.4 Population trend of Gansu Province. *Source* Gansu Development Yearbook-General Survey (2010)

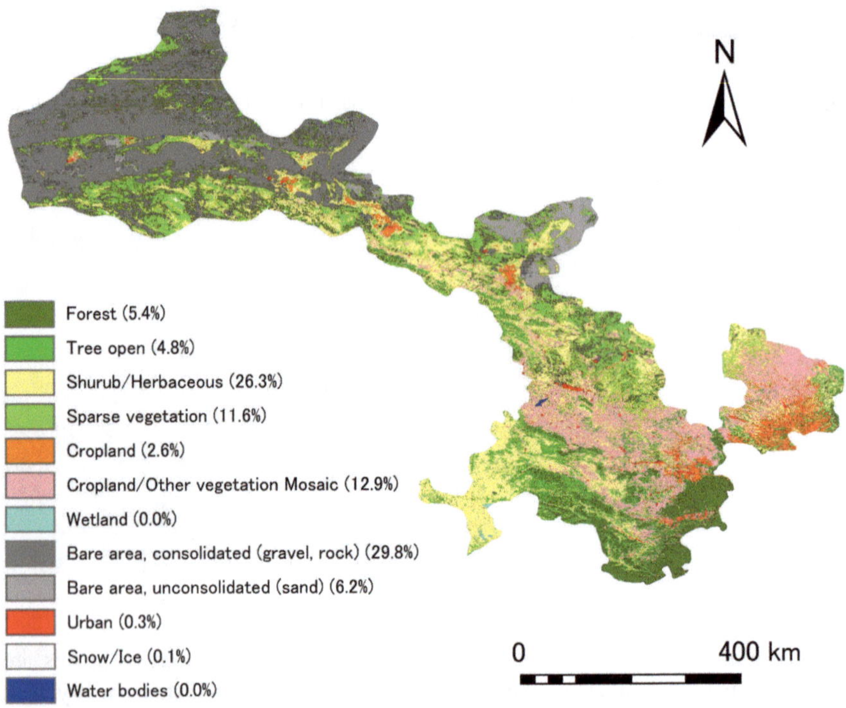

Fig. 4.5 Land cover of Gansu Province. *Source* Global Map ver.2

Fig. 4.6 Loess Plateau, Huining County (October 5, 2012)

use for China as a whole.[2] Major land cover categories in the province are Bare Area, Shrubs, and Sparse Vegetation. A large portion of the western part of the province past Wushaolin Pass is occupied by Bare Area. The northeastern area is dominated by sand stands, and the northwestern area is dominated by gravel and rock stands. The climate of these areas is desert-like, and the area is suitable for neither farming nor human habitation. Cropland occupies only 2.6% of the province. Except for some small farms that depended on rainfall, cropland of the region was limited to the river basin. Cropland has expanded in the region during the second half of the twentieth century due to the development of irrigation facilities that draw water from oases or rivers. Wheat, cereals, oilseed, beets, and other crops are cultivated here. Grassland of the area has been damaged by drought and soil salinization, and farming on the grassland has retreated in recent years. Cao et al. (2013) indicate that personalization of pasture rights of by the central government was another cause of grassland degradation.

Land use in the Loess Plateau, located in the southeastern part of the province, is a mosaic of Cropland and Other Vegetation. The Loess Plateau is deficient in irrigation facilities, and therefore cannot support widespread agriculture. Cropland is distributed over the Loess Plateau and river basin, mainly of wheat and corn. The soil in this region is easily eroded by rainfall, but the plateau can maintain a terraced field well (Fig. 4.6). Urban land only occupies 0.3% of the whole, and the principal local cities, such as Lanzhou and Zhangye, are mentioned in Fig. 4.5.

The following chapter confirms the land use distribution in Gansu Province based on a field survey from 2011 to 2013.

[2]Wu and Guo (1994) considered the land use distribution for China as a whole in the 1990s by dividing the entire country into four districts and 17 subdistricts. According to this geographic division, Gansu Province is a boundary area in three regions: the Northern, Southern, and Northwestern districts (Fig. 4.33).

4.4 Land Use Changes in the Western Area of the Province

4.4.1 Agricultural Land Uses and Recent Changes

The Western area of the province receives scanty annual rainfall, and the average temperature is rather high, so the physical conditions of the area are severely limiting for agriculture, thus limiting the capacity for human population. However, this area is characterized by a rich history, due to its location on the Silk Road, which linked the East to the West in ancient times. This area was also the boundary between the Han Tribe and Nomadic pastoral peoples. There are many historical monuments across the western area of the province. The Great Wall of China was constructed to protect the Han ethnic group from invading people groups. Gate Number 1 was built by the Ming Dynasty in the barren Gobi and specified as a national monument for cultural protection (Fig. 4.7). The gate attracts many tourists from around the world.

Almost all of the land in the western part of Gansu is occupied by the Gobi Desert. Agricultural land is therefore climatically limited in the flatlands, but small urban and agricultural developments have historically surrounded oases. Even so, there are plenty of cornfields in the flatlands (Fig. 4.8). The soybean could be grown in places that seem to be bare land. In some places, we can see harvested corn being dried in the sun (Fig. 4.9). Figure 4.10 shows the cultivating area of corn for Gansu in 2007. Corn is widely grown in the province, and its cultivating area has been steadily increasing due to the Reform and Opening Up policy (Fig. 4.11). The total corn cultivation area in the western area is much less than in the eastern area, but corn is an important crop in this semiarid region.

Arid climate conditions dominate the region, with several deserts, including the Gobi, located in Gansu. Potato and corn are cultivated using precipitation even in the Gobi. However, planting areas for commercial crops such as vegetables, fruits, and cotton are grown in places where water conditions are appropriate, such as at the foot of a mountain slope or in oases. Figure 4.12 displays the ratio of irrigation

Fig. 4.7 Great Wall No. 1 Gate of Ming Dynasty (September 29, 2012)

Fig. 4.8 Cornfield near Wuwei City (September 29, 2012)

Fig. 4.9 Place corn in the sun near Zhangye (September 30, 2012)

in cultivated areas, illustrating how irrigation maintains agricultural land in the western area of the province.

A farm village is located in the southern part of the province, about 50 km from Zhangye City (Fig. 4.13). In the figure, you can see the Jilianshan Mountain Range (altitude about 4000 m) with some early snow cover. Underground water from the mountain range can be used at the foot of a mountain slope. There are many old greenhouses producing vegetables and grapes in this village.

Commercial crops, such as various fruits, are located along the road and the surrounding areas of Guazhou and Dunhuang City, both located in the westernmost area of the province (Fig. 4.14). Figure 4.15 shows a time series for several crops in the area. Cotton has been increasing rapidly in recent years, as well as harvested grape. Cultivation of other fruits is also carried out in this area. Stalls selling melons can be found along the desert road to Guazhou and Dunhuang (Fig. 4.16). Grazing of animals is another important activity in this region. While greenery is sparse, meadows can be found at the foot of a mountain slope where sheep grazing is common.

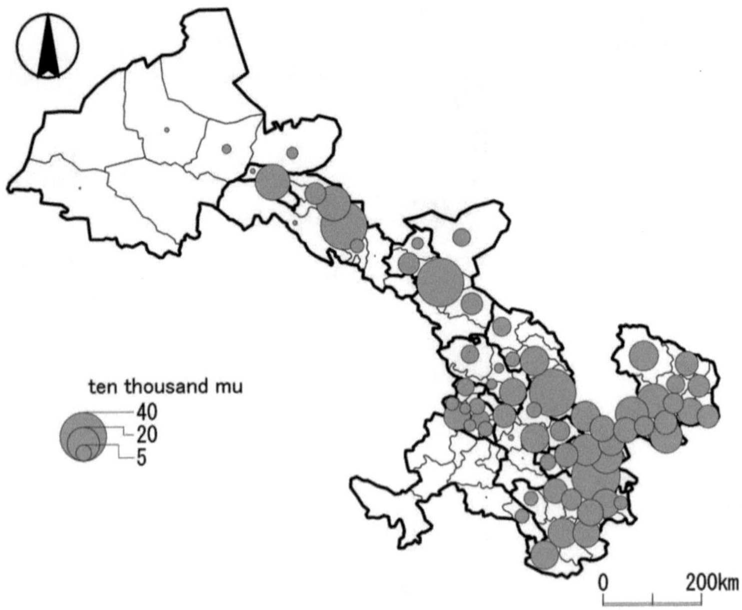

Fig. 4.10 Cultivating area of corn in 2007. *Source* Gansu Rural Year Book (2008)

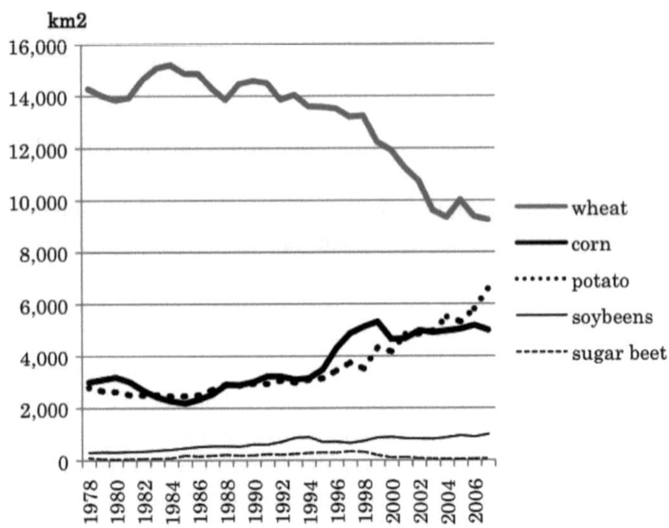

Fig. 4.11 Cultivating area trend of some crops. *Source* Gansu Rural Year Book (2008)

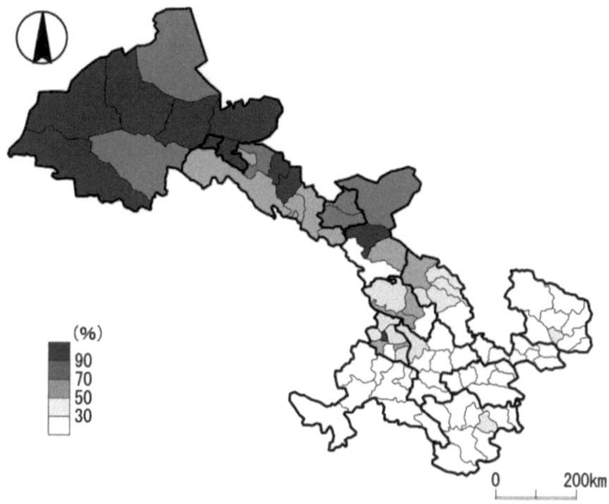

Fig. 4.12 Ratio of irrigation in the cultivated area (2011). *Source* Gansu Development Yearbook-General Survey (2012)

Fig. 4.13 Rural village on the foot of Jilianshan mountain range (September 30, 2012)

Fig. 4.14 Cotton field, Dunhuang City (October 1, 2012)

4.4.2 Cities and Energy Producing Industry

Historical cities along the Silk Road remain important in the area even today. The cities of this area are located at the crossings of the Silk Road and south-to-north rivers supplied by the mountain range. These cities have experienced development and decline with high vulnerability to the natural environment combined with a

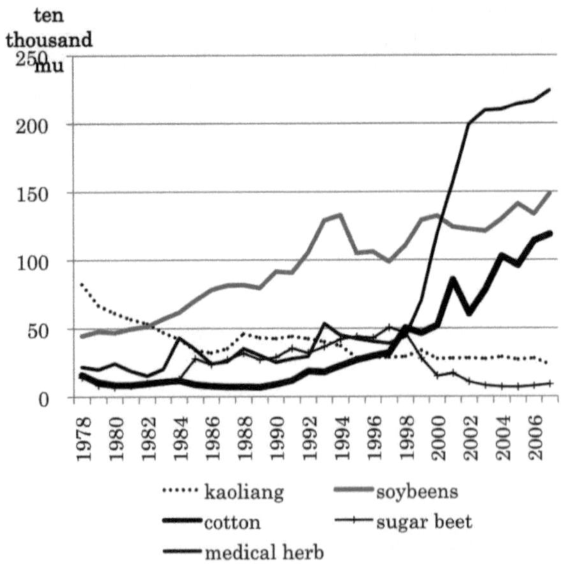

Fig. 4.15 Time serial trend of several crops. *Source* Gansu Rural Year Book (2008)

Fig. 4.16 Stall selling melon, Guazhou County (October 1, 2012)

complicated ethnic mixture and changes in the political system (Xie et al. 2007). The Jiayuguan-Gate in Jiayuguan City, built during the Ming Dynasty, is famous worldwide and attracts many tourists (Fig. 4.17).

Figure 4.18 shows the GDP of secondary industries of Gansu in Chinese Yuan for the year 2011. Several cities, including Jiuquan and Jiayuguan, play a significant role in secondary industries of the region. The urban areas of Jiuquan and Jiayuguan are surrounded by the Gobi, and are famous for both industry and historical sightseeing. The famous Jiuquan Steel Company is located in Jiuquan City and has many buildings (about 20 stories tall) for their residents (Fig. 4.19). The atmosphere here is polluted, especially compared to the beautiful, clear air in other places of the province (except for in Lanzhou City). Jiayuguan is also famous for its manufacturing industry, and new projects in energy and space exploration.

Fig. 4.17 Jiayuguan-gate (October 3, 2012)

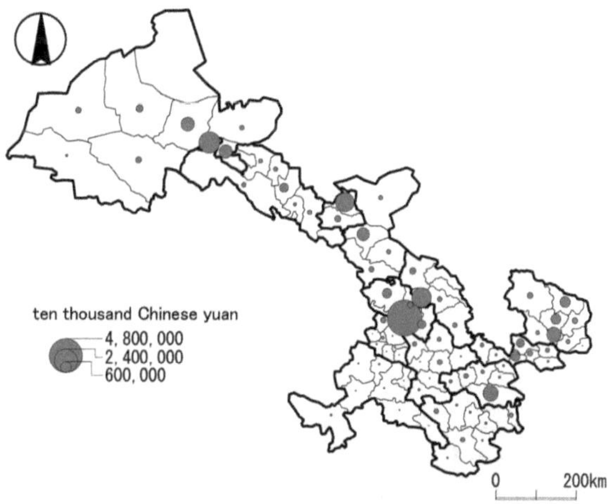

Fig. 4.18 GDP of the secondary industry (2011). *Source* Gansu Development Yearbook-General Survey (2012)

Fig. 4.19 Jiuquan City (October 1, 2012)

Fig. 4.20 Newly industrialized area of Zhangye City (September 30, 2012)

Fig. 4.21 Oasis in Zhangye City (September 30, 2012)

There are numerous manufacturing plants located here, e.g., "East Car Jiuquan New Energy Base" and "China Aerospace."

Zhangye City is also famous historically as an oasis city along the Silk Road. There are many buildings of about ten stories or so, though there are few high-rise buildings. Various residential houses are five–six stories in the built-up area. The northwest urban area of Zhangye is a newly developed industrial zone (Fig. 4.20). The surrounding area of the swamp was preserved as a park for the citizens, and the marsh has been extended to the north (Fig. 4.21). Villa style houses are constructed along the swamp.

The western area of Gansu Province has a high potential for a wind power and solar energy generation, but is isolated from the wind turbine production district. Renewable energy is a primary method for developing economic sustainability in the western area of China (Cyranoski 2009; Wu and Xu 2013). The new industries of solar and wind power generation are actively advancing near the cities in the western area of the province. Approximately 20 wind power generators are located at the foot of mountain slopes (Fig. 4.22). To the west of Dunhuang lies the Gobi, which holds a strong historical legacy in and around built-up areas (Fig. 4.23). However, the construction of new industrial estates is also advanced around here. Many sunlight panels are set up next to industrial estates, with one corner of the site becoming the new energy-supplying district (Fig. 4.24). Several steel towers are also under construction in Guazhou and Dunhuang (Fig. 4.25).

Fig. 4.22 Solar-power plant near Guazhou County (October 1, 2012)

Fig. 4.23 Mugaoqu, Dunhuang City (October 2, 2012)

Fig. 4.24 Solar power plant, western suburb of Dunhuang (October 2, 2012)

Fig. 4.25 Steel tower under construction (October 1, 2012)

4.5 Land Use Changes on the Loess Plateau

The Loess Plateau is a famous topographical landscape in the eastern area of the province. It extends to the middle and upper river basin of the Huanghu River and occupies over 400,000 km^2. The soil of the Loess is mostly silty sand, and is very easily eroded. There are many gullies from deep erosion, and it is not suitable for agricultural use. There is a long history of agriculture in this area, managed because of the physical environment providing little rainfall and sufficient temperature and sunlight. In the Loess Plateau, the Gansu government is promoting a resident measure against poverty by various methods. For instance, a lower dependence on cereal production is being pursued by increasing market crop cultivation and livestock production. Another involves rearrangement of farm villages to make fertile land available for cultivation (Sharna et al. 2008).

There is very little vegetation on the slopes and top of the Loess Plateau in Baiing City of Huining County. However, the gentle slope of the plateau is used extensively for terraced fields, and potato and corn are chiefly growing in the arable land. Recently, there was a slope rich in greenery due to the tree-planting program (Fig. 4.26). Many fruit trees, for instance the jujube tree, were adopted for the tree-planting program (Fig. 4.27). Afforestation has been executed widely at the national level since the 1990s as a means of nature conservation and protection from natural hazards. The Program for Conversion of Cropland to Forest is a public afforestation project that began in 1999. The eastern area of the Loess Plateau has experienced the most afforestation in Gansu (Fig. 4.28). Production of medicinal herbs has also been growing rapidly. The eastern part of the province is a major producing area of medicinal herbs for Chinese herbal medicine in Lanzhou City.

We can easily observe afforestation sites in the area surrounding Lanzhou City. The trees are placed on the shallow terrace along a contour line of the slope. The expressway lying on the west edge of the Loess Plateau has many small fluctuations (Fig. 4.29). The slopes neighboring mountains have thin green meadows near the bottom. Corn and other crops are sometimes planted on the flat land. In the field, there are many places without any vegetation, and the soil becomes barren.

Fig. 4.26 Loess Plateau, Huining County (October 5, 2012)

Fig. 4.27 Tree planting, Huining County (October 5, 2012)

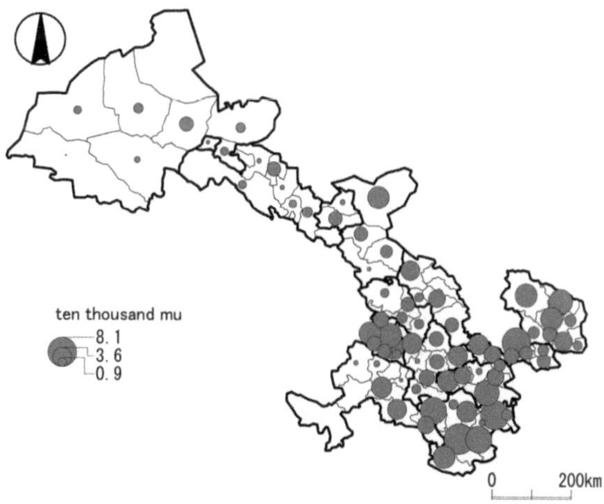

Fig. 4.28 Afforestation area (2007). *Source* Gansu Rural Year Book (2008)

Fig. 4.29 Mountain slope on the northern suburb of Lanzhou City (September 29, 2012)

Potatoes, etc., can be cultivated here. The tree-planting program has recently advanced, along with terraced field extensions. Nonetheless, vegetation is scarce in the Loess Plateau. However, forest area increases were also affected by political control of deforestation in many provinces. The per capita GDP had a significant influence on the changes in forested areas of China (Zhang et al. 2006). The inland area, including Gansu Province, has not yet experienced a drastic increase in forest because of a primary focus on economic development. There is also a shortage of proper silviculture after afforestation. Although these afforestation activities are performed in various places, it is not clear whether they have yet accomplished the goal of a full-grown forest.

4.6 Land Use Changes in Urban Areas

4.6.1 Urban Development of Lanzhou City

Lanzhou is the capital of Gansu Province, and it is the biggest city located in the westernmost in the inland area of China. This is another famous historical city along the Silk Road. The urban area of Lanzhou extended to include the valley plain of the Huang-Ho River in the latter half of the twentieth century (Fig. 4.30). The population of Lanzhou City has been increasing consistently since the 1970s, and exceeded a registered population of 3 million in 2002 (Fig. 4.31).

Urban development of the city has been affected by national policies, such as the Three Line Construction in the 1950s and the Western Area Development Plan in recent years. The city has accepted some major manufacturing companies and their employees from other regions, mainly the Coastal Region. The population of the city doubled in about 10 years, and the built-up area has come to include almost the entire river valley basin of the Huanghu River.

Economic development of the city has further depended on the tertiary industry as a capital of the province. However, it is important to note that secondary industries triggered the economic development of the city, especially the manufacturing sector. Figure 4.32 shows that heavy industry accounts for nearly the

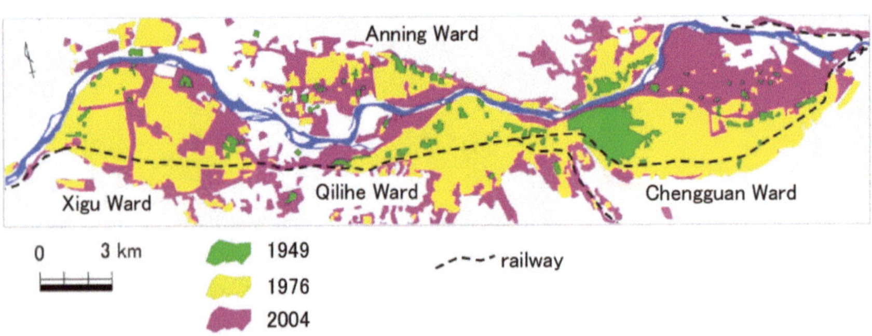

Fig. 4.30 Expansion of built-up area of Lanzhou City. *Source* Maps of each year

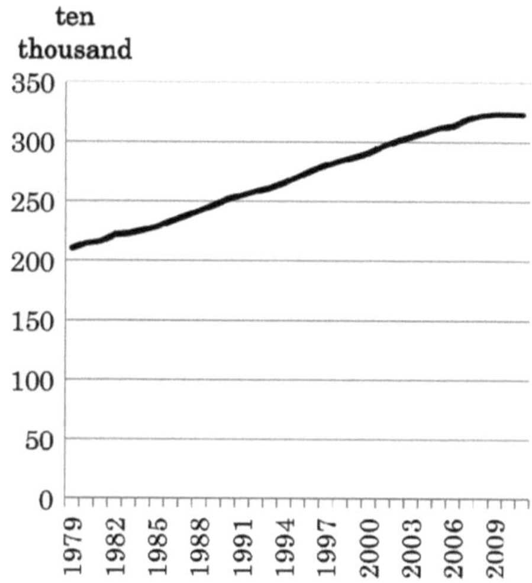

Fig. 4.31 Population increase of Lanzhou City. *Source* Lanzhou Statistical Yearbook (2012)

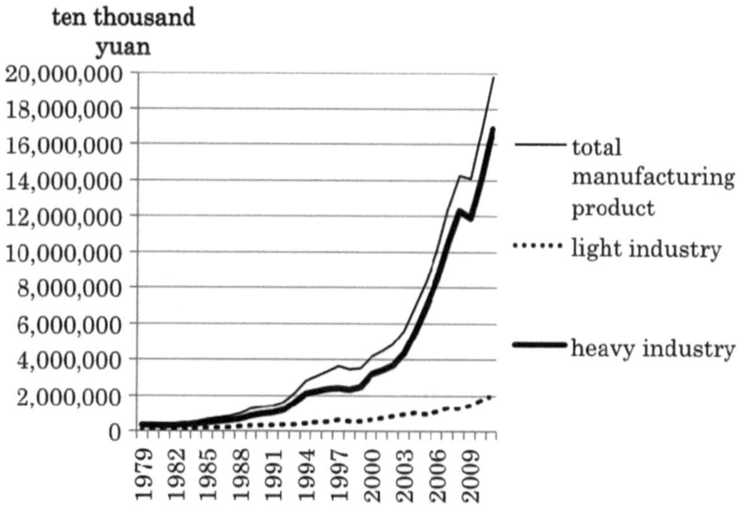

Fig. 4.32 Manufacturing product of Lanzhou City. *Source* Lanzhou Statistical Yearbook (2012)

entire production capacity. Lanzhou has been fostered as a core city for regional development of the inland area by national policy. The national working unit managed by the central government played a significant part in the jobs market, but its role has drastically decreased since the latter half of the 1990s.

Land creation is a peculiar feature of urban development of this area. Many areas in the eastern Gansu Province are covered in Loess Plateau. Since level land is scarce, land reclamation offers a development site in the urban facility by removing a peak to fill a valley. According to Li et al. (2014), 700 summits were removed, and 250 km^2 of flatland were developed in Lanzhou City. Figure 4.32 shows a large-scale cemetery garden under development.

Fig. 4.33 Large-scale cemetery garden (October 26, 2013)

Fig. 4.34 Downtown of Lanzhou City (October 6, 2012)

4.6.2 New City Development

There were many factories and residential houses developed after New China on the ground of the river valley basin in Lanzhou (Chai 1991). Because these factories and residential house have aged, it is necessary to redevelop or maintain the facilities. After the 1990s, the construction of new residential homes was advanced with an aim of redevelopment in older urban areas. There are many high-rise residential buildings in the downtown area (Fig. 4.34). In Lanzhou, little construction of independent villa houses has occurred, although some of the commercial housing has been constructed to target upper income citizens and outside investors.

A modern urban development is under construction in the area, about one hour away from the city by expressway. The Lanzhou New District is a large development project occupying 400 km^2 (Fig. 4.35). This development district is adjacent to the Lanzhou new airport that opened a few years ago. We can easily identify that the planning area mainly occupies the area designated for agricultural land. Therefore, a substantial amount of agricultural land will be converted to urban land use.

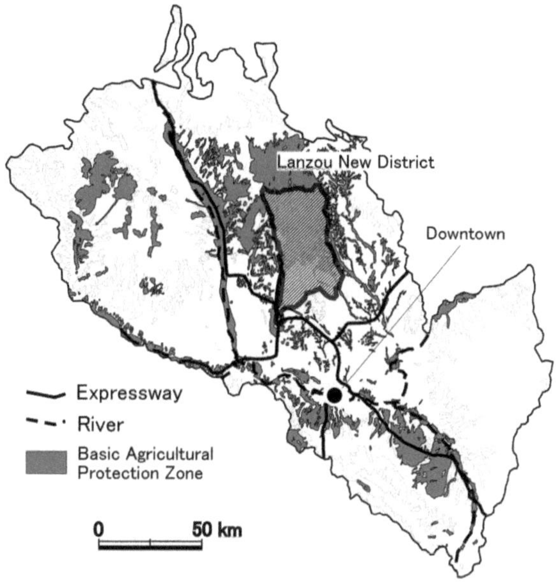

Fig. 4.35 Location of Lanzhou New District

Fig. 4.36 Automobile plant in Lanzhou New District. (November 28, 2011)

The Lanzhou Electric Equipment Company, located in the aging built-up area, is now planning to move its plant to the Lanzhou New District. Transfer plans are available to other major manufacturing companies. The Lanzhou Electric Equipment Company has already acquired land in the new district. The company will start to manufacture wind power generation equipment in cooperation with a German enterprise in the Lanzhou New District. The existing production line of the factory is also scheduled to move to the new district. Figure 4.36 shows an automotive plant of the South Korean company. Compared with an eastern Chinese city, the inland cities rarely depend on international markets for economic development (Wang et al. 2011). A lot of investment and economic development has occurred in recent years by introducing capital mechanisms both in and out of the countryside in Gansu Province.

4.7 Factors Affecting Land Use Changes in Gansu Province

The major factors driving land use changes in Gansu Province are shown in Fig. 4.37. There are several types of land use change in the province based on field research and other materials, and they can be summarized into roughly four types: urbanization, industrialization, commercialization of crops, and afforestation. Many residential houses and business buildings, mainly supplied by the private sector, are expanding the urban areas, causing urbanization. In Gansu Province, industrialization goes hand in hand with urbanization, with national regional development policy affecting industrialization. Solar and wind power plants have been recently established in numerous places as a way to support further urbanization and industrialization. The cultivated areas of grain crops such as wheat and corn have been decreasing, while commercial crops such as cotton, vegetables, and fruits have been increasing. Native vegetation is scarce due to climatic and historical conditions. Tree-planting projects have been undertaken, mainly in the eastern part of the province, and with the goal of producing healthy afforested areas.

The physical character of Gansu Province provides a basic background for producing these land use changes. Gansu Province is located in the remote inland area of China and lags behind the eastern coastal area in economic development.

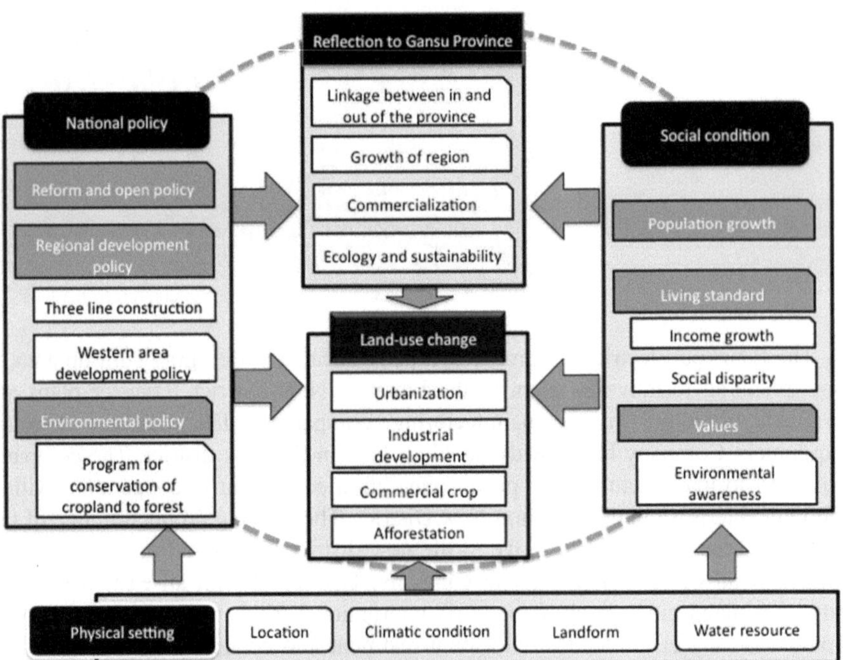

Fig. 4.37 Driving forces affecting land use changes of Gansu Province

Climatic and geomorphologic conditions peculiar to the inland area provide constraints on agriculture and afforestation activities. Insufficient water resources in the province are a critical factor for overall human activity. National policy and social conditions of the area have affected land use changes. A drastic change in the national economic policy occurred with the introduction of the Reform and Opening Up policy of 1978, and further development policies regarding the inland area have affected the socioeconomic activities of Gansu Province. On the other hand, environmental policy has also been recently implemented in response to calls for environmental protection. Improvements in living standards that accompany population increases and economic development are driving economic activity of the province. Environmental pollution has appeared as a negative effect from economic growth, and has fostered consciousness of environmental protection in the local people, pressing them to change their values. These internal and external conditions in the province induced economic activity of the administration, companies, and people of the province, leading to a change of consciousness, and driving land use change in the province.

There are some constraining factors in the future of land use in Gansu Province, located in an inland area of China where economic development and population increases continue. Piao et al. (2010) pointed out that the Chinese economy has a vulnerability to climate change, even though China is supporting 22% of world population with only 7% of the global cultivated area. An intensification of human activities causes changes in the natural environment. The streamflow of the Huanghu River has decreased greatly since 1960 (Piao et al. 2010). In the inland area of China, where there is little precipitation, a change in streamflow has a significant influence on local industry and social life. Although climate change is also a cause of the flow reduction in the upper stream region of the Huanghu River, Cuo et al. (2013) indicated that the intensification of human activities before 1960 had a substantial influence on the streamflow near Lanzhou City. Ta et al. (2006) showed that many human activities, including land cultivation and economic construction, were major causes of a mid-twentieth century sandstorm in the northwestern area of China. The sandstorm days in spring have a high correlation with annual cultivation area in the Huxi Corridor. Furthermore, Li et al. (2014) reported that land creation for developing urban building lots not only destroys forests and farmland, exposing wild animals and plants to danger, but also causes air and water pollution, soil erosion, and land subsidence.

These reports have suggested that the intensification of human activity has been the driving force of the economic development of the area and has given a large load to the local natural environment of Gansu since the 1980s. Zhou et al. (2012) pointed out that the agricultural activity by humans in the Bronze Age caused a decrease in soil fertility and an increase in soil salinization. Their activity further brought about bee colony abandonment, as indicated by a pollen analysis of the stratum of the Bronze Age in the Huxi Corridor. In the Loess Plateau during the Han Dynasty, agricultural development created farmland through deforestation corresponding to a population increase. Sterilization of farmland and desertification progressed because of insufficient manure supplies and deforestation (Hippo and

Ryu 2005). Thus, Gansu Province has a long history of severe battles between the natural environment and human inhabitants. If the history of this area is taken into consideration, it is necessary to have a continuing dialogue always regarding changes to the natural environment for economic development. Land use policy must be carefully considered for the inland area of China.

This chapter is the revised edition of Doi (2014), which summarized the result of SLUAS.

References

Cao J, Yeh ET, Holden NM, Yang Y, Guozhen D (2013) The effects of enclosures and land-use contracts on rangeland degradation on the Qinghai–Tibetan plateau. J Arid Environ 97:3–8

Chai Y (1991) Urban internal structure of China's cities: a case study of Lanzhou City. Jpn J Hum Geogr 43(6):16–35 (in Japanese with English abstract)

Chen R, Ye C, Cai Y, Xing X, Chen Q (2014) The impact of rural out-migration on land use transition in China: past, present and trend. Land Use Policy 40:101–110

Cuo L, Zhang Y, Gao Y, Hao Z, Cairang L (2013) The impact of climate change and land cover/use transition on the hydrology in the upper Yellow River Basin, China. J Hydrol 502:37–52

Cyranoski D (2009) Beijing's windy bet. Nature 457(22):372–374

Doi H (2014) Land use changes in Inland area of China—a case of Gansu Province. SLUAS science report—towards sustainable land use in Asia (V), pp 29–44

Editorial Board of Geographic Atlas of China (2009) Geographical atlas of China. Sino Maps Press, Beijing (in Chinese)

Editorial Board of Gansu Development Yearbook (2010) Gansu development yearbook-general survey 2010. China Statistics Press, Beijing (in Chinese)

Editorial Board of Gansu Rural Year Book (2008) Gansu rural year book 2008. China Statistics Press, Beijing (in Chinese)

Hippo Y, Ryu K (2005) The arrangement of water resource in northwest of China and counter measures against desertification around western corridor of Koga and on Loess plateau. Keizaigakukiho (Rissho University) 54(3/4):13–61 (in Japanese)

Lin GCS, Ho SPS (2003) China's land resources and land-use change: insights from 1996 survey. Land Use Policy 20:87–107

Li P, Qian H, Wu J (2014) Accelerate research on land creation. Nature 510:29–31

Long H (2014) Land use policy in China: introduction. Land Use Policy 40:1–5

Piao S, Ciais P, Huang Y, Shen Z, Li J, Zhou L, Liu H, Ma Y, Ding Y, Friedlingstein P, Liu C, Tan K, Yu Y, Zhang T, Fang J (2010) The impacts of climate change on water resources and agriculture in China. Nature 467(2):43–51

Sharna N, Murray U, Shen Y, Li L, William B (2008) Farming systems of the Loess Plateau, Gansu Province, China. Agr Ecosyst Environ 124:13–23

Statistical Bureau of Lanzhou City (2012) Lanzhou statistical yearbook 2012. China Statistics Press, Beijing (in Chinese)

Ta W, Dong Z, Sanzhi C (2006) Effect of the 1950s large-scale migration for land reclamation on spring dust storms in Northwest China. Atmos Environ 40:5815–5823

Wang D, Gong J, Chen L, Zhang L, Song Y, Yue Y (2012a) Spatio-temporal pattern analysis of land use/cover change trajectories in Xihe watershed. Int J Appl Earth Obs Geoinf 14:12–21

Wang J, Chen Y, Zhang Y, Cao Y (2012b) Land-use changes and policy dimension driving forces in China: present, trend and future. Land Use Policy 29:737–749

Wang S, Yang Z, Liu H (2011) Impact of urban economic openness on real estate prices: from thirty-five cities in China. China Econ Rev 22:42–54
Wu Y, Xu R (2013) Current status, future potentials and challenges of renewable energy development in Gansu province (Northwest China). Renew Sustain Energy Rev 18:73–86
Wu Z, Guo H (1994) Zhongguo Tudiliyong (land use of China). Science Publishing, Beijing, 422 pp (in Chinese)
Xie Y, Ward R, Fang C, Qiao B (2007) The urban system in west China: a case study along the midsection of the ancient Silk Road—He–Xi Corridor. Cities 24(1):60–73
Zhang Y, Tachibana S, Nagata S (2006) Impact of socio-economic factors on the changes in forest areas in China. For Policy Econ 9:63–76
Zhang Z, Wang X, Zhao X, Liu B, Yi L, Zuo L, Wen Q, Liu F (2014) A 2010 update of National Land Use/Cover Database of China at 1:100000 scale using medium spatial resolution satellite images. Remote Sens Environ 149:142–154
Zhou X, Li X, John D, Zhao K, Pia A, Sun N, Yang Q (2012) Land degradation during the Bronze Age in Hexi Corridor (Gansu, China). Quatern Int 254:42–48

Chapter 5
The Change of the Traditional Urbanization and Its Future Development Focus in China

Qi Lu

Abstract The modernization of China has lasted more than 170 years since 1840. As the important component of modernization, urbanization of China had expressed different styles in the past but unsuccessful before the reform and open policies in 1978. After Mao's era, China restarted its modernization movement, and therefore its urbanization has been accelerated. The national urbanization rate has been quickly increased from about 17% in 1978 to 52.7% in 2012 (right now, it might be around 56%). But, the accelerated urbanization has led to many problems, such as the high population density in big cities, the unexpected high price of house compared to the very low-income families, the very heavy traffic jam of the cities, the larger social disparities between the rich and the poor and between urban and rural, the most serious air pollution and the tense relation between demand and supply in various resources and so on. After the 18th National Congress of the Communist Party of China held in Beijing from Oct. 15 to 18, 2012, the modernization and the related affairs were re-emphasized as usual. A so called 'new type urbanization' became a popular word to describe the current and coming urbanization. The author thinks it may be incorrectly used by Chinese. Why people say it a new type urbanization. It is just because we had new leaders and a new premier for the party and the government in power. But, urbanization is urbanization in its nature. It is only expressed in different features and in development ways in different periods. Is there something new for the coming urbanization? This paper plans to answer the question through the following discussions: 1. the understanding of urbanization; 2. a brief recall of the historical urbanization process in China; 3. the lessons of the past urbanization and the present situation of the urbanization in China; 4. the problems resulting from the acceleration urbanization; 5. The development of middle- or small-scaled cities in the coming urbanization in China.

Q. Lu (✉)
Institute of Geographical Sciences and Natural Resources Research, CAS, A 11 Datun Road, Chaoyang District, Beijing, China
e-mail: Luq@igsnrr.ac.cn

Y. Himiyama (ed.), *Exploring Sustainable Land Use in Monsoon Asia*, Springer Geography, https://doi.org/10.1007/978-981-10-5927-8_5

Keywords The understanding of urbanization · Historical urbanization
Present urbanization · Problems in urbanization development of middle scaled city
China

5.1 The Understanding of Urbanization

Logically speaking, a correct understanding of something plays a very important role in the relevant practices. A correct concept not only provides a right thinking in making policies and establishing reasonable institutions, but also leads to a smooth way for the practical development. In its nature, urbanization is a necessity of economical and social development and the improvement of life standard for us human beings, continuous land use change and a long continuous process in history. However, this new concept came into being to describe the current and the coming urbanization in China after the replacement of the former leadership of the Party and the government by the new leadership in 2012, i.e., the New Type Urbanization. It seems the concept appeared not following basic change of the urbanization in nature, but after the political change, and also soon becomes a popular concept in China. So, it is easy to see the concept in the public opinion of media (both official and academic opinions), but, what does it mean exactly? We know that the new leadership of China may adopt some different policies (as Hukou institution) in the current or future urbanization development for some cities at certain levels. But, is it a basically different thing from the urbanization in the history of China or the urbanization trend all over the world from the land use point of view? Does it mean something different from the former urbanization in nature? This concept may easily mislead to the understanding in the nature of urbanization. Historically and theoretically speaking, urbanization is a continuous process in history all over the world, and it may present different performances (rapid or slow development) in the different historical periods or in different countries due to the complicated and various physical and social factors. So, we should not have the definition to describe the development of this process by using the concept of old type or new type urbanization.

5.2 A Brief Recall of the Historical Urbanization in China

China is a country with a very long history. Its urbanization process lagged behind the developed countries in the modern history though. However, the interesting phenomenon is its urbanization process in ancient time was much advanced if calculated by Chinese way (urban population/total population). From Tables 5.1 and 5.2 below, we can see it clearly.

If compared to some big cities in the countries of the western Europe in the sixteenth or seventeenth centuries, we may find that even at the beginning of modernization of Europe, the scale of the urban population in some highly

Table 5.1 The changes of the estimated urbanization rate since Warring States in China

Time	Rate of urban population (%)
Warring States (300 BC)	15.9
Western Han (2 AD)	17.5
Tang Dynasty (745 AD)	20.8
Southern Song Dynasty (1200 around AD)	22.0
Qing Dynasty (1820 AD)	6.9
Modern Period (1949 AD)	10.6
Modern Period (1978 AD)	17.9
Modern Period (2012 AD)	43.9

Sources Zhao, Treaties on the urban development of China, p. 84, (Chinese edition); New Star Publisher, 2006

Table 5.2 The estimated urban population of some big cities in Tang Dynasty in China

14th year of Tian Bao reign (755 AD)	Urban population
Changan (today's Xian)	600,000
Luo Yang (today's Luoyang)	300,000
Seats of prefectures	4,400,000
Seats of counties	5,690,000
Total urban population	10,990,000

Sources Zhao, Treaties on the urban development of China, p. 63, (Chinese edition); New Star Publisher, 2006

modernized cities such as Paris and London or some other cities was smaller than that in the big cities of China such as Changan (Xian today, the capital of Shannxi Province) in the medieval period. See Tables 5.3 and 5.4.

But, with the development of the European modernization development, things became different. Compared to these countries, the urbanization of China was not only lagged behind these European countries and some other developed countries, but also lagged behind some developing countries in the modern history. See Tables 5.5 and 5.6 which provide the comparison.

If compared the average annual growth of urbanization rate of China to some developed countries in the world during the period from 1890 to 1933, we can see the growth of China's urbanization rate was the lowest, only 0.05% per year, but

Table 5.3 The urban population of the big cities in Western Europe in sixteenth or seventeenth centuries

Name of the cities	Urban population
Paris (1594)	180,000
End of seventeenth century	500,000
London (1602)	250,000
London (1700)	674,350

Sources Population data for Paris and Landon is from Songbate, Luxury and Capitalism, pp. 29–30 (Chinese edition, 2000)

Table 5.4 The urban population in some other cities in Europe in sixteenth–seventeenth centuries

	1530	1563	1575–1577	1629	End of sixteenth cen.	1560	1622
Florence	60,000						
Venice		168,626	195,863				
Roma							
Lisbon				110,800			
Seville					100,000		
Antwerp						104,927	
Amsterdam							104,927

Source Songbate, Luxury and Capitalism, pp. 29–30 (Chinese edition, 2000)

Table 5.5 The rates of urbanization in the world (1800–1997) unit %

	World	Developed area	Developing area
1800	5.1	7.3	4.3
1825	5.4	8.2	4.3
1850	6.3	11.4	4.4
1875	8.8	17.2	5.0
1900	13.3	26.1	6.5
1925	20.5	39.9	9.3
1950	29.0	52.5	16.7
1975	38.4	68.6	27.2
1997	46.0	77.3	38.4

Sources 1, Zhou Yixing, Urban Geography, p. 78, 81, 1995 (Chinese); World Bank, 1988/1989 World Development Report, pp. 190–193 (Chinese edition, 1999)

Table 5.6 Rates of urbanization population in some countries of the world (1890–1938) unit: %

	1890	1900	1913	1920	1928	1933
Britain	29.9	32.8	34.6	37.3	38.2	39.2
USA	15.3	18.7	23.1	25.9	28.7	32.8
Germany	11.3	15.5	21.0	35.7	34.4	30.2
France	11.7	13.3	14.8	15.1	15.3	15.0
Russia	3.6	4.8	7.0	3.1	7.1	20.2
Italy	9.0	9.6	11.6	13.2	16.1	18.2
Japan	6.3	8.6	12.8	11.6	15.6	28.6
China	7.7 (1893)					10.6 (1949)

Source Kennedy, The Rise and Fall of the Great Powers, p. 193, (Chinese edition), 2006, the Publishing Company of International Culture

0.52% per year for Japan more than 10 times), 0.44% per year for Germany (8.8 times), 0.41% per year for USA (8.2 times), and 0.39% per year for Russia (7.8 times) (See Table 5.7). However, another estimate shows that in 1820, the urbanization rate of China was only 6.9% (Zhao Gang. p. 81). If calculated by this data, the average annual growth for China was rather low, only 0.01% per year during the period from 1820 to 1893.

Low share of the urban population in megacities is another obvious characteristic in modern China. Take Beijing for example, there is about 20 million population right now, accounting for about 1.4% of the whole population of China, but, there concentrates about 30% of the Japanese population in Tokyo (Table 5.8).

China's economic development also shows the same trend in its modern period. According to a research result, in the early time of modernization period, the share of manufacturing value of China occupied 32.8% of the world in 1750, but declined to 6.2% in 1900. The same proportion changes for the developed countries are just opposite. It was 23.2% for Europe in 1750, but 62.0% in 1900, 0.1% for USA in 1750, but 23.6% in 1900, 3.8% for Japan, but 2.4% in 1900. See Table 5.9.

5.3 The Lessons from the Past and Present Urbanization in China

If compared China's modernization with the modernizations in Russia and Japan, we may find in the far east areas, these three countries started their modernizations almost at the same time. For Japan, it started its Meiji Restoration in 1868 (the purpose is "to become a rich and strong country", "to get every good things from the world"). Meanwhile, it was the period of 'the Great Reform' in Russia. Russia also started engaging in its modernization (it declared the abolishment of the slavery system and many major changes in the economy and society) (Black et al., The Comparison of Modernizations in Japan and Russia, Chinese version, 2000, p. 42). But, China failed in its modernization and urbanization in early modernization period.

Table 5.7 The average annual growth of urbanization rate in some countries of the world (1890–1933) unit %

	1890	1933	Annual growth
Britain	29.9	39.2	0.22
USA	15.3	32.8	0.41
Germany	11.3	30.2	0.44
France	11.7	15.0	0.08
Russia	3.6	20.2	0.39
Italy	9.0	18.2	0.21
Japan	6.3	28.6	0.52
China	7.7 (1893)	10.6 (1949)	0.05

Source 1. calculation is based on Table 5.2 from Kennedy, The Rise and Fall of the Great Powers, p. 193, (Chinese edition, 2006); 2. 7.7 and 10.6% are from Zhao, Treaties on the urban development of China, p. 84, 2006, (Chinese)

Table 5.8 The comparison of urban population rate and its concentration in nineteenth century

	Britain (1801)	Japan (1868)	China (1893)
Total population(1) 1000 persons	8892	31,500	426,000
Urban population(2) 1000 persons	2725	5200	32,662
(2)/(1)	30.6%	16.5%	7.7%
Population in the largest cities (3) 1000 persons	948	1000	1000
(3)/(2)	39.9%	19.2%	3.0%

Sources Zhao, Treaties on the urban development of China, p. 158, 2006, (Chinese)

Table 5.9 Relative shares of manufacturing values of the world (1750–1900) (%)

	1750	1800	1830	1860	1880	1900
Europe	23.2	28.1	34.2	53.2	61.3	62.0
USA	0.1	0.8	2.4	7.2	14.7	23.6
Japan	3.8	3.5	2.8	2.6	2.4	2.4
China	32.8	33.3	29.8	19.7	12.5	6.2

Source Kennedy, The Rise and Fall of the Great Powers, p. 144, (Chinese edition, 2006)

In the recent 60 years, China still has been on the way to its modernization and urbanization. But it had also wasted about 30 years (from 1949 to 1978) for its modernization and urbanization, owing to some wrong thinkings and policies for urbanization (for example, the separation of urban-rural household registration system) and many revolutionary activities such as Cultural Revolution and so on. The best period of modernization in China is the past 30 years around (from 1978 to the present), and we really have achieved a lot mainly in material aspects (the so-called 'four modernizations' are all emphasizing material aspects). But, in my opinion, as far as the institutional modernization is concerned, the modernizations of social, political, and cultural aspects are still lagging behind if checked by the standards of modernization in its nature. One important thing is we have failed in improving the relationship between urban areas and rural areas. The ridiculous phenomenon is, the disparity between urban areas and rural areas has been enlarged since 1978.

In the distinguished past more than 30 years from 1978, China has experienced great changes both in its urbanization and modernization. We have kept a very high economic growth, and the GDP was increased to the second place in the world, and the number of cities and the urban population has been greatly increased compared

Table 5.10 Increase of cities in China by the three regions (1978–2002)

Region	Total	East	Central	West
2002	660	287	247	126
1978	193	69	84	40
2002/1978	3.42	4.16	2.94	3.15

Source City Statistical Yearbook of China, (2003)

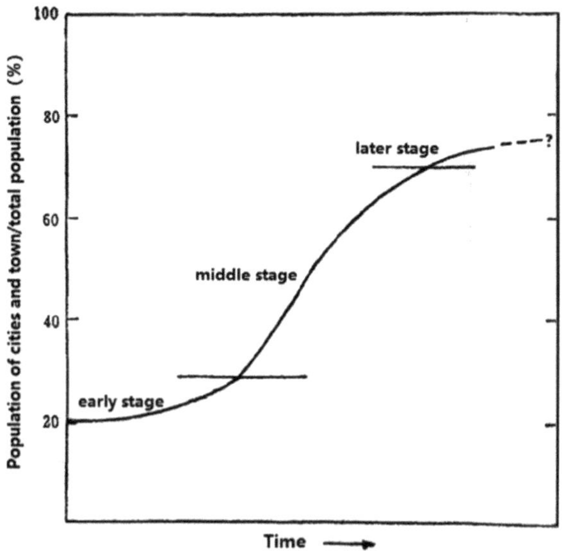

Fig. 5.1 Urbanization theory

to the period of centralized planning economy and the urbanization development of China entered the acceleration development period (see Table 5.10; Fig. 5.1 below), therefore, the general situation of the distribution all over China and the different levels of the cities can be seen in Figs. 5.2 and 5.3.

Figure 5.1 shows a very famous urbanization theory raised in 1979 by American geographer Ray M. Northam. According to this urbanization theory, urbanization process can be divided into three stages, i.e., Early stage, Middle stage, and Later stage, if the urbanization rate is below 25%, the urbanization development is in a very slow process. If it is in the stage between 25 and 70%, the urbanization development is in the accelerated period, so, we can say the urbanization development in China is in the acceleration process in recent years. Figures 5.2 and 5.3 show us the dense distribution of the cities and the expansion of urbanized areas from 1990 to 2010.

According to the information from the official media, in 2012, the urbanization rate in China reached 52.57%, which equals the average level in the world. The regional urbanization rates in three main regions are 60% for East China, 47% for Middle China and 43% for West China (New Beijing Daily, June 27, 2013, A 12). Except East China, the rates in Middle China and West China are, respectively, lower than the average world level 5.57 and 9.57%.

Comparatively, from 1949 to 1978, the average annual growth of urbanization rate is 0.25 point each year, 1978 to 2006, the average annual growth of urbanization rate is 0.9 point each year (2006 China Statistical Abstract, p. 38). However, from 2006 to 2012, the annual growth of urbanization rate is 1.6 point, much higher than that from 1949 to 1978. The changes of the average annual growth of

Fig. 5.2 Global night light distribution satellite map of China, by NASA's Earth Observation

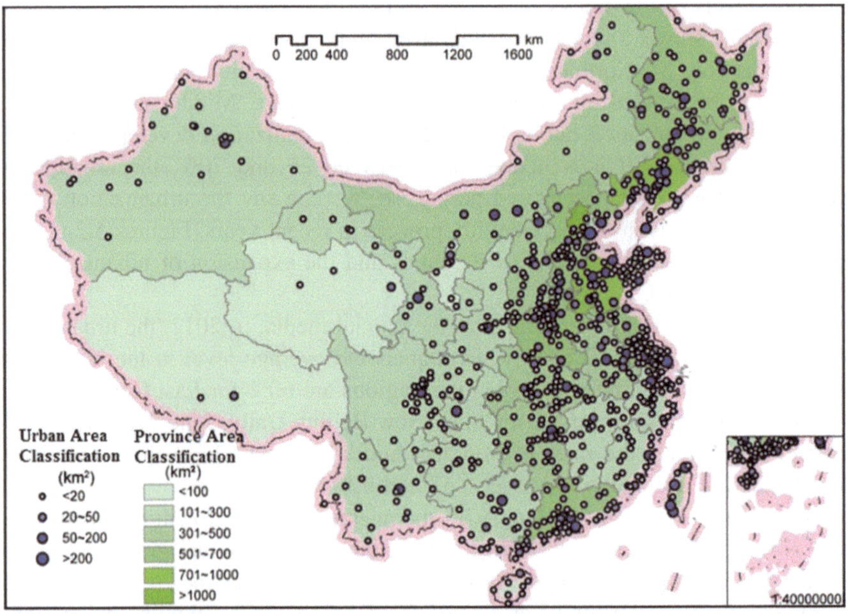

Fig. 5.3 Satellite and remote sensing mapping about the expansion of urbanized areas from 1990 to 2010

urbanization rate in China also clearly prove the characteristic of China's urbanization development.

Though the urbanization rate in China reached 52.57% in 2012, the gap between the urbanization in China and that in the highly developed countries such as Britain (89.1%), France (82.5%), America (94.7%), Japan (77.9%), and Germany (81.2%) is still large (2001–2002 The Report for the Urbanization Development of China, 2003, p. 20).

The present urbanization rate of China is not only lower than that in the developed countries, but also lower than that in some developing countries. According to the report, around 2000, the urbanization rates of some developing countries as Bulgaria (69%), Jamaica (55%), Jordan (73%), Latvia (73%), and Philippine (56%) (2001–2002, Report of Urban Development of China, 2003, p. 32).

5.4 The Problems Resulting from the Acceleration Urbanization

China's urbanization has been accelerated gradually after the open and policy reform. However, China's urbanization development has resulted in many problems during the past 30 years. If we cannot resolve these problems and find a reasonable way for the future, the coming urbanization may become more difficult. We have quoted some information from the official media about the major problems as follows:

(1) **Huge income from selling land for the governments**. The governmental financial income from selling land has been a problem in the land management for a long time and it is still existing today and may exist in future. For instance, at the very beginning of 2014, the land market of Beijing is still huge. Some information reveals that until Jan. 16, 2014, the volume of land sale business was 30 billion yuan (New Beijing Daily, Jan. 20, 2014, A 21), which is about 0.5 billion USD. By another piece of news, at the end of 2014, the income from selling land for Beijing government was 190 billion yuan in total that year (the first financial daily, Dec. 1, 2014), exceeding the income in 2013.

In 2013, only the four cities at first level, namely Shanghai, Beijing, Guangzhou, and Shenzhen, have received more than 500 million yuan from land sale in total, which was much higher than that of about 200 million yuan in 2012, increasing ratio was 153.5% (New Beijing Daily, Dec. 26, 2013, B 04). According to a new information, in the recent 17 years, the total land sale income all over China for the government has been 2700 billion yuan (the first financial daily, Feb. 16, 2016).

(2) **Endless increase of housing price**. Under the most strict control policy on housing price rising, the housing price has continued to go up. For instance, in 2013, the price of the high-end commercial housing in Beijing experienced

from about 40000 yuan/per square meter to about 60,000 yuan/per square meter, for some more luxurious houses, it reached 100,000 yuan/per square meter to 150,000 yuan/per square meter (New Beijing Daily, Jan. 17, 2014, B 14), which is about 15,000 USD to 23,000 USD/per square meter.

According to the official statistics, the housing price in the 69 cities among the 70 cities went up in 2013. For example, the price in November 2013 of Shanghai, Beijing, Shenzhen, and Guangzhou, respectively, went up to 21.9, 21.1, 21.0, and 20.9% compared to that in November 2012 (New Beijing Daily, Dec. 19, 2013, A 07).

Revealed by a survey body, if a middle-class family in Shanghai plans to buy a house with 100 m^2, it takes 30 years, 28 years for Beijing and Guangzhou, 20 years for Chengdu, but it takes 11 years for Stockholm, 10 years for Sydney, less than 10 years for Zurich (New Beijing Daily, Mar. 20, 2012, B 12). Chinese people may take much longer time to buy a house than the people in the cities in some other foreign countries.

If we compare the housing price changes with the annual income changes of the residents in Beijing from 2003 to 2010, we may find that in 2003, with the annual income they could buy 5 m^2, in 2004, 5.6 m^2, in 2005, 5.2 m^2, in 2006, 4.4 m^2, in 2007, 3.5 m^2, in 2008, 3.6 m^2, in 2009, 3.5 m^2, in 2010, 2.8 m^2 (New Beijing Daily, Mar. 5, 2013, A 09).

(3) **Air pollution of cities**. Air pollution of cities has been a serious problem for a long time. In May 2013, the Ministry of Environment Protection issued the monthly bulletin about the condition of air quality within the 74 cities in China, the results showed that about 60% of the cities were up-to-standard. The up-to-standard days in Beijing, Tianjin, and Hebei province declined from 62.1% in April to 25% in May. Of the 10 serious pollution cities, 7 cities were in the North China Plain (including Beijing, Shijiazhuang, Tangshan, Xingtai, Handan, Baoding, and Hengshui), air quality which was up-to-standard in the cities in North China Plain was only 36.5% (New Beijing Daily, June 20, 2013, A 24).

In July 2013, the Ministry of Environment Protection issued the bulletin about the condition of air quality within the 74 cities in China again, the results showed that about 70.7% of the days were up-to-standard, but the air quality which was up-to-standard in the cities in North China Plain was only 36.5% (New Beijing Daily, Aug. 24, 2013, B 08).

Many reasons contribute to the air pollution of cities. But the main causes for the urban pollution in China are: (1) expansion of industries; some reports show that industrial pollution may take more than 300 years to happen to the developed countries, but only 30 years for China; (2) the connection of regional urban expansion and the reduction of the pollution dilution in rural areas; (3) booming increase of cars, for instance, the annual increase of cars in Guangzhou is about 150,000 per year (Southern Weekly, April 3, 2008, front page). In 2007, fog and haze weather days for Shenzhen were 231 days, but 164 days in 2006; in December of 2007, fog and haze weather days for Guangzhou were 22 days (Southern Weekly, April 3, 2008, front page).

In 1987, a report by the world environment and development committee named 'Our Common Future' was published and a new concept of sustainability was put forward. What is the sustainability, the concept says: "the sustainability means one ability which can meet the demands of people in modern time and at the same time it cannot threaten the ability which can meet the demand and desire of the future generations" (the world environment and development committee 'Our Common Future' Jilin People's Publisher, 1997, p. 48). From this, we can see how far the urbanization of China is from sustainability.

(4) **Traffic jam**. Traffic accidents have been increasing in China more and more, and most of them happened in the urban areas. In 1994, there were 255,000 traffic accidents happened to China and caused the death of 66,000 people in China; in 1998, 346,000 traffic accidents and caused the death of 78,000 people; in 2000, there were 617,000 traffic accidents and caused the death of 94,000 people; in 2004, there were 568,000 traffic accidents and caused the death of 79,900 people (Xueqiang, Urban Geography, 2009, pp. 326–327, Educational Publisher).
(5) **The huge population scale**. After the open and reform policies, there has been a boom migration wave from rural to urban in China, more and more rural population has surged into the urban areas for better income. Take Beijing, for example, in the recent 10 years in average, the annual increase of the permanent residents have been 600,000 people. At the end of 2012, the total population was 20.7 million which exceeded the officially planned goal of 18 million for 2020 (New Beijing Daily, Jan. 21, 2014, A 10). At the end of 2013, it was 21.15 million (New Beijing Daily, Feb. 14, 2014, A 05).
(6) **Serious shortage of water resources**. High population pressure leads to high pressure on natural resources on land and water. According to the open data issued by the official body, Beijing Water Management Bureau, right now, the annual average water resource is 2.1 billion cubic meters, but the total water demand is in average 3.6 billion cubic meters, and the per capita water resource is less than 100 cubic meters which is less than one-tenth of the national level and 1/77 of the world level (New Beijing Daily, Jan. 21, 2014, A 10).
(7) **The expansion of built-up area**. The built-up area has been expanding quickly after the reform. Take Beijing, for example, in Qing Dynasty, the total built-up area was 62 km^2 (Zhu Zuxi, The Building of Beijing, p. 1); in 1949, it was about 100 km^2, annually expanded 1.26 km^2; from 1949 to 1997, the expanded rate was 8.1 km^2 (Qi etc. Geographical Study, 20 Vol. 6, 2001).

The area of construction land use in Beijing in 1982 was 1792.9 km^2, accounting for 10.7% of the total land area of Beijing, in 1992, the total area of construction land use was 2281.74 km^2, accounting for 13.9% of the total land area of Beijing, but in 2008, the total area of construction land use was 3377.15 km^2, accounting for 20.6% of the total land area of Beijing (Qi and Shasha, SLUAS Science Report 2010, pp. 1–26).

(8) **The social problems**. Particularly the migration population scale is huge. Nowadays, there are at least 200 million migrants in China from rural areas to urban areas. In 2012, there were 260 million migration laborers in total with very low income, in average 2290 yuan per month (New Beijing Daily, May 28, 2013, A 17). It is easy for these migrants with low income and unstable work to make various social problems.

(9) **Big income gap between rural and urban**. According to a recent report by CASS, in 2010, the per capita income for urban family was 3.23 times than that for the rural family, in 2011, it was 3.13 times (New Beijing Daily, Dec. 19, 2012, A 18), there is only a slight reduction.

(10) **The unreasonable structure of the growth of land use scale and population scale**. According to a report, the land use scale for urbanization has been exceeded the scale expansion of permanent population in urban areas. From 1980 to 2011, the land area expansion of urbanization has been 9.2 times, but the growth of population scale has been 3.5 times (New Beijing Daily, June 27, 2013, A 12).

Since the megacity's urbanization caused many social and environmental problems and these problems have already limited the future urbanization development, China needs a different way to realize its urbanization. So, it seems that to develop middle and small cities may become the focus of the coming urbanization. But, we also need to pay a very high cost.

5.5 The Development of Middle- and Small-Scaled Cities in the Coming Urbanization in China

In 2013, urbanization again became an important task for the new leadership of the Party and the government, the urbanization goal of China is to realize 70% of the urbanization rate in 2030, if so, the total urban residents will exceed 1 billion. It means more than 300 million people will move to the cities from countryside (New Beijing Daily, June 8, 2013, B 03).

Urbanization means three things for China in future: a lot of land occupation, the huge cost of investment, and getting rid of the problems in the urbanization before. Land occupation by urbanization means appearance of about 15 megacities like Beijing. The present construction land area of Beijing is about 3000 km^2, so land occupation for the coming urbanization all over China might be about 45,000 km^2, accounting for about 0.5% of the total land area of China.

Some data shows, owing to the rapid development of urbanization, more and more arable land has been occupied, from 1978 to 1986, about 6 million migrants (mu is area unit in China. 1 mu is 0.06667 ha) of arable land loss per year; from 1996 to 2003, about 8 million migrants of arable land loss each year (Zhang Xinan edited, "the Analysis Report of the Land Resource Security in China," 2005, pp. 108–112). The food security of China may be threatened due to the serious arable land loss.

According to a recent study report by CASS, we can imagine that the cost of the coming urbanization will be extremely high. By the estimates, the per capita public cost of urbanization is 130,000 yuan/a person and the total population for urbanization is 390 million yuan, so the total investment by 2030 is about 51,000 billion yuan (New Beijing Daily, July 31, 2013, A 22). It is almost equal to the annual GDP and it may need China to put 3000 billion yuan around each year in the coming 17 years.

In the urbanization process, farmers need to pay some personal cost including life cost and housing cost themselves, in average estimate, it is about 120,000 yuan in total per year (New Beijing Daily, July 31, 2013, A 22) the overall cost of China may be 48,000 billion yuan, if the public cost of 51,000 billion yuan is added, the grand total is 99,000 billion yuan. The annual social cost in total may be 5800 billion yuan per year.

In the past, the Chinese government thought that the 'urban sickness' only resulted in the developed world (we said the capitalist world), and the so called 'urban sickness' could not happen to the socialism country like China. But, in the urbanization development process, China itself has seriously faced this sickness. Today, we have to admit this urban sickness in China.

The Third Plenary Session of the 18th CPC Central Committee raised the goal of urbanization. After the Session, the Central Committee held a special meeting on the urbanization. This meeting put forward the tasks for the coming urbanization and discussed "The National Planning for the New Type Urbanization" (New Beijing Daily, Dec. 15, 2013, A 03).

One important thing is to strengthen the reform of household registration in the coming urbanization, according to the decision of the Central Committee, the reform focuses on the four respects: complete openness of the household registration of the small cities; gradually and in-order to open the system of the middle cities; formulation of the conditions for the settlement of the big cities; strict control of the population scale of the megacities (New Beijing Daily, Dec. 18, 2013, A 08).

But, if the immigrants from the countryside are not allowed to enter some big cities, the reform policy of the household registration may be still an unreasonable policy. However, if we open the household registration to all cities at any levels, the problems for the megacities may become more serious. This is the basic contradictory question for China's coming urbanization.

To avoid the 'mega-city sickness' and resolve the present problems in the big cities and urbanized areas, to strengthen the construction of the vice-central cities, new centers or smaller cities may appear as another important focus for the future urbanization of China. For example, a development planning for the vice-central cities of Beijing in 2020 will be formed by Beijing government according to the concerned news (New Beijing Daily, Jan. 7, 2014, A 07).

Based on the spirits of the decision of the Central Committee, the Ministry of Security soon formed an official document for the household registration reform,

Fig. 5.4 The photos about the expansion of middle or small cities in the north areas of Beijing

Fig. 5.4 (continued)

saying a new household registration system will be raised in 2020 (New Beijing Daily, Dec. 18, 2013, A 08). But, what will happen, we have to see.

We really do not know what will happen to the urbanization in future, we do not know if all the problems could be resolved as well, but, the government has really given us some promises again for more healthy urbanization. So, we have to keep eyes on what will happen to the developments of megacities and middle-small cities or towns in the future.

Within the recent decades, the expansion of middle or small towns is booming all over China, which is swallowing more and more arable land. The following photos give us some examples of the expansion of middle or small cities in the north areas of Beijing. Except the township Yudaokou, Zhangbei, Guyuan, Longhua are all the county towns, they are all in Hebei Province, around 200 km^2 away from Beijing (see the photos a–o of Fig. 5.4).

1. **The expansion of Zhangbei**: Zhangbei, the county seat of Zhangbei County, this county was a poverty-stricken county, more than 200 km away from Beijing. Now, the situation has been changed due to rapid development. These photos taken from the newly expanded areas, some new buildings are still under the construction (Fig. 5.4a–d).
2. **The expansion of Guyuan**: Guyuan, the county seat of Guyuan County, this county was also a poverty-stricken county, but now the situation has become much better. In this county, tourism has been rapidly developed. These photos taken from the city's newly developed areas, the pattern of the streets and the buildings are almost similar to the new areas of Zhangbei (Fig. 5.4e–h).
3. **The expanded features of Longhua**: Longhua, the county seat of Longhua County, this county was also a poverty-stricken county. This county belongs to Chengde, less than 200 km from Beijing. In this county, tourism has been not developed as in the other counties. These photos also taken from the newly expanded areas, the streets and the buildings look new (Fig. 5.4i–l).
4. **The boom construction of small town**: Yudaokou is located along the highway, it is a township belonging to Weichang County. Yudaokou is near to the famous royal hunting place (Mulan Hunting Area), so new hotels and restaurants distribute everywhere along the two sides of the highway. In the future, Yudaokou may grow a small city (Fig. 5.4m–o).

References

Achievements and problems of modernization process and environmental management of China in the past 50 years. Edited by Shinichi Kawai, New Challenges and Perspectives of Modern Chinese Studies. University Academy Press of Japan, 2008

Black CE et al (2000) The comparison of modernizations in Japan and Russia, Chinese edition. Commercial Publisher

Edited by the Institute of Land Survey and Planning, China, The expansion of the urban construction scale of the key cities in China—the atlas of the monitor remote sensing. Geological Press, 2013

Kennedy P (2006) The rise and fall of the great powers, Chinese edition. Publishing Company of International Culture, Beijing

New Beijing Daily, 2012, 2013

Lu Q (2001) A brief comment on the impacts of social and human factors on the LUCC of Beijing in the recent 100 years. Geogr Study 20(6)

Lu Q, Lu S (2012) Economic development, population growth and land use changes in the last 30 years in Beijing. Chief-editor: Yuko Himiyama, SLUAS Science Report

Songbate W (2000) Luxury and capitalism, Chinese edition. Shanghai Translation Publisher, Shanghai

The Chinese Mayors Association (2003) 2001–2002 report for the urbanization development of China. Xiyuan Publisher, Xiyuan

The State Statistical Bureau (2003) City statistical yearbook of China. The Statistical Publisher

World Bank (1999) 1988/1989 world development report, Chinese edition. Chinese Financial Publisher, Beijing

Xu X et al (2009) Urban geography. Educational Publisher

Zhou Y (1995) Urban geography. Educational Publisher

Zhao G (2006) Treaties on the urban development of China, Chinese edition. New Star Publisher, Beijing

Part III
Land Use Change in India

Chapter 6
Landuse Sustainability of Agricultural Zones

Arun Das, Koichi Kimoto, M. Ravi Kumar, R. Umakanth, Dhritiraj Sengupta and H. R. Vishwanth

Abstract Karnataka state is the second most drought-prone region in India next only to Rajasthan. Even then, the traditional cropping land use pattern in the state has kept the people and the agriculture alive. But the recent onslaught of the high-yield variety seeds and western crops are posing a serious threat to the sustainability of the region. Located on the southeastern part of the peninsular India, Karnataka state enjoys the both wet and dry conditions from west to east. The climatic condition of the state has manifested different agriculture regions, which has paved the way to grow more than 200 varieties of crops. The mixed- and multi-cropping land use pattern is the tactful crop selection game played by the farmers to overcome the failure of monsoons. The entire cropping system of Karnataka depends on rain. Physiologically Karnataka has been classified into four divisions, such as, Coastal, Hilly, Northern, and Southern Plains. The coastal and hilly region experiences high rainfall and temperature, which are apt for agriculture, followed by the southern region. The northern plains are situated amidst of the land-locked condition, which experiences high temperature with scanty rainfall along with an overall dry condition. Despite these conditions, the northern region is the most sustainable compared to other regions. The gist of this article is to explain the sustainable factor existing in the land use pattern of northern dry region in comparison to the other three physiographic divisions.

Keywords Land-use · Land cover · Agro climatic zones · Food security
Sustainable agriculture

A. Das (✉) · M. Ravi Kumar · R. Umakanth · D. Sengupta · H. R. Vishwanth
Department of Geography, University of Mysore, Mysore, India
e-mail: arundasmysore@gmail.com

K. Kimoto
Kwansei Gakuin University, Nishinomiya, Japan
e-mail: k.kimoto@kwansei.ac.jp

Y. Himiyama (ed.), *Exploring Sustainable Land Use in Monsoon Asia*,
Springer Geography, https://doi.org/10.1007/978-981-10-5927-8_6

6.1 Introduction

India being at the center of this scenario is exerting extreme pressure on its land resources. Feeding 1000 million populations within the available agricultural land is a challenging task. Land being a limited resource has not been used judiciously. Land and water resources and the way they are used are central to the challenge of improving food security across the world. Demographic pressures, climate change, and the increased competition for land and water are likely to increase vulnerability for food insecurity, particularly for Africa and Asia (Diouf 2011). The spell of prevailing Monsoon climate in India is uncertain over space and time. This again has put most of the agricultural land into drought condition. Land resources such as soil, geological structure, and underground water are under rampant usage. Excessive mining and quarrying has turned valuable land in to wasteland. Most of the states in India experience these conditions. Among the 29 states, Karnataka is the second drought-hit region of India next only to Rajasthan.

6.1.1 Background

6.1.1.1 Note on Study Region

Karnataka is second largest state next only to Andhra Pradesh in south India. The state possess 191,791 km^2 of area, which accounts to 5.83% to the total geographical area of the country. Karnataka state comprises a total population of 61.1 million as per 2011 census. It ranks ninth largest populous state in the country. Being a part of the peninsular India it is positioned in southwestern parts of Deccan plateau adjacent to the Arabian Sea. The land stretches north to south in an elongated form (18° north–11° south), whereas the west to east extent stretch (300 km) is Narrower (750 km) (Fig. 6.1). The state of Karnataka which lies in southwestern part of India, experiences varied climatic condition, due to its topography. Out of the total geographical area, 70% of the land is under dry condition, where maximum population lives. The entire state has been classified into 10 agro climatic zones by the agriculture university (Fig. 6.2; Table 6.2). Normally the southwest monsoon rainfall is the prime source of raising crops in the state. During the retreating monsoon, the state experiences scanty rainfall. The rainfall rate decreases from the west to the east. The central and eastern part of the state has been classified as the dry regions (Kumara and Hosamani 2012). Quantum of rainfall in these regions is inadequate, but sufficient to produce only dry crops. Only a few patches of land along the river channel path are fed with canal irrigation. The varied land and climate of the state has paved way for cultivation of variety of crops (Acharya et al. 2011). More than two hundred types of crops are cultivated in different parts of Karnataka. The cropping pattern in the state is also unique, compared to the other states of India. Mixed- and multi-cropping system is commonly seen in all parts of

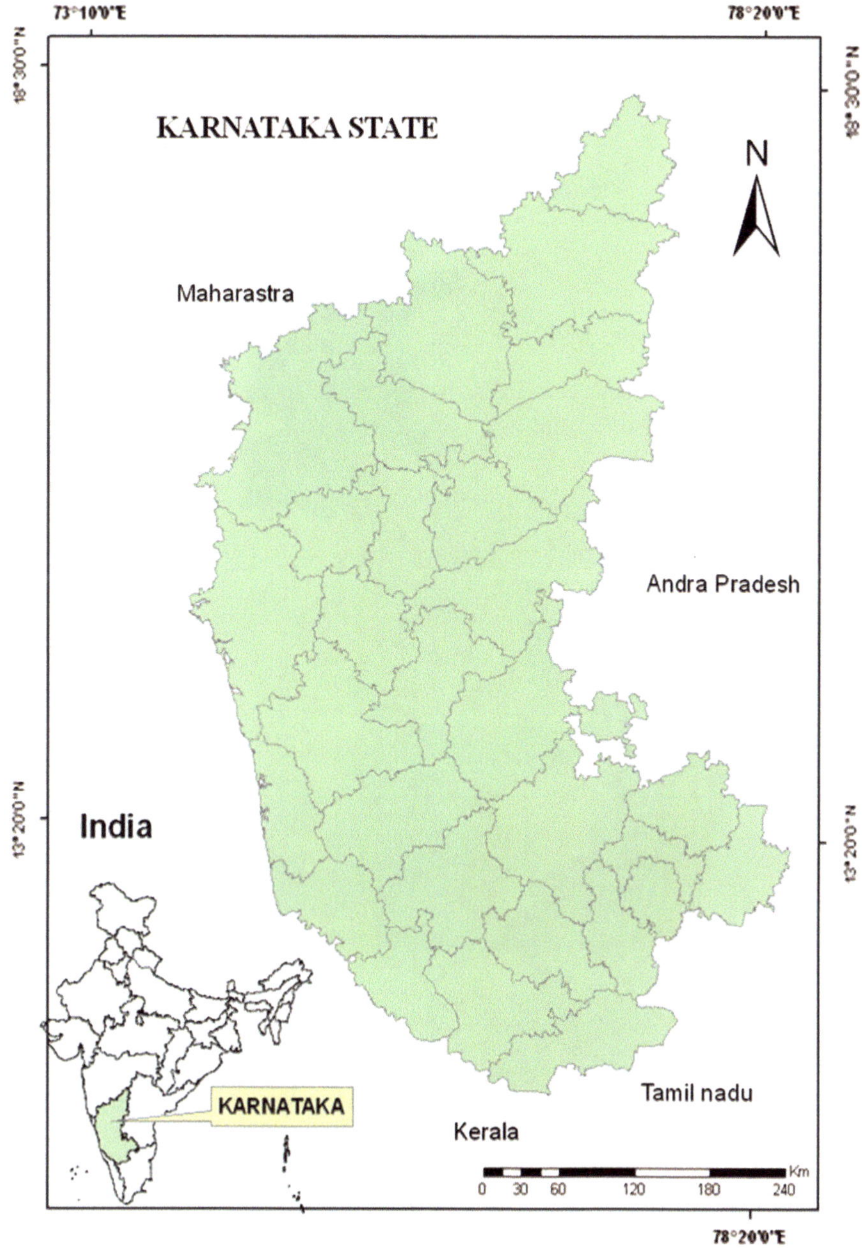

Fig. 6.1 Location of Karnataka State

agricultural regions of Karnataka. The state has always taken a lead ahead of the other States in India; in many respects as far as Agricultural Policy initiatives are concerned. It became the first State in the country to have unveiled its own

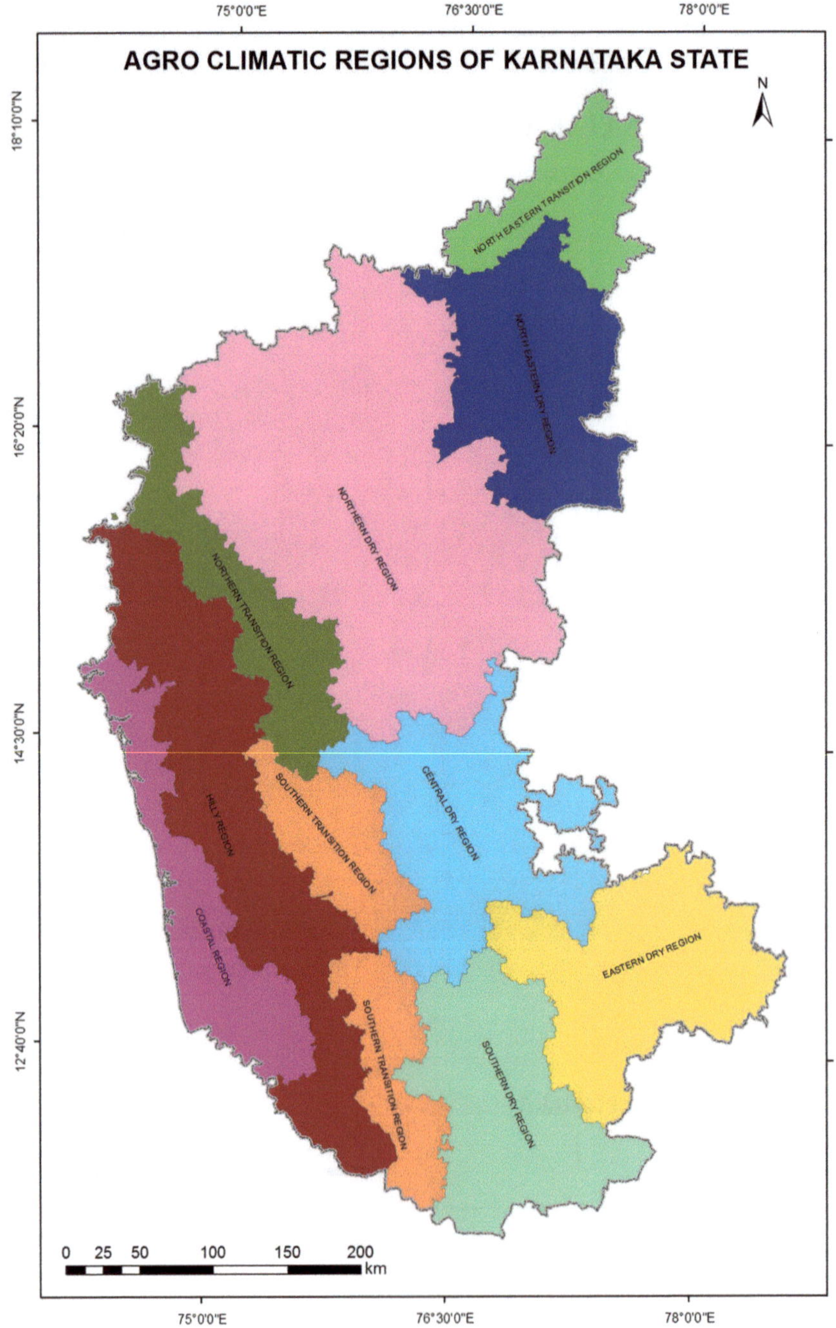

Fig. 6.2 Agro climatic regions of Karnataka State

Agricultural Policy as early as 1995, in order to demonstrate that agriculture is a subject enshrined in the state list under the Constitution of India (Khashempur 2006).

6.1.1.2 Global Impact on Local Food Security

Influence of international food habits through various means of electronic and print media partly developed a change in cropping pattern from tradition-based to health-based crop cultivation. At present each house has got television connection. In Brazil, for example, television is today the main source of entertainment, even for lower income families, who spend 5–8 hours a day in front of the television (Sawaya et al. 2004). The advertisement of multinational companies on protein-rich foods, such as Kelloggs, Barley, and Oats has tuned the minds of each household mother to feed her children with these types of foods.[1] The result of this, the very food habits and the food security of Indian farmers are at stake. The growing demand for cereals such as maize, barley, oats, have reduced the importance of indigenous crops, causing a direct negative impact on the traditional crops. On the other hand there is no feasible geographical condition existing in the state to produce barley or oats. Only few patches of land may have the said condition to produce barley and oats, but yield wise it is not cost effective. The end result in the near future will lead to a different food habit, which will compel to import food grains from other countries, sidelining the local crop value. In near future, most of the crops will vanish from the state which is currently producing 200 varieties of crops. Certainly, this will end up in putting the agriculture system and the food security of the country at great risk.

6.1.1.3 Farmers Blunder in Crop Selection Method

The decision of crop selection is done by the land owner after due consultation with fellow farmers on observation made about the trend of market for each crop. Seldom, farmers are also indirectly influenced to grow a particular crop due to subsidy provided by government to promote the production of those crops, e.g., Palm and Cotton. The subsidy on seeds and fertilizers directly influences the crop selection (Yu 2015). Second, providing loans to grow certain crops is also another means of influencing farmers on selection of crops. Banks provide agricultural/crop loans to any farmer for growing crops that are economically viable. The banks award loans only after due evaluation of the credit-worthiness of the farmer and the farmers ability to repay the loan without defaulting. It is, therefore, essential to demonstrate to banks the commercial viability of alternate crops, before they advance loans (Prasad 2007).

[1]Consumption, health and nutrition in urban areas: A case study of Brazil.

6.1.1.4 The Question of Sustainability Between Climatological Zones

The common assumption among the people of Karnataka is that, the coastal belt has a higher productive capacity, followed by mountains regions and transitional zones. Whereas the Dry belt zones are assumed less productive. From these point of view, there is a false belief existing among the people that, southern districts are productive, than northern districts. With these assumptions in mind, agro-cultural land use scenario was analyzed; the fact that came in to light was astonishing.

6.2 Methodology

Agro climatic suitability assessment is gaining weightage as an important basis for sustainable agricultural development planning for rainfed agriculture (Sarkar 2008). Keeping in mind the importance of this technique, a systematic method was adapted to analyze the sustainability of the study region on the grounds of Agro climatic regions. At first level, a general agricultural land use of Karnataka has been explained, followed by a zonal-level land-use scenario of major types of crops, such as, cereals, pulses, oil seeds, and commercial crops. Further, the analysis has been done to examine the condition of land use and sustainability between rained dry land and irrigated crop at an agro climatic zonal level crop analysis. This method has been adapted mainly to contemplate the assumption existing about the varying crop-producing capacity between these two regions. The zonal-level agricultural land use and zonal level crop wise irrigated and unirrigated analysis present a clear picture about the status of each zone with respect to its sustainability. From this analysis, the reader will get a better understanding whether the dry regions are more sustainable for the future, in comparison with the irrigated regions. Hence, the following analysis will be substantiating information towards understanding the reality between northern and southern dry zones of Karnataka.

6.3 General Land Use of Karnataka

Utilization of land for various purposes in Karnataka has maintained a stabilized trend over the decades. According to the standard land use classification done by agricultural Department, Government of India, the entire geographical land has been classified into nine types, Such as, Forest Land Cover, Non Agriculture Land Use, Barren land, Cultivable Waste, Pasture Land, Trees and Groves, Current Fallow, Other Waste, Net Sown Area (Fig. 6.3; Table 6.1). Among the different land uses, net sown area covers 53.4% of the total geographical area,[2] which is a

[2]Directorate of Economics and Statistics, Department of Agriculture and Co-operation, Govt of Karnataka, India, Repot of the Expert Committee on Agricultural Statistics.

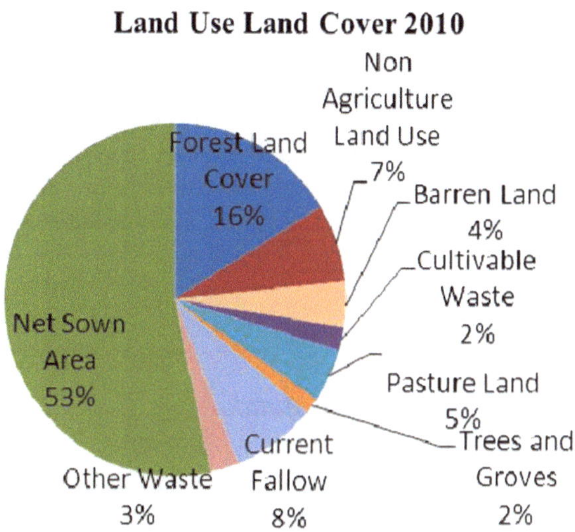

Fig. 6.3 General land use of Karnataka

Table 6.1 Land use cover 2010

Sl. no.	Type of land use	Area in hectares	In percentage
1	Forest land cover	3,071,833	16.12
2	Non agriculture land use	1,375,047	7.21
3	Barren land	787,628	4.13
4	Cultivable waste	412,913	2.16
5	Pasture land	923,383	4.84
6	Trees and groves	289,715	1.52
7	Current fallow	1,499,755	7.87
8	Other waste	515,973	2.70
9	Net sown area	10,173,589	53.40
10	Total	19,049,836	100

clear sign of agriculture-oriented state. Few decades ago the forest cover was 35%. Over the time, population pressure has reduced the area of forest cover to 16.1%. At present, the per-capita net sown area accounts to 0.166 hectares. Pertinently it implies, over stress on agriculture land due to the growing population. The pressure on agriculture land could have been overcome through optimum utilization of land under fallow, which accounts to 7.8% at present. The land under barren accounts to 4.13%, wherein there is a limited scope to rejuvenate this in to crop land. Apart from focusing just only on the production of crops, there is also a greater need for enhancing land under forest, at least from 16 to 23% (Nayak et al. 2012; Sundar 2013). Another major drawback of forest cover in Karnataka state is its irregular spatial distribution. Out of the total geographical area under forest 90% of forest area is concentrated along the west coast and western mountains of Karnataka, leaving behind, rest of the state without any green cover. It is also necessary to bring land under forest to maintain international standards of ecological and environmental balance (Dutta and Deven 2009; Gadgil et al. 1983). The national forest policy enunciated in 1982 advocates maintenance of fully one-third of the geographical area of our country under forest cover. Efforts need to be made to reach this level of vegetation (Ministry of Forest, Govt of India 2005). The general land use of Karnataka has been depicted through maps (Figs. 6.4, 6.5, 6.6, 6.7, 6.8, 6.9, 6.10, 6.11 and 6.12).

6.3.1 Cropping Periods in Karnataka

Broadly the cropping periods of Karnataka has been classified in to two periods, kharif and rabi crops. The climatic pattern of Karnataka has four seasons, i.e., Dry Summer Season (March, April, May), Wet Summer Season (June, July, August), Wet Winter Season (September, October, November), and Dry Winter Season (December, January, February), As the monsoon rains begin in the month of June, the agriculture activity will be at its peak, which prevails from June and September. During these months the dry rainfed farmers grow only one crop and leave their land fallow after the harvest in the month of September or in October. Whereas, the farmers in the valley course get irrigation through canal and they produce second crop between September and January. During these months the farmers of both irrigated and dry land grow rainfed crops such as Ragi, Jowar, Bajra, Maize in dry and transitional zones. At the same time, the farmers of hilly and coastal zones grow paddy as rainfed crop. In the month of August, soon after the south west monsoon rainfall recedes, farmers practicing irrigation gets water from the reservoirs. During these period (August, September, October, November) through canal irrigation farmers grow wet crop such as paddy and wheat, which are harvested by the mid of January. Farmers owning both wet and dry land, fix the timings of the crops in anticipation of rainfall condition. Usually the dry summer spell in southern Karnataka prevails during February, March, April, and May. The convectional current rainfall which occurs in the month of April is highly suitable for preparing the land for the upcoming monsoon rains. As soon as the second convectional rainfall occurs in the last week of April or first week of May they broadcast dry cereal crops such as ragi, Jowar, Bajra. Most of the crops chosen are suitable to produce within the duration of 4 months. In its initial stages of growing, at an interval of 15 days convectional rainfall is required to keep the crop growing until the setting of monsoon, which begins in the first week of June. Once the on time monsoon rain begins it gives a great relief for the farmers until harvest. These crops are harvested during at the end of south west monsoon rainfall, i.e., August or September. A farmer owning irrigated land will prepare land for wet crop usually at the end of southwest monsoon period. The water from the reservoir are released almost at the end of southwest monsoon period, with the help of canal source of

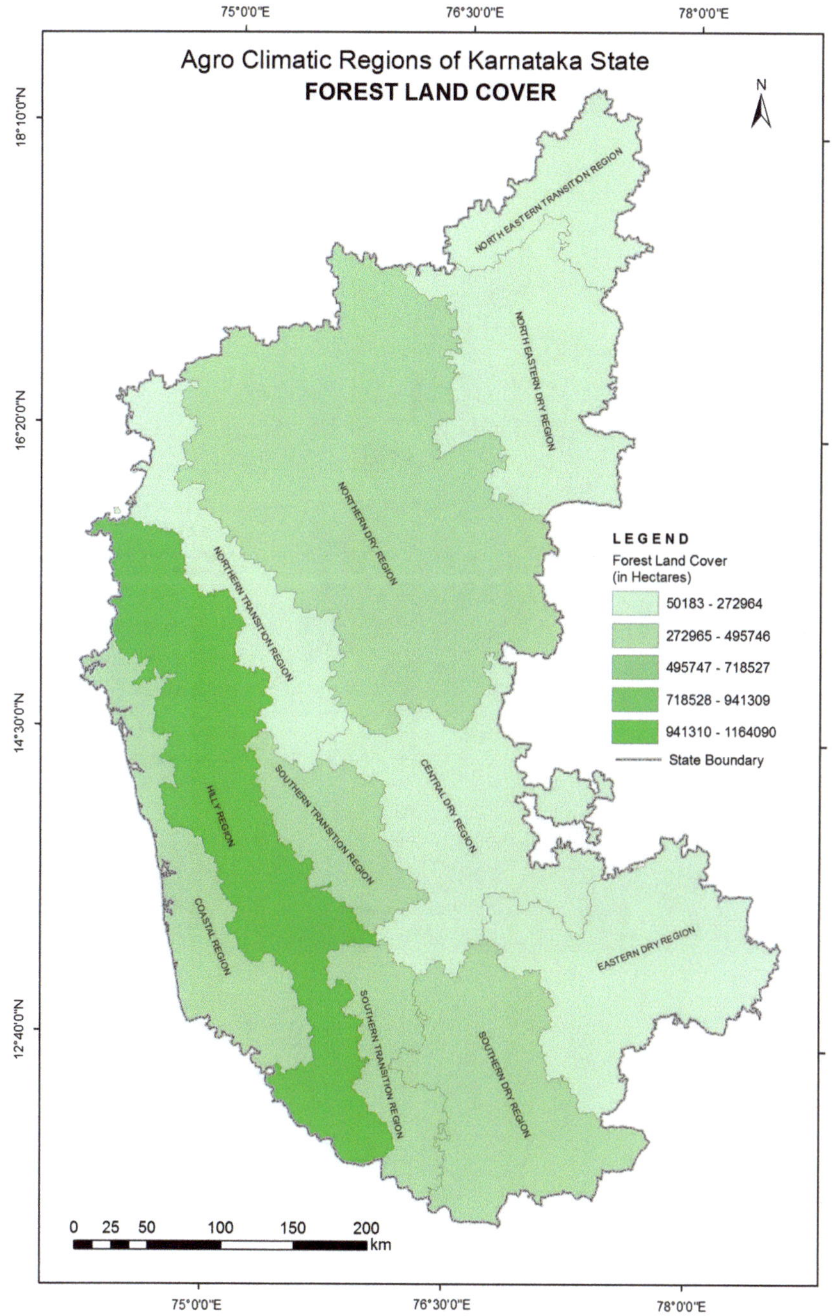

Fig. 6.4 Forest land of Karnataka State

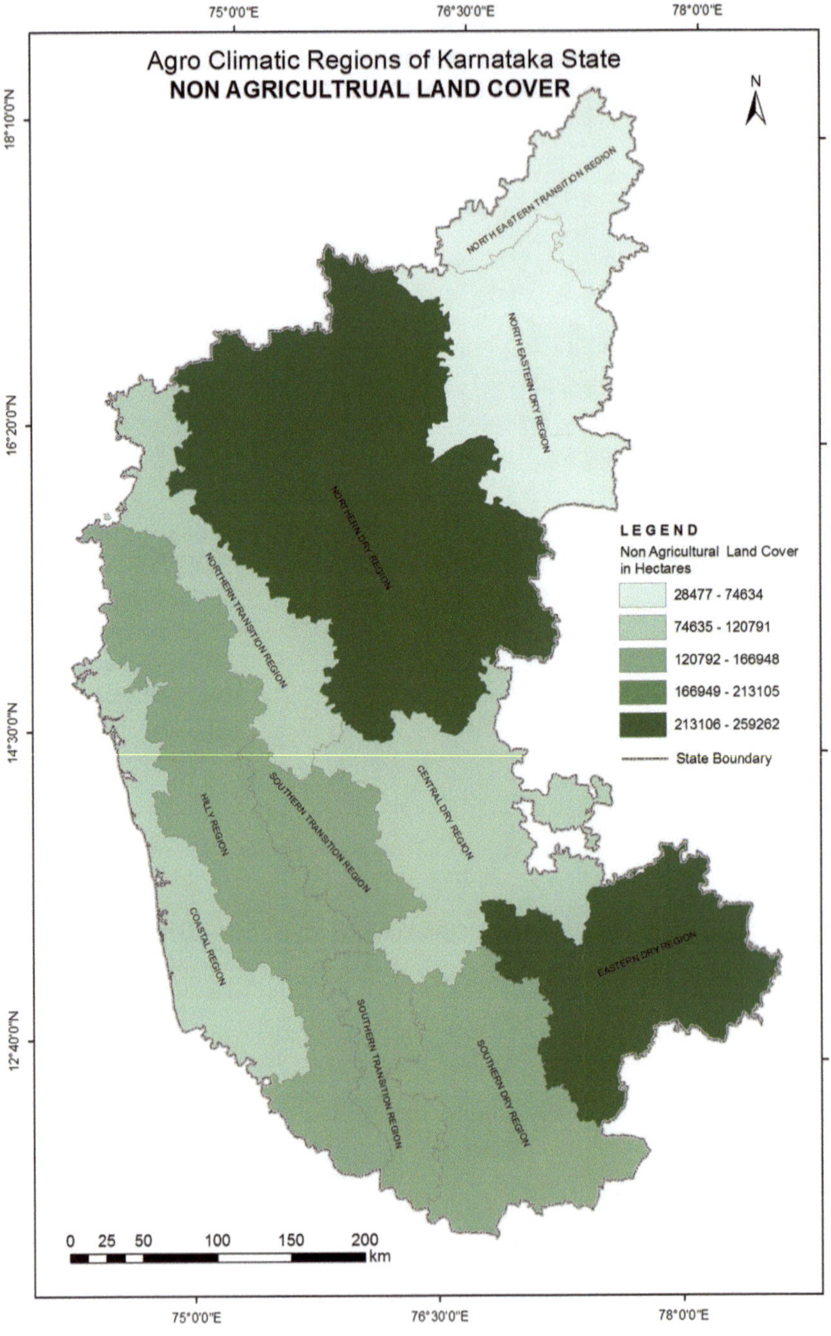

Fig. 6.5 Non agriculture land use of Karnataka State

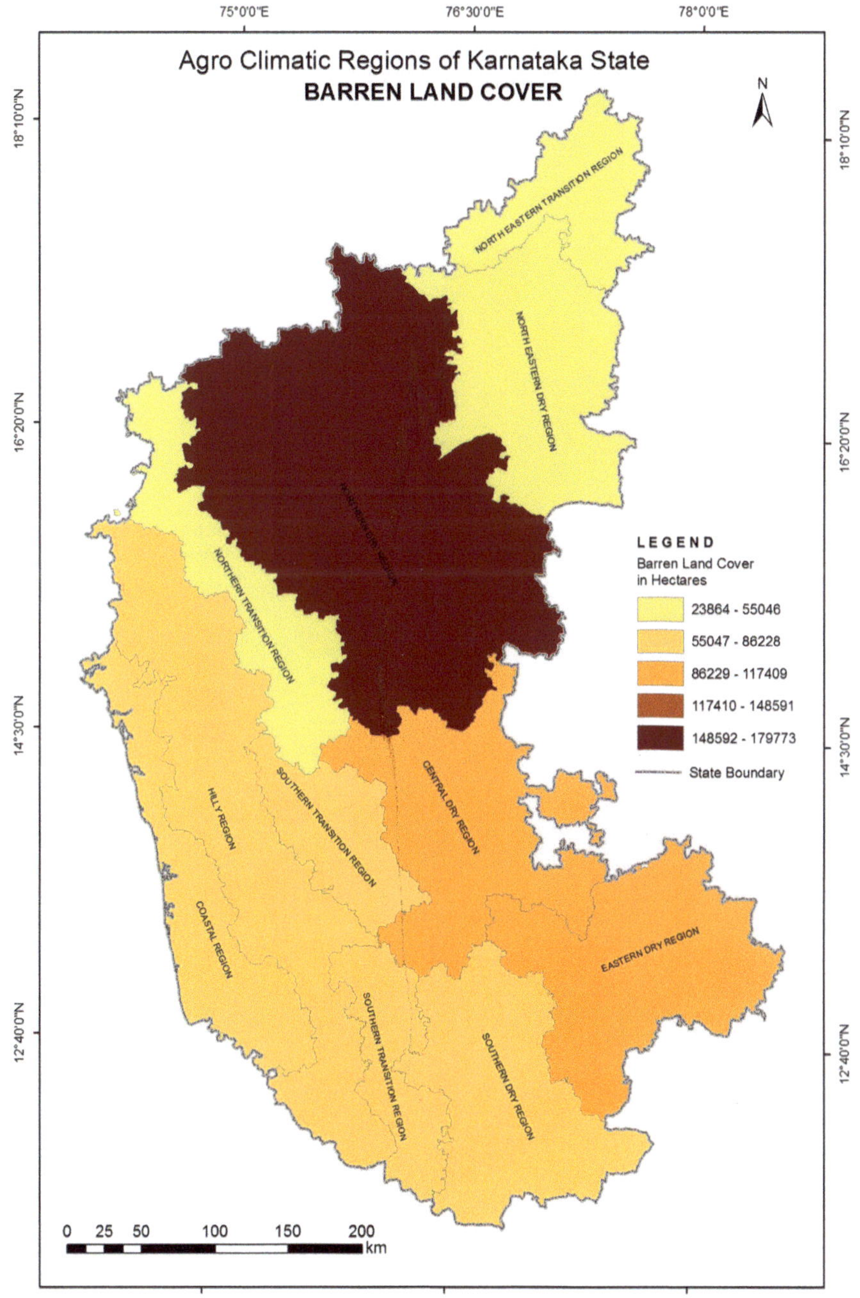

Fig. 6.6 Barren land of Karnataka State

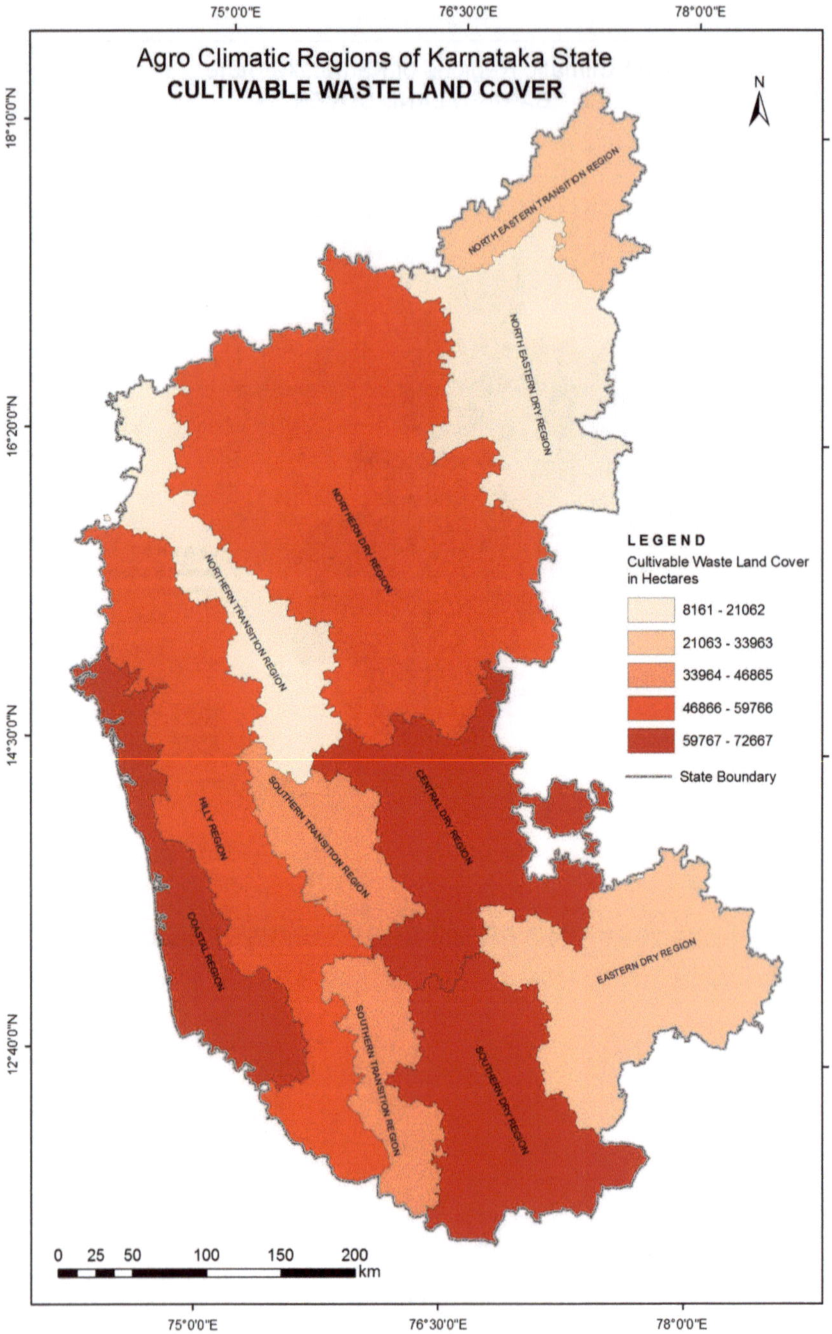

Fig. 6.7 Cultivable waste land of Karnataka State

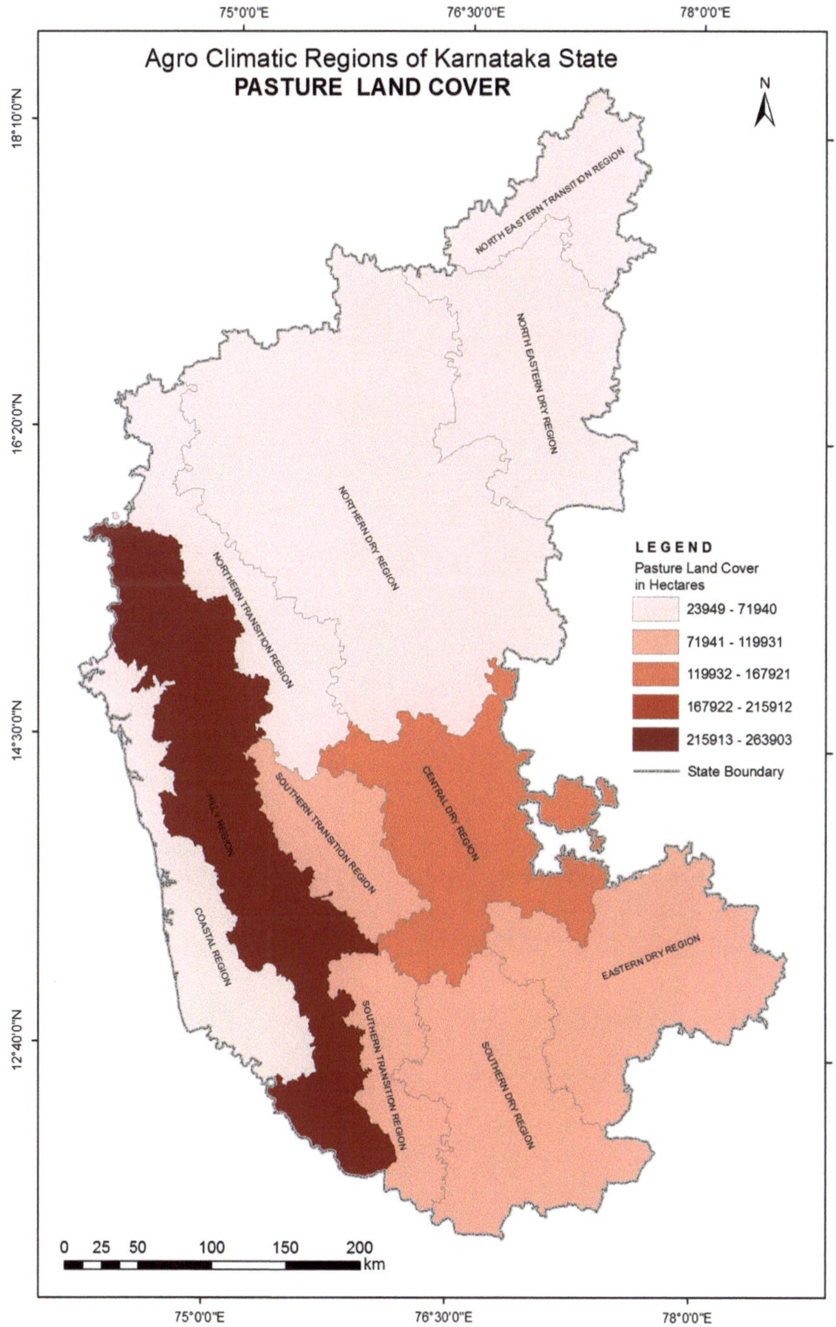

Fig. 6.8 Pasture land of Karnataka State

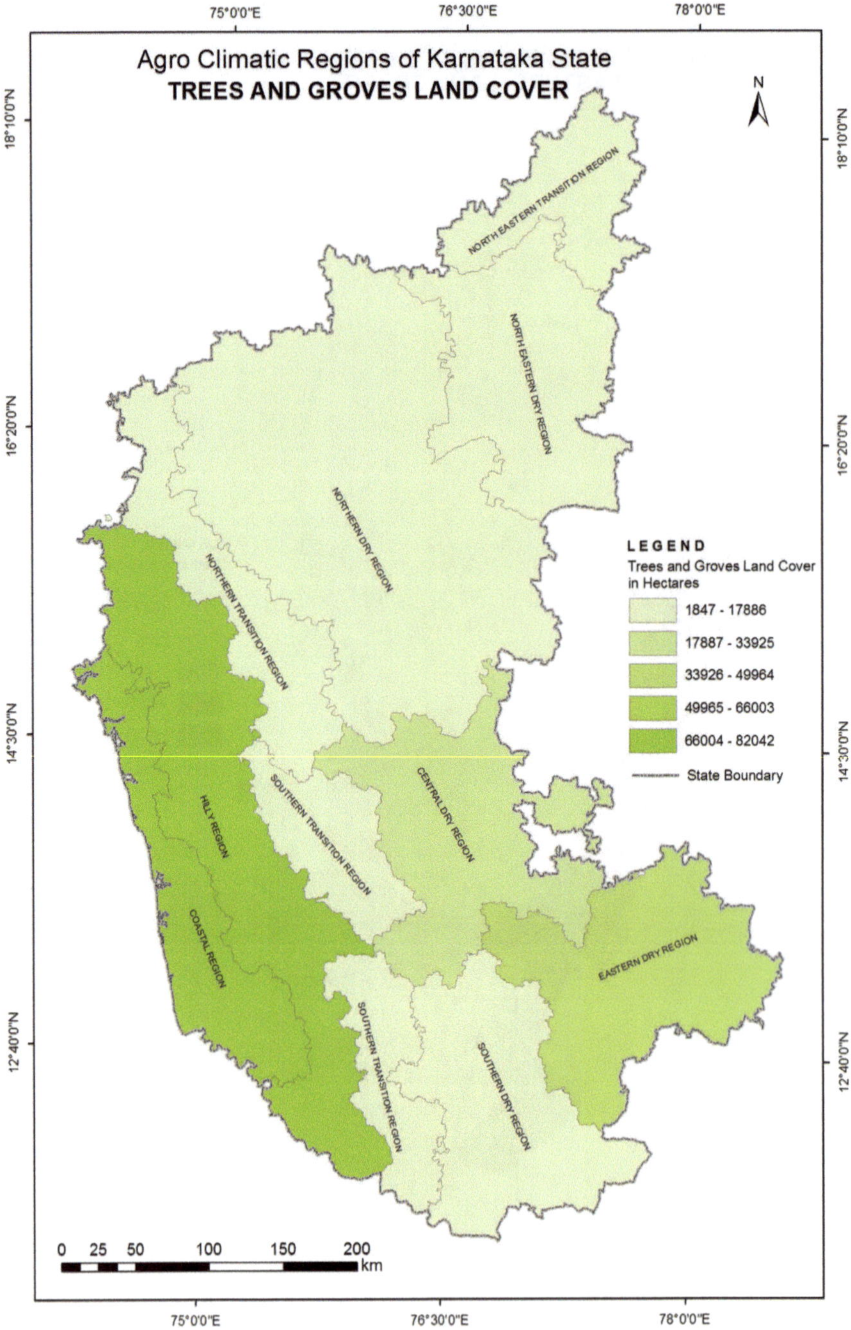

Fig. 6.9 Trees and groves of Karnataka State

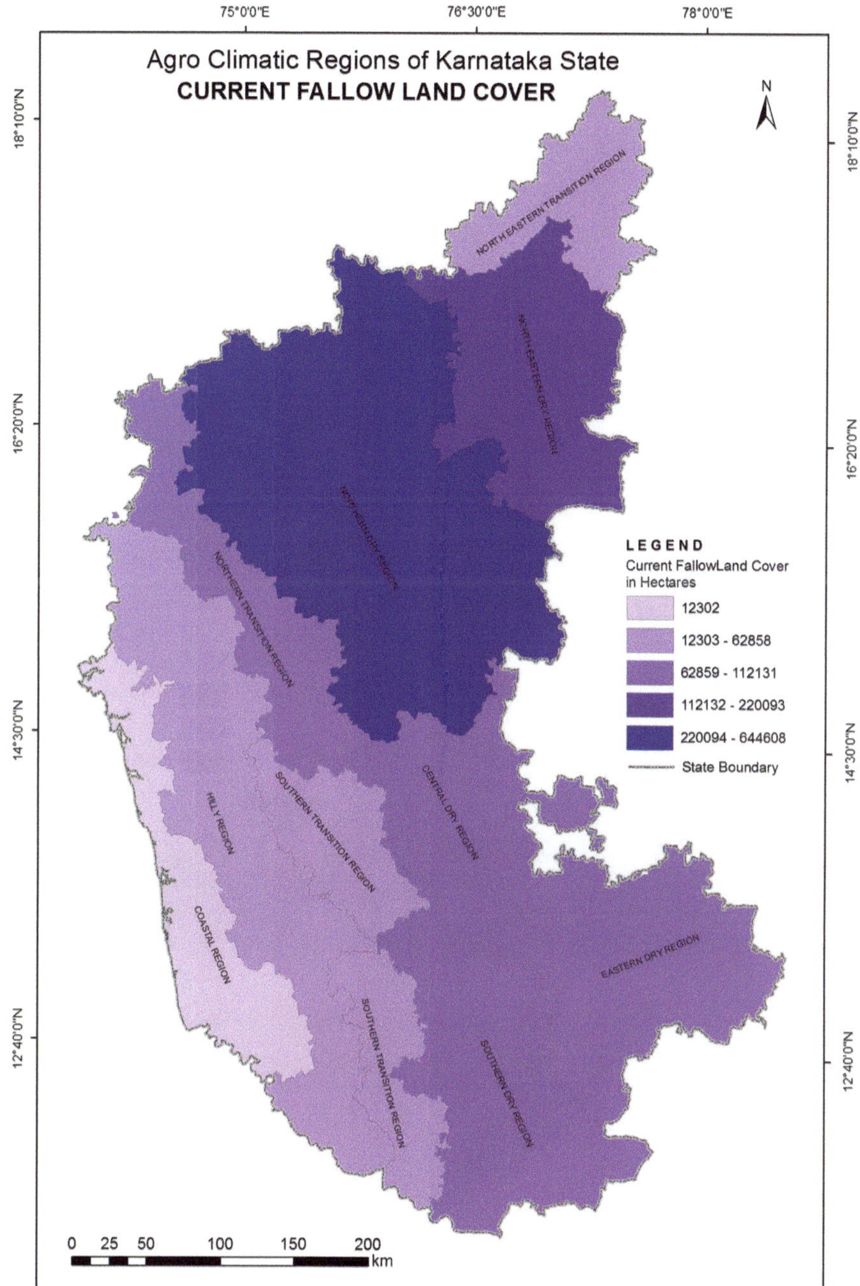

Fig. 6.10 Current fallow land of Karnataka State

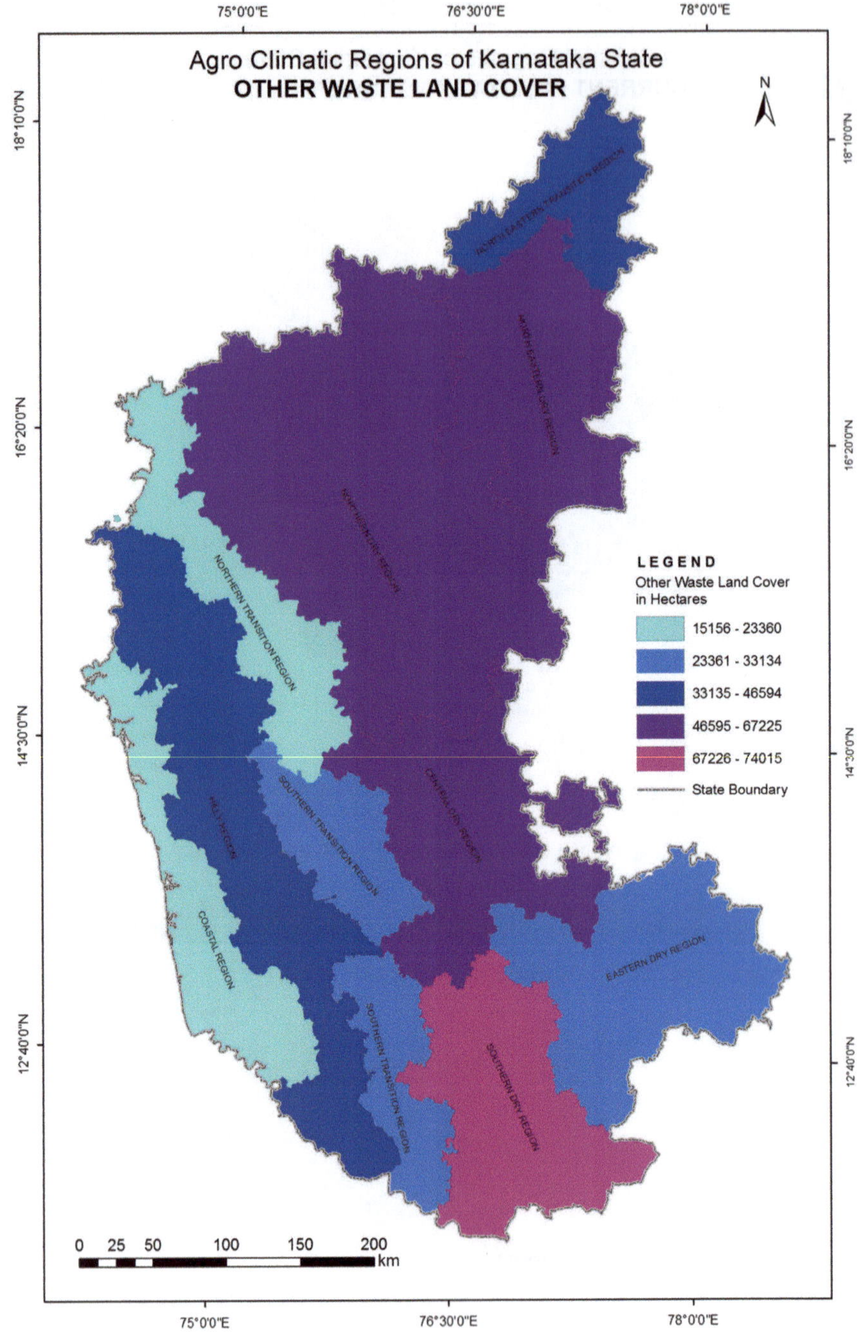

Fig. 6.11 Karnataka State other waste land

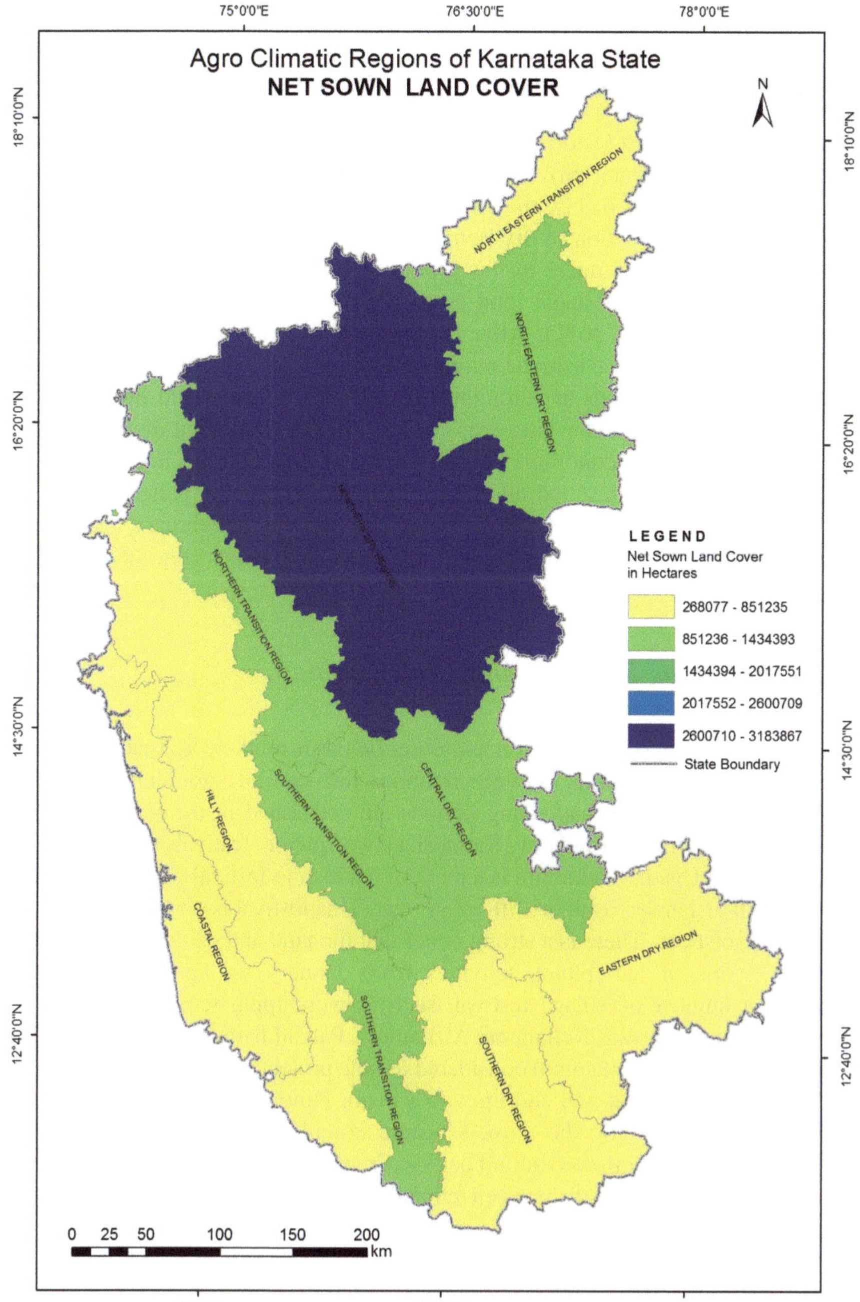

Fig. 6.12 Karnataka State net area sown

irrigation, wet farmers grow paddy in both southern and northern Karnataka. If there is a failure in monsoon rainfall, in such situation, the water released is adequate to grow only one crop. In case of one crop, the irrigated land farmers grow pulses and oil seeds as second crop. The moisture retained in the land after the monsoon spell between October, November, December are suitable to grow pulses and oil seeds. If the water provided for the second crop is not sufficient enough to produce wet crops like paddy, the farmers produce Ragi or Jowar or maize in the irrigated land. The situation of dry rainfed land scenario is different from dry land irrigated farmers. The farmers in this type of climatic condition during monsoon period produce cereal or staple food crops which are the most essential for their survival (Sharma et al. 2011). After harvesting the cereal crops they choose to produce pulses and oil seeds in the retreating monsoon period, i.e., from September to November. These crops are drought resistant and can withstand high temperature, they are also called as winter crops or rabi crops or retreating monsoon crops. Because of varying dry condition, farmers choose variety of crops in different parts of the state, which has resulted into multi- and mixed-cropping pattern.[3] Both dry and irrigated land will be left fallow during the months of February and March, which happens to be the resting period for farmers and cattles (Fig. 6.13).

6.3.2 Religious Festivals and Crop Timings

According to the field interrogation and observation in southern Karnataka, there is a close relationship existing between religious festival and agricultural activities. The commencement and harvesting periods are celebrated in the form of festival (Sivananda 2000). In southern Karnataka two major festivals are practiced; Sankranthi and Ugadi. Sankranthi is a mark of harvesting festival which is observed in the month of January 14th or 15th every year. This festival is observed on the day of summer solstices. There is a strong belief that the land and cattles should be free from all forms of agriculture activities from January to April. During these 2 months a number of village festival, such as worshiping female deities like 7 sisters (Chamundeshwari, Kellamma, Aladamma, Pattalddamma, Suralli Maramma, Godnalli Maramma, Kellamma) considered as sole protector of farmers community. All these village festival will end up with a major Hindu New Year festival called Ugadi. Soon after Ugadi, the farmers restart agricultural activity as soon as the conventional or pre-monsoon rainfall begins. From this one can understand that, the interrelationship that exists between climatic conditions, religious festivals and agriculture (Fig. 6.13; Table 6.2).

[3]Govt of Karnataka, India, Report, Prospective land use plan for Karnataka 2025.

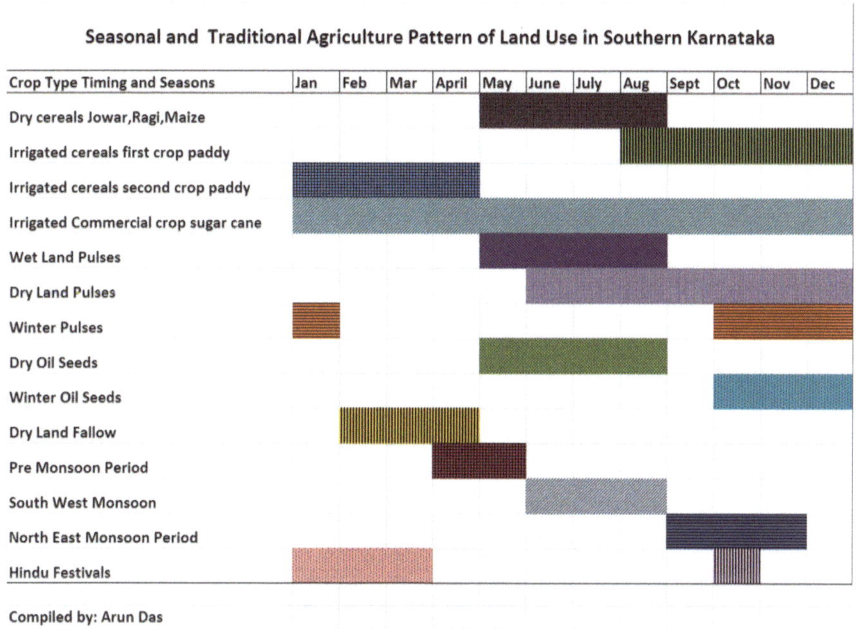

Fig. 6.13 Season and traditions based agriculture land use pattern of Karnataka (complied by Arun Das)

Table 6.2 Agro climatic regions of Karnataka

Sl. no.	Agro climatic regions	Area in hectares	In percentage
1	Northern dry zone	4,783,642	25.11
2	Central dry zone	1,943,830	10.20
3	Eastern dry zone	1,808,217	9.49
4	North eastern dry zone	1,762,604	9.25
5	Southern dry zone	1,647,612	8.65
6	Southern transition zone	1,309,847	6.88
7	Northeastern transition zone	871,036	4.57
8	Other waste	515,973	2.70
9	Net Sown Area	10,173,589	53.40
10	Total	19,049,836	100

6.3.3 Agro Climatic Regions of Karnataka

Based on the prevailing climatic conditions in Karnataka, the state has been classified into ten agro climatic regions (National Remote Sensing Agencies 2004). Each region has a unique characteristic feature with respect not only to climate, but also with respect to geomorphology, soil, vegetative cover, hydro-geology, and slope (Ramesh Kumar 2002). The categorized agro climatic regions are as follows (Fig. 6.2; Table 6.3).

Table 6.3 Agricultural land use of Karnataka. *Source* Karnataka at Glance, Govt of India, 2010)

Sl. no.	Crops	Area in percent
1	Cereals and millets	46.25
2	Pulses	14.05
3	Oil seeds	21.14
4	Commercial crops	8.04
5	Plantation and horticulture crops	10.52
6	State total	100

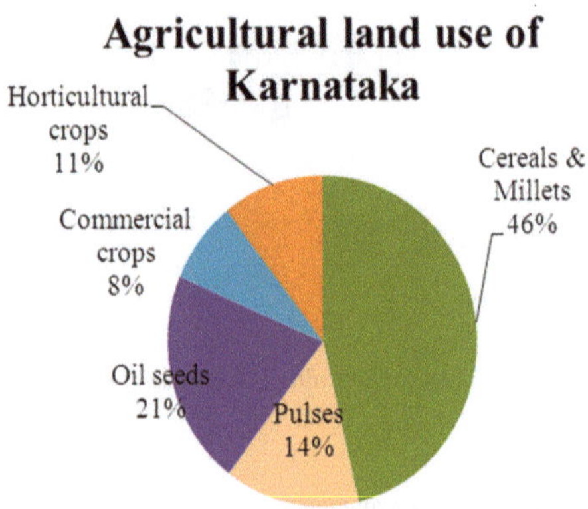

Fig. 6.14 Karnataka State net area sown

6.3.4 Agricultural Land Use of Karnataka

As stated in the above paragraph, Karnataka is a state of various climatic conditions with varied crop-growing regions. Agricultural department of Karnataka has classified crop-growing regions of the state into six categories, such as, cereal crops, pulses, oil seeds, commercial crops, plantation crops, and horticulture crops. Out of the total area sown, Cereal crops accounts 46.25% followed by Pulses 14.05%, Oil seed 21.14%, Commercial crops 8.04%. Plantation crops and Horticulture crops put together 10.52% (Fig. 6.14; Table 6.3).

6.4 Crop Wise Agriculture Land Use of Karnataka

6.4.1 Cereals

Among the various types of cereal crops cultivated in Karnataka, six types of crops are largely produced and another five types of cereals are produced at minor scale. The six major cereal crops produced are paddy which covers 28% of total Geographical area under cereal crop, followed by Jowar 26%, Maize 20%, Ragi 14%, Bajra 5%, wheat 5%, and other millets <0.5%. The major share of the cereal crops are seen in north Karnataka, which accounts almost 70% of total area under cereal crop. This clearly indicates that, though north Karnataka is experiencing semi-arid condition, it is possessing sustainable agro climate condition (Fig. 6.15; Table 6.4)

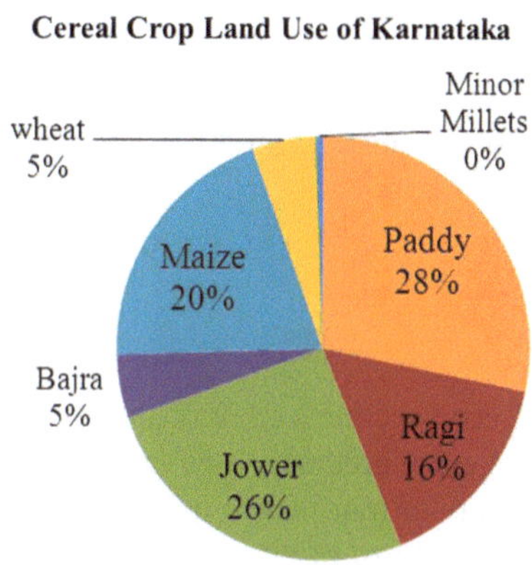

Fig. 6.15 Cereal crop land use of Karnataka

Table 6.4 Zonal level land use scenario of Cereal crops (area in hectares) *Source* Karnataka at Glance, Govt of India, 2010

Sl. no.	Crops	Dry zone	Transitional zone	Hilly zone	Coastal zone
1	Paddy	566,268	269,616	310,204	172,239
2	Jowar	1,549,584	404,599	10,000	0
3	Bajra	342,367	30,369	0	0
4	Maize	718,159	144,527	6224	0
5	Ragi	789,994	175,615	11,150	29
6	Wheat	191,469	47,665	125	0
7	Other cereals	60,489	20,152	616	0

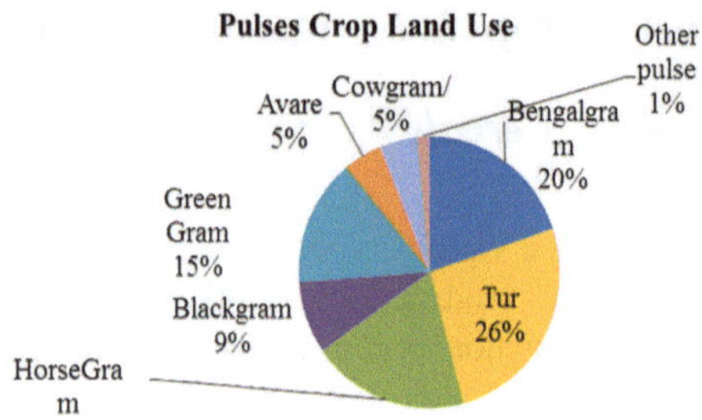

Fig. 6.16 Pulses crop land use

6.4.2 *Pulses*

Pulses are next important food crop next only to cereal crops. It is used for preparing of curry and side dishes. Pulses are as much as important like cereal crops, because without pulses there is no meal. But only by pulses there cannot be a meal. As such, pulses account 14% of the total cropped area of Karnataka. Although a variety of pulses are produced, the major pulses cultivated in all parts of Karnataka are tur gram 26%, bengal gram 20%, horse gram 13% (Fig. 6.16).

6.4.3 *Oil Seeds*

The total area under oil seeds accounts 21% of the cropped area of Karnataka. In general, oil is one of the major ingredients used in the preparation of all varieties of food. On an average, a family of four people uses 2 kg of oil per month. Karnataka is no exception to this, the land use of oil seed crop is different, compared to other type of food crops. Most popular edible types of oil used in Karnataka are groundnut, sesame, and sunflower. Total area under cultivation of groundnut accounts to 33 another 33 is under sesame whereas, sunflower occupies 26%. These three crops itself accounts 93% of the total area under oil seed cultivation. Remaining 7% is shared by sunflowers, soybean, linseed, castor, and other oil seeds. The recent health conscious and recommendation by doctors for its low cholesterol, groundnut oil has been replaced by sunflower (Sundeep and Manchanda 2012). Another reason for increasing land use of sunflower production is due to, its adaptability to harsh climatic condition and resistant to decease. Due to these features sunflower is cultivated both in summer and winter season, especially where day time temperature during winter ranges between 20 and 25 °C and summer day time temperature ranging between 30 and 35 °C. It is also a profitable crop compared to pulses. The area devoted for oil seeds production, is a clear indication of a profit and the climatological feasibility to produce oil crops (Fig. 6.17).

6.4.4 Commercial Crops

Major commercial crops cultivated in Karnataka are cotton, sugarcane, and tobacco. Commercial crops account to 8% of the total cropped area of Karnataka state. This clearly indicates that the state do not have a favorable ambiance for the commercial crops. Only Tobacco is exported for national market, where as sugarcane is marketed within the state. Cotton accounts to 62% of total area under commercial crop, and tobacco account to 7%. Cotton is mainly cultivated in northern Karnataka, whereas tobacco is concentrated mainly in southern Karnataka (Prasad 2007). In the case of sugarcane, it is an irrigated crop mainly cultivated in the southern Karnataka, which accounts 31% of total commercial area of the Karnataka state (Fig. 6.18).

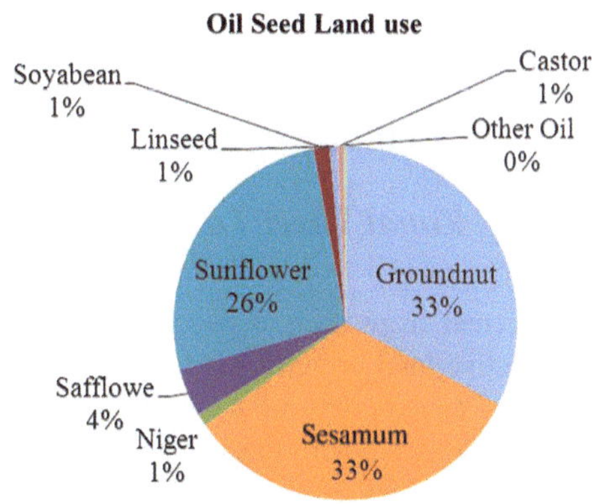

Fig. 6.17 Oil seed land use

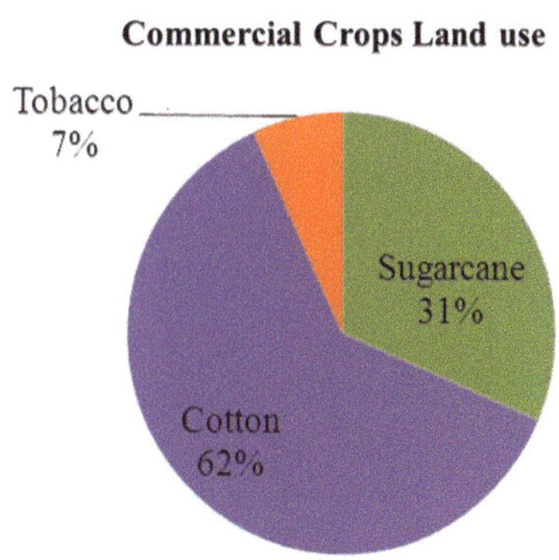

Fig. 6.18 Commercial crops land use

6.4.5 Plantation and Horticultural Crops

The total area covered by plantation and horticultural crops accounts to 11% to the total cropped area of the state. More than 50% of the total area under plantation and horticultural crop is covered by plantation crops, which accounts to 53% followed by condiments and spices 21%, vegetables 15%, fruits 11% and <0.5% by flowers. Among the plantation crops the main types of crops are coconut, areca nut, cashew nut, coffee, rubber, betel vine. In the case of condiments a variety of crops such as chilly, ginger, turmeric, cardamom, coriander, garlic, pepper, and tamarind are main crops. With respect to fruits, popularly grown crops are mango, banana, grapes, papaya, pomegranate, citrus, ber, guava, and sapota. All varieties of horticultural crops are cultivated in all parts of Karnataka whereas the planation crops are such as areca nut, cashew nut, coffee, and rubber are restricted in hilly regions due to topographical factors (Das 2013). These crops need abundant moisture without flood or stagnation. Areca nut is grown as the main crop in coastal region (Fig. 6.19).

6.5 Agricultural Land Use Under Each Zone

6.5.1 Zonal Level Land Use Scenario of Cereal Crops

The cereal crops such as paddy, jowar, ragi, maize, wheat and other cereals forms distinctive land use coverage under different zones. The interesting fact is that, all types of cereal crops are concentrated in dry zone (Karnataka at a Glance 2005), in comparison to other zones, such as transitional zone, hilly zone and coastal zone. Second, the wet crop such as paddy and wheat also cover 566.26 thousand and

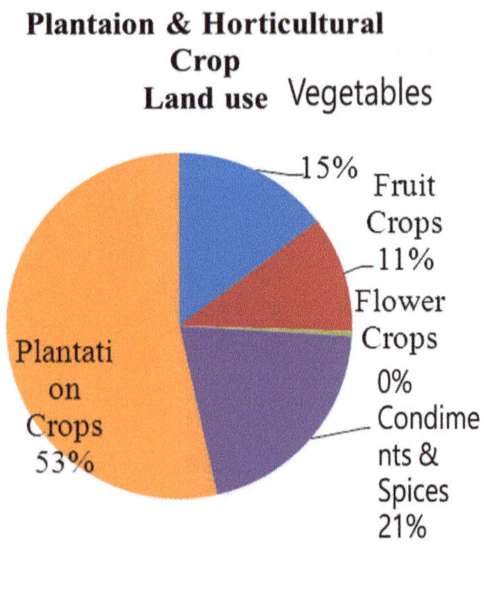

Fig. 6.19 Plantation and horticulture land use

191.46 thousands of hectares of land under paddy and wheat cultivation respectively. Similar to dry zone, transitional zone possess all types of cereal crops cultivation, but at lower scale compared to dry zone, whereas hilly and coastal zones are dominated by only one type of cereal crop, i.e., paddy. From the graph (Fig. 6.19), it is very much clear that, dry zones are highly productive followed by transitional zone, hilly zone, and coastal zone. Another interesting factor which emerges out of the analysis is, it also states that region with high rainfall and high temperature are less productive than region experiencing harsh climatic condition with low rainfall and with high temperature (Fig. 6.20; Table 6.4).

6.5.2 Zonal Level Land-Use Scenario of Pulses

The scenario of pulses land use in Karnataka depicts a distinctive feature as per zone. All types of pulses are produced in dry and transitional zones, with higher concentration in dry zone. With respect to hilly zone and coastal zone, very meager area is under pulses, which is <7000 hectares (see table and graph). In the coastal zone, only three types of pulses are cultivated at very small area of <3000 hectares, other than black gram which is cultivated over 10,221 hectares. From the table shown above, it is evident that the land use of pulses are more in dry and transitional zones, which varies between 303 thousand (0.3 million) and 20 thousand (0.02 million) hectares. Once again, it is proved that, dry and transitional zones are highly sustainable and productive compared to hilly and coastal zones in the case of pulses (Fig. 6.21; Table 6.5).

6.5.3 Zonal-Level Land-Use Scenario of Oil Seed Crops

Oil, which is one of the important ingredients in Indian food, is produced in most of the states of India, but all varieties of oil seeds are not produced. In Karnataka, most of the oil seeds are produced for domestic and commercial purposes (Yu 2015). It is

Fig. 6.20 Land use of cereal crops

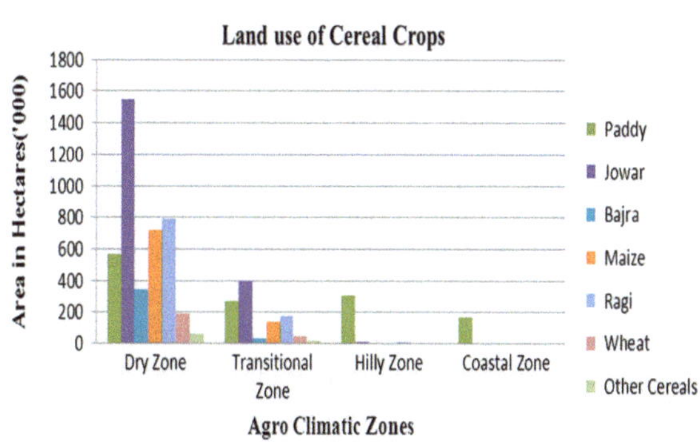

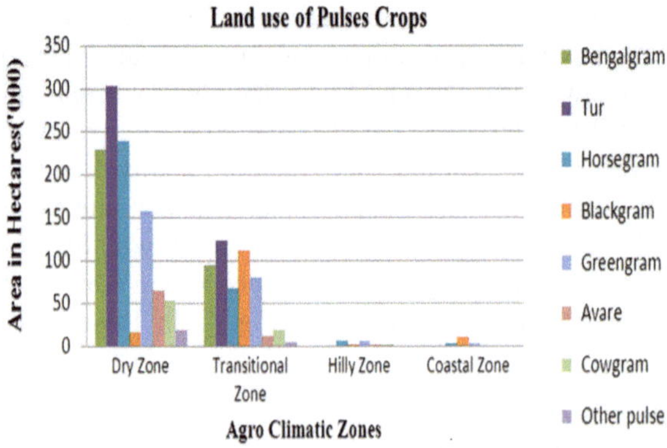

Fig. 6.21 Land use of pulses crop

Table 6.5 Zonal level land cover scenario of pulses crops (area in hectares) *Source* Karnataka at Glance, Govt of India, 2010

Sl. no.	Crops	Dry zone	Transitional zone	Hilly zone	Coastal zone
1	Bengal gram	229,572	95,147	1277	0
2	Tur	303,903	123,758	858	0
3	Horse gram	239,405	68,524	6184	3983
4	Black gram	16,762	111,891	1756	10,221
5	Green gram	158,480	80,590	6963	3730
6	Avare	64,561	12,795	2322	5
7	Cow gram/cowpea	53,966	19,892	2040	1122
8	Other pulse	19,411	4484	112	62

evident from the cropping pattern that the dry and transitional zones are the land under all varieties of oil seeds which is a good indicator of the sustainable characteristics of land compared to hilly and coastal regions (Fig. 6.22; Table 6.6).

6.5.4 Zonal-Level Land-Use Scenario of Commercial Crops

The black soil dry regions of northern Karnataka have been proved to be most congenial for the production of cotton. Nearly 345 thousands (0.35 million) hectares of land is under cotton cultivation which is in the dry zones of Karnataka. Under the commercial crop categories, cotton is the predominant crop having land use more than one time; Compared to other commercial crops both in dry and transitional zones. Sugarcane is produced with help of irrigation. It is seen as second next important commercial crop in dry zone, transitional zone, and also in hilly zone. Tobacco cultivation is seen to be higher in transitional zone compared to dry zone and transitional zone. It covers an area of 55 thousand hectares, whereas it

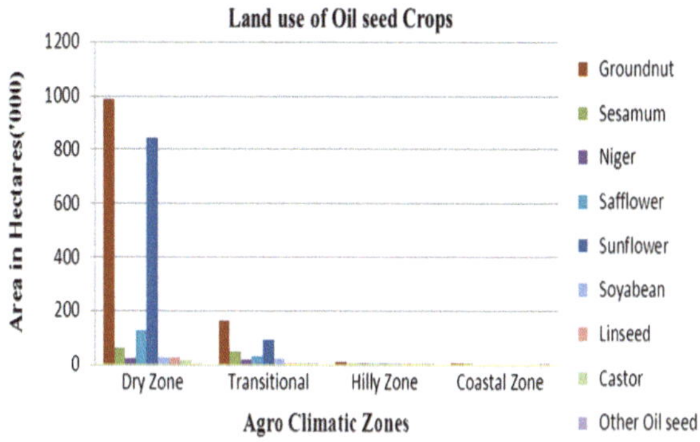

Fig. 6.22 Land use of oil crops

Table 6.6 Zonal level land use scenario of oil seeds crops (area in hectares) *Source* Karnataka at Glance, Govt of India, 2010

Sl. no.	Crops	Dry zone	Transitional zone	Hilly zone	Coastal zone
1	Groundnut	985,959	163,813	9062	6614
2	Sesamum	62,345	47,644	418	1015
3	Niger	23,434	18,022	427	0
4	Safflower	128,811	29,450	133	0
5	Sunflower	844,623	90,795	3240	0
6	Soybean	27,195	24,357	2019	0
7	Linseed	25,121	1520	391	0
8	Castor	17,616	3622	128	0
9	Other oil seeds	5352	1742	381	599

is only 11.4 thousand hectares in dry zone and 271 hectares in hilly zone. The congenial climatic condition prevailing in southern transitional zone of Karnataka is the major reason for higher concentration of tobacco land use. From the above-mentioned data, it is once again proved that dry and transitional zones are highly sustainable with respect to commercial crop compared to hilly and coastal zones (Fig. 6.23; Table 6.7).

6.5.5 Zonal-Level Land Use Scenario of Plantation and Horticultural Crops

Plantation crops such as, coconut, areca nut, cashew nut, coffee, rubber, betel vine are the major crops of hilly and coastal zones (Kannabiran et al. 2012). Since coconut has been categorized as a plantation crop and it is best grown in dry regions of southern Karnataka. The area under plantation crop in the dry zone is shooting up to an extent of 262.45 thousands (0.26 million) hectares. The plantation crop under

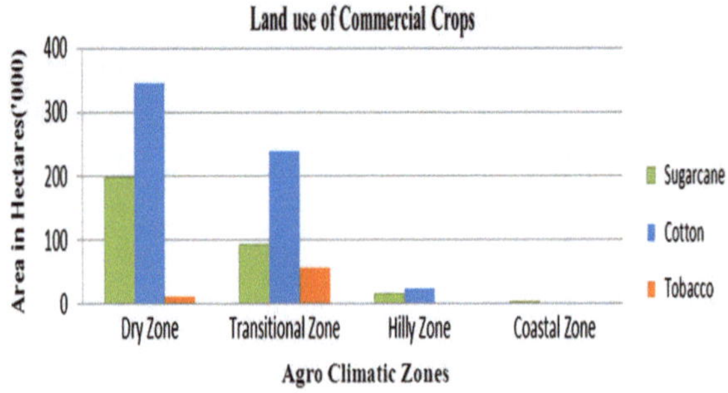

Fig. 6.23 Land use of commercial crops

Table 6.7 Zonal level land use scenario of commercial crops (area in hectares) *Source* Karnataka at Glance, Govt of India, 2010

Sl. no.	Crops	Dry zone	Transitional zone	Hilly zone	Coastal zone
1	Sugarcane	197,698	91,878	16,120	2767
2	Cotton	345,873	237,819	22,663	0
3	Tobacco	11,487	55,093	271	0

transitional zone occupies 68.18 thousands hectares of land under coconut cultivation. In hilly region, the planation crop occupies 204 thousand hectares of land, out of which 80% is under coffee, 10% under rubber, another 10% under areca nut. The plantation crop in coastal zone is mainly occupied by Areca-nut, which accounts to 118.3 thousand hectares. The scenario of vegetables, fruits, and condiment crops are more are less similar to the other food crops. Taking into account, the area under, each of these crops, although dry and transitional zones possess larger area it does not imply that, the coastal and hilly zones are unfavorable or non-conducive to these crops. The share of total cropped area under each zone is larger compared to other three zones, imperatively dry zone possess larger area with respect to plantation and Horticultural crops, see the given table and graph. Once Again, It is proved that, sustainability with respect to plantation and horticultural crop his equally shared in all the regions of Karnataka (Fig. 6.24; Table 6.8).

6.6 Zonal Wise Irrigated and Rainfed Agricultural Land Use Scenario

6.6.1 Cereals

Paddy cultivation in India is concentrated in the southern states like Andhra Pradesh, Tamil Nadu, Karnataka, and Kerala. The second most important paddy-producing region is located in the Gangetic and Indus plains of India. Third northeast hilly region and northern Himalayan regions produce paddy. Paddy being

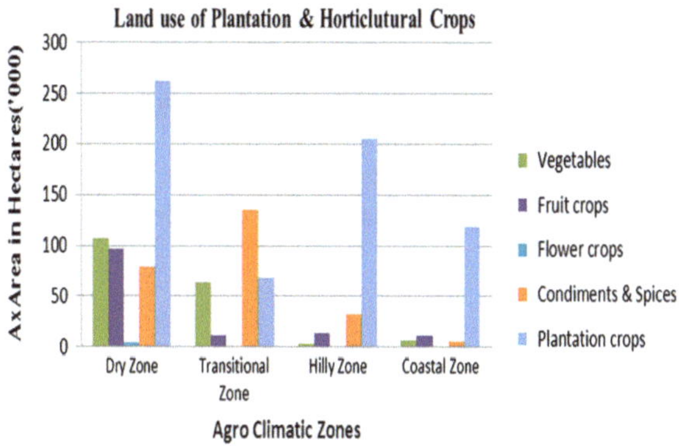

Fig. 6.24 Land use of plantation and horticulture crops

Table 6.8 Zonal level land use scenario of plantation and horticultural crops (area in hectares) *Source* Karnataka at Glance, Govt of India, 2010

Sl. no.	Crops	Dry zone	Transitional zone	Hilly zone	Coastal zone
1	Vegetables	106,774	63,581	3153	6276
2	Fruit crops	96,836	11,740	13,320	10,828
3	Flower crops	4443	651	159	159
4	Condiments and spices	78,847	134,726	32,180	5382
5	Plantation crops	262,452	68,181	204,708	118,334

a crop grown in high temperature with high moisture is concentrated in the high rainfall regions and irrigated tracks in Karnataka. Among the cereal crops paddy is the largest wet crop. Among the four zones, the natural climatic conditions suitable to grow paddy is seen in coastal, hilly, and transitional zones. Whereas in the case of dry zone, paddy cultivation is seen as an irrigated crop due to inadequate rainfall. Among the cereal crop, jowar is the largest cereal crop produced in Karnataka, which is followed by ragi, maize, bajra, and wheat. The transitional, hilly, and coastal zones are not as congenial as of dry zone for producing jowar. In fact, jowar is produced in the transitional zone as an irrigated crop for an extent of 500 thousand hectares. Through this it is very much clear that, jowar cultivation is concentrated in the dry zones of Karnataka rather than other zones. The unique climatic conditions prevailing in the northern Karnataka with black soil has favored jowar cultivation, which is also the staple food of this region. Similarly ragi is concentrated in dry and transitional zones with less area under hilly and Costal zones. The fourth important cereal crop maize also seen as a rainfed and irrigated crop in dry zone and it is rainfed in transitional zone, whereas in hilly and coastal zones it is very less due to adverse climatic condition. The scenario of bajra land use, is almost similar to maize, where dry and transitional zones occupy the major share of bajra cultivation. Wheat cultivation requires high temperature along with higher rainfall and humidity. The concentration of wheat cultivation is seen in dry and transitional zones, both as rainfed and irrigated crop. This clearly states that, the dry region is more sustainable than rainfed regions. Nearly 75 thousand hectares of land in dry zone is under irrigated wheat crop, whereas it is 115 thousand hectares under rainfed condition. Transitional zone also possess larger area under rainfed condition and nearly 18 thousand hectares of land under irrigated condition.

6.6.2 Pulses

Pulses are largely grown as rainfed crop. The graphs shown below neatly depict the concentration of all types of pulses in dry and transitional zones and also as a rainfed crop. The hilly and coastal zones possess a weak land use scenario with respect to pulses due to high rainfall.

6.6.3 Commercial Crops

Among the commercial crops cotton and tobacco are cultivated as a rainfed crop. These crops are also concentrated in dry and transitional zones. In the case of tobacco, transitional zone possess larger area under rainfed condition compare to dry zone rainfed tobacco. With respect to hilly and coastal zones, it is marginal. Under commercial crops, sugarcane is the only crop which requires high amount of moisture and temperature condition. Although it is found in all zones of Karnataka, but it is concentrated in the dry and transitional zones as an irrigated crop. Another reason for less area under sugarcane is due to huge capital investment and it is also for being an annual crop. Compared to sugarcane the other plantation crops are more profitable in hilly and coastal zones, Hence sugarcane cultivation is seen to a limited extent in these two zones.

6.6.4 Oil Seed Crops

Among the oil seeds, groundnut ranks first in area as well as production. Due to higher demand and profit, groundnuts are produced on large quantity only after the harvest of cereal crops, i.e., during the period of September–December. According to agricultural statistics, groundnut is produced as a rainfed crop to an extent of 768 thousand hectares of land. In places where water is limited for irrigation either through canal or from tank or bore wells, groundnut is cultivated as irrigated crop. The scenario in the transitional zone is almost similar to dry zone, where rainfed area is greater than irrigated area. In the other two regions, i.e., hilly and coastal, groundnut cultivation is very much limited because of moist soil condition which leads to rotting of nuts. Next to groundnut, sunflower occupies major share with respect to land and production. Cultivation of this crop also is in timing with groundnut. The regions which are having hard and black soil, sunflower is suitable. The irrigated condition is not suitable for sunflower cultivation, as this crop requires high temperature with moderate rainfall. Because of this reason, sunflower cultivation is totally absent in coastal region and about 3000 hectares of land is seen in hilly region. With respect to safflower, it is third major oil seed crop in Karnataka. Another interesting aspect of this crop is that, it is grown only in rainfed regions under very harsh soil and climatic condition. The total area of this crop accounts to 125 thousand hectares in dry regions. In transitional zone, it is around 29 thousand hectares of area. This crop is totally absent from production in hilly and coastal zones. There are many other variety of oil seed crops cultivated in the dry regions of Karnataka, under rainfed condition such as sesamum, niger, linseed, castor oil, soya, mustard. These crops are cultivated in a limited area in the transitional zone. Whereas in hilly and coastal zone it is almost negligible. With respect to oil seed land use scenario, once again the dry region proves a high level of sustainability existing in this region rather than the coastal and hilly region.

6.6.5 Plantation and Horticultural Crops

The plantation and Horticultural crops are although conducive to be produced in coastal and hilly zones. Due to the limited land available and other high profit earning crop such as coffee and other condiment crops, vegetable, fruits and flowers are grown to a limited extent. But in the case of dry and transitional zones vegetables, fruits, and flowers are also cultivated due to the conducive climatic conditions. Whereas in the case of plantation, the hilly zone possesses coffee, coastal zone possesses areca nut and dry zone possess coconut as a plantation crop. The condiments such as chilly, ginger are major crops in dry and transitional zones. Whereas, pepper, cardamom, and ginger are major condiment crops in coastal and hilly zones. The land use distribution of plantation and horticultural crops also substantiate that dry regions are highly resistive and sustainable compare to other regions.

6.7 Conclusion

The agricultural production of land is determined by physiographic and agro climatic regions, whereas, the carrying capacity of land is determined by land productivity. Ultimately the land determines the carrying capacity and sustainability. The sustainability factor is a crucial element, which remains as a camouflage, more or less as a positive agent. But when greater attention and focus is leveled toward, which type of land possesses greater carrying capacity and sustainability, the actual fact will get revealed. In the case of Karnataka state, in order to understand the land production and sustainability factor, the ten agro climatic regions have been further merged into four broad categories, without curtailing or overlooking the climatic effects on each region. The four broad classifications are, dry zone, transitional zone, hilly zone, and coastal zone. Dry zone comprises of North Eastern zone, Central dry zone and Southern dry zone. Similarly transitional zone comprises of north eastern transitional zone, south transitional zone, and northern transitional zone. The dry and transitional zones are clubbed together to get a broader prospective related to agricultural land use. Overall scenario of cereal crop cultivation in Karnataka depicted through the figures clearly indicates that, both wet and dry cereal crops are highly concentrated in dry and transitional zones. These two regions are also most congenial to produce all types of cereal crops under good rainfall conditions. Because of these conditions virtually the dry and transitional zones are possessing higher land-carrying capacity with high level of sustainability compared to coastal and hilly zones. The recent suicide incidents in Karnataka substantiate the decline sustainability of irrigated wet crop cultivation against the dry and rainfed crop sustainability. More than 110 farmers committed suicide who happens to be irrigated land owners. The assured income from irrigated land has given scope to borrow money from money lenders leading farmers to fall in a debt trap. This ultimately pushes them into a serious step of committing suicide. In contrast to the farmers of the irrigated land, the situation of dry land farmers who are practicing multi-crop cultivation and are also actually gambling with monsoon rainfall could able to get profit without committing suicide. The entire analysis with respect to the sustainable factor clearly reflects that the dry land farmers are careful in financial management and are also proved as successive farmers.

Acknowledgements This chapter is the revised edition of SLUAS Science Report Vol. 5 pp. 127–147. **Landuse Sustainability of Agricultural Zones** was an endeavour of many people who have contributed directly and indirectly. I personally wish to thank all my students for their efforts in generations of remote sensing data and also in the preparation of maps and charts. This research work would not have been carried out without the financial support from the JSPS Japan. Prof.Himiyama being the LULCA project director recognised and encouraged me for five years. The final out come are the two papers published in this book and also many more papers. I owe my indebtedness to Prof. Koichi Kimoto who introduced me to Prof. HImiyama.

References

Acharya SP, Basavaraja H, Kunnal LB, Maharansett SB, Bhat ARS (2011) Crop diversification in Karnataka: an economic analysis. University of Agricultural sciences, Dharwad

Das P (2013) Cropping Pattern (Agricultural and Horticultural) in different zones, their average yields in comparison to National average/critical graph/reasons identified and yield potential. Indian Council of Agricultural Research, New Delhi

Diouf J (2011) The state of the worlds land and water resources for food and agriculture

Dutta and Deven (2009) Encroachment creating alarming situation, Guwahati

Gadgil M, Narendra Prasad S, Ali R (1983) Forest management and forest policy in India: a critical review

Kannabiran K et al (2012) Legal Clinics and Adivasi rights: a report of national workshop. Econ Polit Weekly 47:21–24
Karnataka at a Glance (2005) Directorate of Economics and Statistics, Govt of Karnataka
Khashempur B (2006) Karnataka agricultural policy. Department of Agriculture and Horticulture, Bangalore
Kumara BR, Hosamani SB (2012) Cost and returns of major cropping system in northern transitional zones of Karnataka. University of Agricultural sciences, Dharwad
Ministry of Forest, Govt of India (2005) Report, Forest Tree Resources in States and Union Territories. N.G Raghu and Associates, Centre for Environmental Law education, India, Soil Resource Management for Land Use Planning
National Remote Sensing Agencies (2004–2005) Department of Space, Govt of India, national land use and Land cover mapping using multi-temporal AWiFS data, project report 2004–2005
Nayak BP, Kohli P, Sharma JV (2012) Livelihood of local communities and forest degradation in India: issues for REDD+, Delhi
Prasad VM (2007) Case study of tobacco cultivation and alternate crops in India, Delhi
Ramesh Kumar SC (2002) Economic land evaluation for sustainable Land management of watersheds in different Agro-Climatic zones of Karnataka. National Bureau of Soil Survey, Bangalore
Sarkar A (2008) Geo-spatial approach in soil and climatic data analysis for agro-climatic suitability assessment of major crops in rainfed agro-ecosystem, Dehradun
Sawaya AL, Martins PA, Jose V (2004) Impact of globalization on food
Sharma BR, Rao KV, Vithal KPR, Amrassinghe UA (2011) Realizing the potential of rainfed agriculture in India, Hyderbad
Sivananda S (2000) Hindu fasts and festivals
Sundeep M, Manchanda SC (2012) Cooking oils for heart health, New Delhi
Sundar N (2013) Violent social Conflicts in India's Forests: society, state and the market, New Delhi
Yu J (2015) Effects of subsidized crop insurance on crop choices, San Fancisco

Chapter 7
Low Carbon Resilient Delhi Megacity for Sustainable Future Earth

R.B. Singh, Subhash Anand and Vidhi Saluja

Abstract The low carbon economy has a minimal output of greenhouse gas emissions into the biosphere, specially refers to carbon dioxide. Urbanization is one of the significant drivers for rising CO_2 emissions. In India, Delhi appears to be a good geographical location for a study of low carbon society because it represents a large urban area where emissions have reached significant levels and are continuing to grow rapidly. This paper aims to prepare a CO_2 emission inventory for Delhi and identifying major sources and sectoral contributions. The primary and secondary survey was carried out to fulfill the objectives of the study. Based on the certain factors, emissions from transport, residential, industrial, and transport sector were calculated. The solar use feasibility index for Delhi was been prepared based on index overlay method. Simulation of proposed solar water heating systems has been done using RET screen International Computer Software. The problem of CO_2 emission is even seen prominently within Indian mega cities. Among various mega cities, Delhi is the most affluent city as it supports a population of 16.75 million. Transport is the leading producer of CO_2 in Delhi, followed by residential, commercial and industrial sector. Being a national capital, Delhi continues to attract the population and to cope with that the residential and industrial development has to take place. So, it is important that the prime attention should be given to make the city environmentally sustainable (use of renewable sources of energy).

Keywords Urbanization · Low carbon · Greenhouse gases · Sectoral Sustainability · Delhi

R.B. Singh (✉) · S. Anand · V. Saluja
Department of Geography, Delhi School of Economics,
University of Delhi, Delhi 110007, India
e-mail: rbsgeo@hotmail.com

S. Anand
e-mail: sanandpv@yahoo.co.in

V. Saluja
e-mail: vidhi.saluja@gmail.com

Y. Himiyama (ed.), *Exploring Sustainable Land Use in Monsoon Asia*,
Springer Geography, https://doi.org/10.1007/978-981-10-5927-8_7

7.1 Introduction

The low carbon economy means an economy with minimum emission of greenhouse gas (GHG) such as carbon dioxide (CO_2), carbon mono oxide, etc. The CO_2 in the atmosphere is a natural temperature regulating mechanism that keeps the earth's lower atmosphere warmer. Urbanization is the one of the significant drivers for increased CO_2 emissions. According to Climate Change Agenda-2012, cities and urban areas consume 75% of world energy and are responsible for 75% of GHG emissions. Global warming caused by increased CO_2 concentration has an alarming impact on the earth's fragile environment.

More than 50% of the world population lives in cities. Global carbon dioxide emissions have risen 19% since 1990. China has largest average annual CO_2 emissions, i.e., 22.5%, whereas India's average annual emission is 5.05% only. India is the fourth largest producer of CO_2 in the world. In India, fossil fuel burning is a principal means of CO_2 emission that results largely from coal burning. Coal fired power plants generate 55% of India's electricity and are responsible for CO_2 emissions. In developing countries like India, climate change would represent an added impact on ecological and economical systems. The problem of CO_2 emission is even seen within Indian mega cities. Among various megacities of India, Delhi has estimated CO_2 emissions of around 15.42 million metric tons (Dhamija 2010). Therefore, in order to achieve sustainability for the future, there is a need to conduct a study on low carbon society because Delhi has a large urban population where emissions have reached significant levels and are continuing to grow rapidly in future.

7.2 Study Area

Delhi is located between 28°37′ N to 28°61′ N and 77°14′ E to 77°23′ E. It borders the Indian states of Uttar Pradesh on East and Haryana on West, North, and South (Fig. 7.1). Delhi lies almost entirely in the Gangetic plains, two prominent features of the geography of Delhi are the Yamuna flood plain and the Delhi ridge. The National Capital Territory of Delhi is spread over an area of 1484 $Km^2$with 16.7 million population is the second largest mega city of India (Census of India 2011). The Delhi has become an area of trade and commerce and industry in the northern region of India.

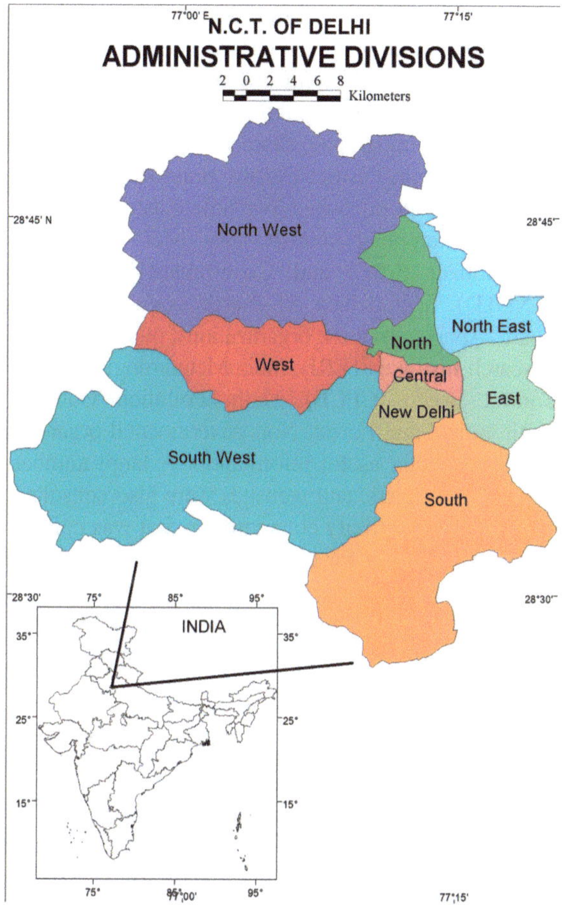

Fig. 7.1 Location Map of Delhi *Source* Census of India 2001

7.3 Data Sources and Methodology

The study is based on primary as well as secondary data sources. The primary data have been generated through structured questionnaires and field observations. 200 respondents were surveyed from households, commercials, industrial, and power sector. For the purpose of household survey, 50 respondents each from Greater Kailash—Higher Income Group (HIG), Vikaspuri—Middle-Income Group (MIG), and slums of Vikaspuri were surveyed. Primary data were also collected for commercial and industrial sector. For the purpose of primary data collection, ten respondents were questioned from Life Style Mall (Rajouri Garden) and ten from Ashoka hotel (Chanakyapuri). Similarly, 30 respondents were interviewed from industries of Okhla and Mayapuri which includes wire, garment, and electronic industries. The questionnaire consisted question on their opinion and awareness about low carbon society, eco-city, green buildings, carbon trading, energy efficiency and conservation, use of non-conventional energy resource, and people's perception on making low carbon society in Delhi.

The relevant and useful information have been obtained from secondary sources both in published and unpublished form. The background information pertaining to an extent, area, demographic characteristics, urbanization and socioeconomic profile have been taken from District Census Handbook of Delhi. Required data, comprehensive information and published available materials related to growth in vehicular population over temporal context have been collected from Centre for Science and Environment (CSE) and Government of National Capital Territory of Delhi (GNCTD). Spatial data on energy consumption was taken from Delhi Electricity Board. Apart from these organizations, other data were collected from The Energy and Research Institute (TERI), India Meteorological Department (IMD), Central Pollution Control Board (CPCB), Delhi Pollution Control Committee and Ministry of Environment and Forest. Non-governmental organization, i.e., Green Peace was also visited to obtain useful information. A large number of published reports, journals, books, newspapers, and websites were also consulted.

A unit of electricity consumption that was collected through primary survey was taken into account for calculating CO_2 emissions using the following formula:

$$\begin{aligned}\text{Burning of 1 kg of Coal} &= 2\,\text{kWh of electrical energy}\\ \text{1 kg of coal emits} &= 2.06\,\text{kg of CO}_2.\\ \text{1 kg of Coal} &= 2\,\text{kWh} = 2\,\text{Units}\\ \text{1 Unit} &= 2.06/2 = 1.03\,\text{kg of CO}_2.\end{aligned}$$

The solar use feasibility index for Delhi was been prepared based on index overlay of all the selected indicators (Fig. 7.2). Simulation of proposed solar water heating systems including their capacity, cost, and financial feasibility has been done using RET (Renewable Energy Technology) screen International computer software.

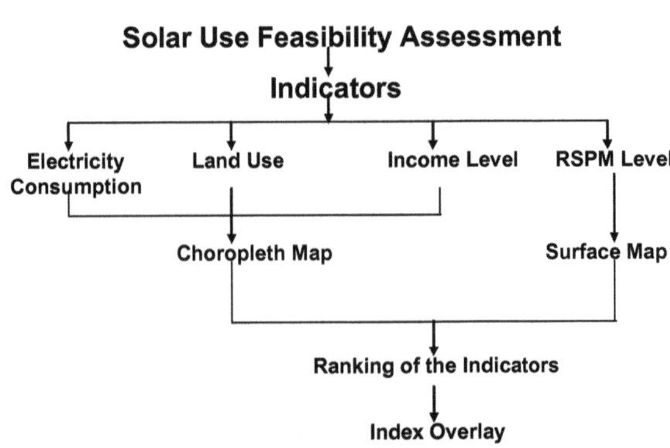

Fig. 7.2 Solar use feasibility assessment indicators using index overlay

7.4 Results and Discussion

7.4.1 Sources of Carbon Emissions in Delhi

a. **Transport sector**

Transport plays a crucial role in the development and constitutes a major share of energy consumption. Transportation primarily relies on petroleum and diesel, which supplies nearly 95% of the total energy use of world transport and accounts for nearly 60% of oil consumption (Bandhopadhaya 2010). Delhi's transport sector emits 12.39 million tons of carbon dioxide (Rediff 2015). Hence, the transport sector has also been largely responsible for the pollution and CO_2 emissions. Delhi has 85 private cars per 1000 population against the national average of 8 private cars per 1000 population (International Business Times 2015). CO_2 levels are greatly exceeding in Delhi and transportation is by far the largest source of the emissions. It accounts for 46% of emissions (Hindustan Times 2009). CO_2emissions from road transport for the year 2007–2008 are 8.66 (Dhamija 2010). Delhi has witnessed an exponential growth in the number of motorized personal vehicles during last few decades. The increase in population translates into increased number of commuters. Simultaneously, there has been a sharp increase in vehicular population in the national capital. In March 1999, Delhi had 3.21 million motor vehicles for a projected population of about 13.4 million. Delhi counted 3.45 million vehicles in March 2001 which has now jumped to 8.47 million in March 2015. The capital has a maximum number of vehicles as compared to any other metropolitan city. Moreover, the total road length has merely increased from 28,508 km in 2001 to 31,373 km in 2009 (Sharma 2012).

Various cars with varying engine size have different contributions to the CO_2 emissions. The post-2000 petrol cars, with engine size more than 1400 cc, emits 143 g/km of CO_2. But post-2005 models of same engine size emit 173 g/km. Fuel economy drops from 16 km per liter to 13 km per ler. Diesel car models with engine size less than 1600 cc of 1996–2000 vintage emit 129 g per km but comparable post-2005 model emits 149 g per km. Fuel economy drops from 20 km per liter to 18 km per liter. While SUV models with engine size less than 3000 cc of 1996–2000 vintage emits 189 g per km, the post-2000 vintage emits 229 g per km and post-2005 models emit 256 g per km. Today, one SUV emits equally to two small petrol cars. Estimated fuel economy drops are dramatic, from 14 km per liter in 1996–2000 models to 10 km per liter in post-2005 models (Centre for Science and Environment 2008).

The cars and two wheelers contribute maximum to the total CO_2 emission load from vehicles in Delhi as much as 60% (Fig. 7.3). During 2002–2007, the CO_2 emissions load from cars has increased by 73% and from two wheelers by 61% (Centre for Science and Environment 2008). The bus-based public transport system carries the bulk of the public transport users in the city. Besides, Delhi Transport Corporation (DTC), private stage carriage operators (blue line buses and RTVs) play a key role in the provision of bus-based public transportation. Buses contribute

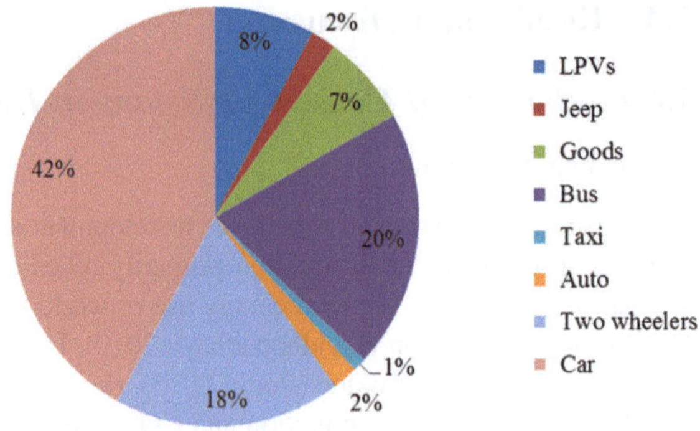

Fig. 7.3 Share of different vehicle segments in the total CO_2 emissions *Source* CSE estimates, 2008

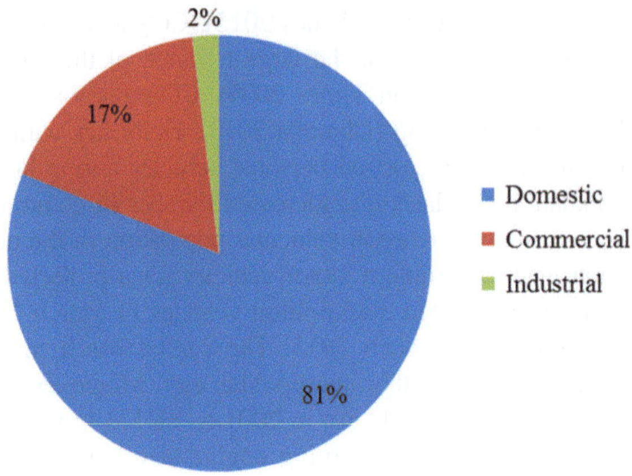

Fig. 7.4 Distribution of consumption of power by different sectors in Delhi (2008–2009) *Source* Government of National Capital Territory of Delhi (2009)

much less of CO_2 load, i.e., 20%. But, it is important to note that buses carry several times more people and consume significantly less fuel per passenger.

The study also reveals that older cars have become more fuel inefficient and emit more CO_2 due to poor maintenance and deterioration. But newer cars, even those produced after 2000 and 2005, are also showing higher levels of CO_2 emissions than the older vintages is quiet disturbing.

b. **Domestic sector**

It is the second main source of CO_2 emissions in Delhi. The pattern of household energy consumption represents the status of welfare as well as the stage of economic development. This in turn, will impact on the types and ways in which energy is accessed, transformed and used. The domestic sector accounts for 81% of power consumption (Fig. 7.4). Household energy consumption is expected to increase in future along with growth in economy and rise in per capita incomes. The residential sector accounts for 45% of total electricity consumption and 46% of the total sanctioned load in Delhi (Department of Environment and Forest 2010). The domestic sector accounts for 34% of total emissions with 5.35 million tons of

Table 7.1 Reasons behind higher CO_2 emission

Reason	Number of respondents	Percentage of respondents
Increasing population	74	49.33
Lack of awareness about Green technologies	36	24.00
Increasing living standard	40	26.67
Total	150	100.00

Source author's own calculation

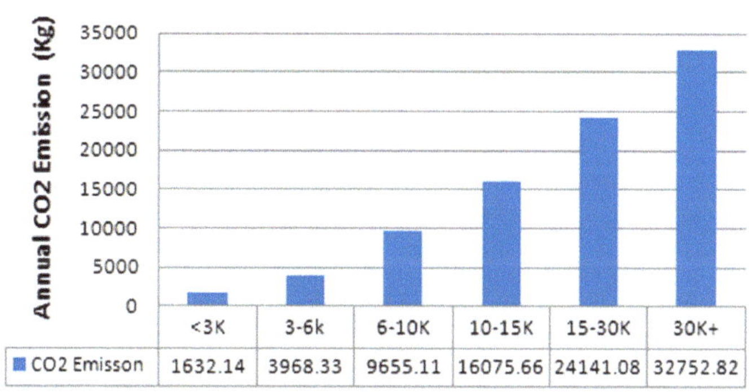

Fig. 7.5 Emissions by various income classes *Source* author's own calculation (K-Thousand Indian Rupees)

emissions from kerosene, diesel (Diesel Generators Sets) and huge consumption of electricity (Dhamija 2010).

It was noted that 49% of respondents mentioned that increase in population is the major cause of rising CO_2 emission. 24% mostly from higher income groups (HIGs) and middle-income groups (MIGs) felt that lack of awareness about green technologies is one of the major causes of CO_2 emission (Table 7.1). About 26% respondents felt that increase in the living standard is the major cause of CO_2 emission. Increase in living standard multifold the usage of electricity as well as vehicles, which in turn results in more CO_2 emission.

For the purpose of study CO_2 emissions were calculated for middle, higher and lower income groups (slums). On the basis of primary survey, it was found that there are high variations in the average annual CO_2 emissions among lower, middle and higher income groups (Fig. 7.5). Increased rate of urbanization will increase the rate of emissions. Since, higher income group (HIG) is living in comparatively bigger houses and have higher standards of living, which finally leads to higher consumption of electricity.

The increasing use of appliances is the main reason for increased emissions (Figs. 7.6 and 7.7). The total connected load in the residential sector in Delhi works to be around 6996 megawatts with the average connected load as 4.8 KW (Department of Environment and Forest 2010). The water heating and cooling

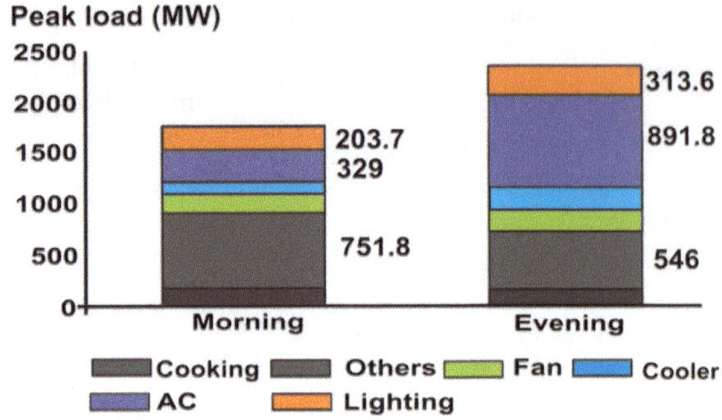

Fig. 7.6 Summer load by end-use

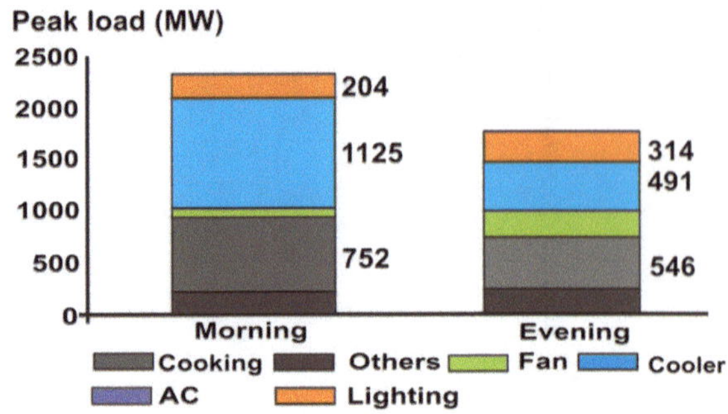

Fig. 7.7 Winter load by end-use

together account for nearly 51% of this load. The share of lighting in the total connected load is around 6%. Of the total connected load for lighting, incandescent bulbs account for about for 40% of the total connected load in the residential sector. Water heating and cooling together account for major share of the electricity consumption. Additionally, it is observed that evening peak in summer is more than morning, while during winter the morning peak is dominant. Space conditioning load in summer and water heating in winter are the major demand drivers, besides lighting load (Department of Environment and Forest 2010).

c. **Commercial sector**

Electricity consumption (2382 million kWh in 2003–2004) in the commercial sector of Delhi contributes to approximately 20% of emissions (Hindustan Times 2009).The total connected load of commercial sector at 1048 MW account for 16% of total connected load (Department of Environment and Forest 2010). The estimated CO_2 emissions from commercial sector of Delhi were 1.87 MMT with power consumption alone contributing 1.69 MMT (Dhamija 2010). The commercial sector comprises primarily of offices, shopping malls, markets, hotels, and restaurants and comprises of a mix of air-conditioned and non-air-conditioned buildings. The prime load centers in the sector are air-conditioning, lighting, and equipment. The major share of electricity consumption is attributed to air-conditioning. In order to calculate, CO_2 emissions in the commercial sector, one mall and one hotel were studied.

City Square Mall

City Square mall of Rajouri Garden is medium sized shopping mall which is built in an area of 50,000 ft^2. This mall contains approximately 80 shops. Each shop normally contains 15–20 compact fluorescent lamp (cfl) lights and is centrally air-conditioned. Average consumption of electricity is 6000 units which emits an average of 8700 kgs of CO_2 per month.

Ashoka Hotel

Ashok Hotel, located at a prime place of diplomatic enclave in posh South Delhi area. It is the biggest hotel with as many as 571 lavishly furnished rooms. All the rooms are equipped with all modern facilities like direct dialing, personal computer with Internet connectivity, channel music, and television. They are centrally air-conditioned with electronic door lock, iron and iron board, Tea or Coffee maker, multichannel music, and interactive television with remote control. The average consumption of electricity is 20,000 units which emit average 26,000 kg of CO_2 per month.

d. **Industrial sector**

Industries are important sources of emissions. Industrial sector contributes 8% due to increase electricity consumption (Hindustan Times 2009). CO_2 emissions from industries were 1.37 MMT with power consumption alone contributing 1.21 MMT (Dhamija 2010). Delhi has changed from a city of trade and governance to a multi-functional city with a judicious mix of small-scale industrial activities as well as wide range of commercial and semi-oriented activities. There are many industrial estates in Delhi. For the purpose of study, Okhla and Mayapuri industrial areas were chosen. On the basis of the primary study, it was found that textiles and metallic industries form a majority and are the major source of CO_2 emission in Delhi.

Okhla, is a neighborhood of one of the old villages in South Delhi district, it is most known as the Okhla Industrial Area (OIA) or Okhla Industrial Estate. It is the area of such estates being developed across India to encourage small-scale industries. It is the hub of textile, engineering, plastics, electronics, wooden, chemical industries, etc. Mayapuri, is a neighborhood around the Naraina village in South West district, it is also known as Mayapuri industrial area or Mayapuri industrial estate (Table 7.2).

Both the industrial areas have almost same type of industries. The average consumption of electronics industries per month is approximately 5000 units which emit 5150 kg of CO_2 per month. Similarly, in the wire industry, the average consumption is approximately 3000 units which emit 3090 kg of CO_2 per month.

Table 7.2 Industry wise electricity consumed and average CO_2 emission

Industry	Electricity consumed (units)	Average CO_2 emission (kg)
Electronic	5000	5150
Wire	3000	3090
Textile	4500	4635

Source author's own calculation

Table 7.3 Year wise coal used CO_2 emission

Year	Coal used in MT	Percentage of CO_2 emissions
1999–2000	7,300,026	15–16.8
2000–2001	612,633	14.82–15.8
2001–2002	541,877	13.6–14.8
2002–2003	704,730	14.8–16.4
2003–2004	622,211	14.2–17.8
2004–2005	590,349	13.8–15
2005–2006	501,321	13–14.1
2006–2007	524,781	13.4–14.5
2007–2008	715,582	15–16
2008–2009	753,978	16–17.8

Source Rajghat Power Station (2009)

Textile industry's average monthly consumption is approximately 4500 units which emit 4635 kg of CO_2 per month.

e. **Energy sector**

Energy is the key driver of most of the world's economies, which is also primarily responsible for the emissions of GHGs into the environment, considered to be responsible for the global warming. Delhi has highest annual consumption of power in the country because of extreme weather conditions and soaring temperatures. Power demand touched all time high of 5925 MW in a single day last July (The Wall Street Journal 2015). Delhi has the highest per capita consumption of electricity among states and union territories of India. The per capita consumption of electricity is about 1615 units (Central Electricity Authority 2010). Rise in population has a direct effect on power demands in Delhi. The power generation in Delhi is undertaken by three governments owned companies, i.e., Indraprastha Power Generation Co. Ltd. (IPPGCL), Rajghat Power House and Pragati Power Corporation Ltd. (PPCL). The city's four gas based power generation plants less than 20% of the city guzzles (The Wall Street Journal 2015).

As a part of secondary survey, Rajghat Thermal Power Station (RTPS) was chosen to analyze the CO_2 Emission. RTPS provides electricity to most parts of Old Delhi. It has two power stations of 67.5 MW each. It is a coal based thermal power station. CO_2 percentage and Coal used in MT were collected from this power station. The average CO_2 emission from last 10 years is 14.9% (Table 7.3).

7.4.2 *Feasibility of Green Technologies*

Energy is an important input for economic development. Therefore, alternative sources of energy have become important and relevant in the twenty-first century. With the intensive use of fossil fuels in Delhi which emits large CO_2, population is facing its adverse impact on the environment. It has therefore, became imperative that cleaner technologies, in terms of higher efficiency, renewable resources, and lower carbon footprints, are adopted to continue economic development in Delhi without impacting the environment adversely. Renewable energy sources like solar and other technologies are coming up with better efficiency and cost effectiveness. Delhi has opportunities to harness solar energy due to its better solar resource potential (Fig. 7.8). Rooftop solar can be a better option to source green energy as it doesn't require additional land. Government of Delhi has announced installation of a solar power plant of 5 MW installed capacity at Delhi. This power plant will provide solar power to meet the energy demands of Delhi Secretariat (Green Clean Guide 2013).

Delhi gets annual solar radiation of about 5 kWh per m^2 per day with an annual mean duration of sunshine hour per day of about 6–7 h (India Meteorological Department 2008). However, global solar radiation during summer increases to about 7 or 8 kWh per m^2 per day and in winter it is around 3–4 kWh/m^2 per day (India Meteorological Department 2008).

The land use map of Delhi has been prepared in accordance with Master Plan of Delhi, 2001. The map has been made by taking commercial areas into consideration. Most of the land is covered by residential area followed by industrial and commercial use. Remaining land is covered by parks, playgrounds and other uses. Mix land use has following reward that adds vitality to city, cuts down commuting distance, saves time and energy, and investment in transport infrastructure to provide the citizens a more integrated daily life cycle. It also helps in improving quality of life in city. India's urban heritage also was built upon mixed land use. Mix land use has worsened the situation of power. Increased vehicular movement has further accelerated the emission of carbon dioxide in Delhi. The demand for electricity is rising in residential and commercial areas. So in the light of these problems, there is a need to develop alternate source of energy in Delhi, and solar energy can be most feasible. It is believed that the parts of Delhi where there are more of commercial areas there are a lot of scope for the application of solar energy. Looking at this map South, North and West Delhi has large number of commercial areas like hotels, malls, etc., where there is enough space for trapping the solar energy (Fig. 7.9).

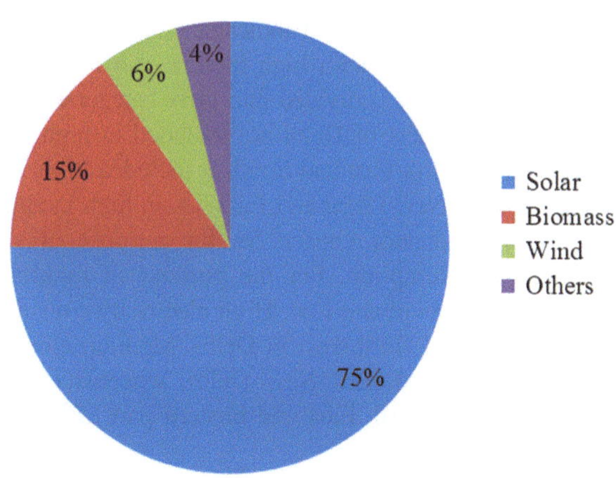

Fig. 7.8 Feasibility of renewable energy

Fig. 7.9 Land use, 2001

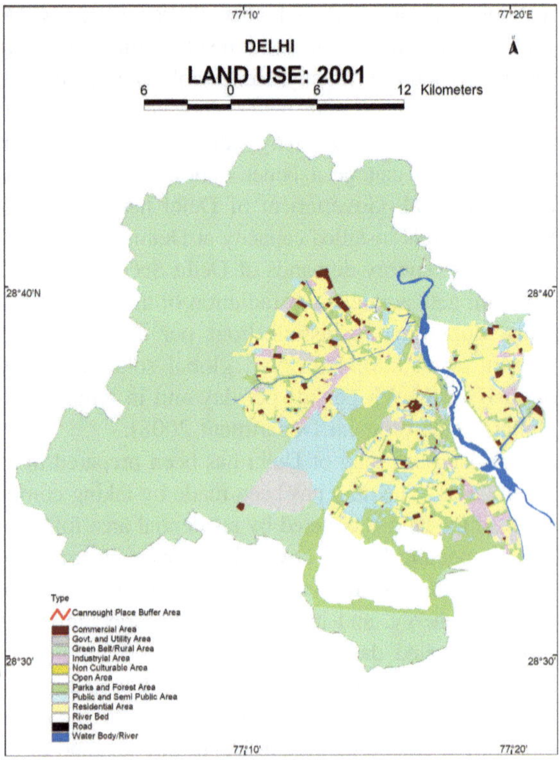

RSPM level is taken into account as it believed that RSPM molecules are heavy which obstructs the incoming solar radiation. The RSPM data was available taken from various monitoring stations, based on the data point map has been prepared and accordingly surface map was prepared. The RSPM level was divided into high, medium and low. There is high RSPM level in the North East, Central, New Delhi and East. Whereas moderate RSPM level is found in North, Central and west and lowest is found in Northwest and South (Fig. 7.10).

Income level is the third indicator chosen to prepare solar feasibility index. It is expected that people with high income can largely use solar energy technologies. Again this indicator is divided into three categories. South West, Central, New Delhi and Eastern districts of Delhi are included in high income level category, Whereas West, North, North East and South is included in medium income level group. South West, Central, New Delhi and East lies in high income group (Fig. 7.11).

The last indicator chosen for the purpose of solar energy use feasibility is electricity consumption. For the purpose of ranking, three categories have been made high, medium and low. High electricity consumption is experienced in South West, South and Northwest in Delhi. Medium electricity consumption is found in Northwest, Central, and New Delhi, whereas lowest electricity consumption is experienced by North East and Eastern part of Delhi (Fig. 7.12).

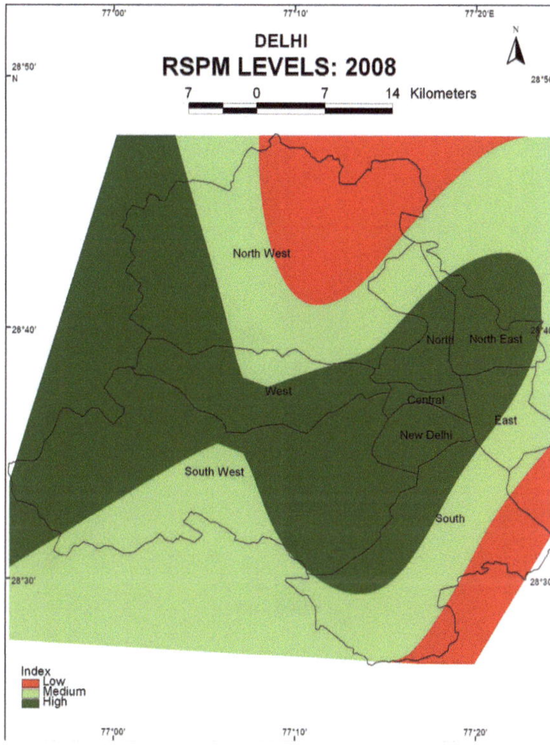

Fig. 7.10 RSPM levels, 2008 *Source* prepared by authors

All the four indicators were overlapped to find out the most feasible site for utilizing solar energy. It was found that South, West and North Delhi has highest solar use feasibility owing to large commercial areas, huge electricity consumption, high income level and low RSPM level (Fig. 7.13). Central, southwest, New Delhi and north east falls under medium category and east Delhi falls under low category.

7.4.3 RET Screen Model for Solar Water Heating System for Various Sectors, Delhi

Solar energy can be harnessed through two routes names solar thermal and photovoltaic, by direct conversion to electricity and heat energy, respectively. Water heating, space cooling, etc., can be replaced through solar thermal and lighting, fanning, etc., can be replaced through solar photovoltaic. Solar water heater uses direct solar energy and can heat water up to 60–80 °C which is sufficient temperature to use for bathing, cooking or washing, etc., and can be stored for 24 h without significant drop in temperature. It can save a huge amount of electricity or fuel bills. The usage of 1000 kWh of 100 l capacity each can contribute to a peak load saving of approximately 1 MW (Dhamija 2010). Also it prevents carbon

Fig. 7.11 Income level, 2007 *Source* prepared by authors

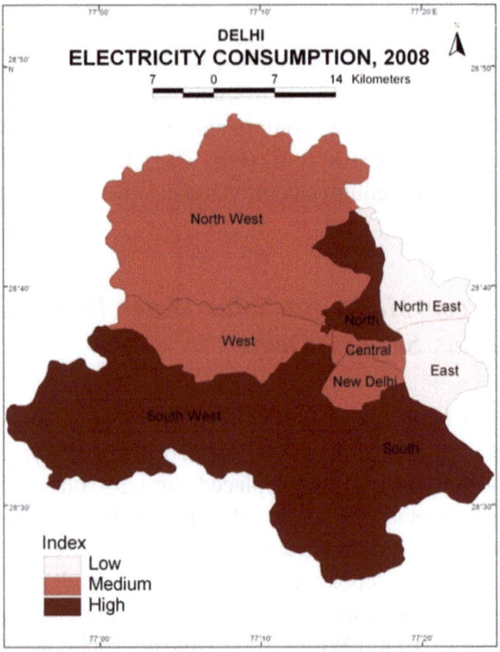

Fig. 7.12 Electricity consumption, 2008 *Source* prepared by authors

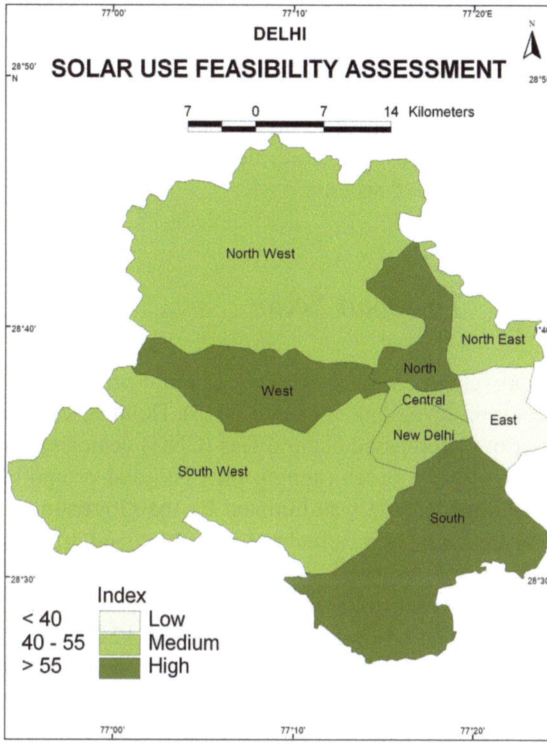

Fig. 7.13 Solar use feasibility assessment *Source* prepared by authors

Table 7.4 Proposed solar water heating systems for different sectors in Delhi

Sectors	Capacity (l/day)	Required no. of collectors	Aperture area (m^2)	Budgeted initial cost (Rs)	Payback period (years)	GHG analysis (tons/year)
Household						
HIG	110	1	1.94	20,000	2	1.09
MIG	80	1	1.94	20,000	2	0.4
Commercial						
Hotels	8650	86	166.84	1,720,000	3.7	38.87
Industry						
Garment industry	250	2	3.88	40,000	1	2.49

Source RET Screen International, 2007

emission for example a 100 l capacity can prevent emission of 1.5 tons of CO_2 per year. On the basis of hot water used in various sectors simulation has been done for the annual reduction in GHGs in Delhi (Table 7.4).

7.5 Policies for Low Carbon Society in Delhi

The emissions from transport, domestic and power sector in Delhi is taken into account, these constitute major sources of GHG emissions into the atmosphere. The policy measures, therefore, need to be directed toward reducing emissions in all these respective sectors.

7.5.1 Transport Sector

Improvement in transport and communication has been recognized as an important sector of development (Singh 2016). Delhi is well on its way to become the largest user of Compressed Natural Gas (CNG) vehicles in the world. By 2012, there were nearly 10,000 buses running on CNG and promoting public transport as a major agenda of change was initiated by the Government of Delhi. The existing diesel operated transport vehicles like tempo and truck should be converted to CNG (Narain and Krupnick 2007).

The government is also promoting the use of battery operated vehicles. There is a refund of value added tax (VAT) of 12.5% on battery operated vehicles along with refund of registration charges. Introduction of e-rickshaws by telecom major Vodafone Essar is a step toward promoting green transportation in Delhi (Shankar 2009). More than 15 year old commercial vehicles have been faced out (Central Pollution Control Board 2008).

The government has also decided to make the emission norms for fuel sold in the National Capital Territory of Delhi to be reduced from 350 PPM sulfur content to 50 PPM sulfur content w.e.f 1st April 2010 (Mashelkar 2002). Total of 500 new pollution control centers have been set up and they are been networked for better analysis of data for emissions. For use vehicles; periodic inspection, maintenance program and emission tax on diesel vehicles was recommended. Other than these, a metro railway has been launched to reduce the dependence on road transport. Rising pollution levels forced Delhi Government to launch controversial odd-even vehicle-numbering formula that hopes to get roughly half of nearly 9 million vehicles off the roads. To curb emissions, the Supreme Court of India has temporarily banned the sale of large diesel cars, such as sports utility vehicles, and others with an engine capacity of 2000 cc or more in Delhi (Rediff 2015).

7.5.2 Policies for Domestic, Commercial, Industrial, and Power Sector in Delhi

Introducing energy efficiency in domestic, commercial, industrial, transport and power sector can address the problem of increasing CO_2 emissions, global warming, energy security and fossil fuel depletion. The various measures taken to improve the energy efficiency were like introduction of Bachat Lamp Yojana Camp. Star rating of the equipments and mandatory use of ISI marked on motors pump sets, power capacitor. It also includes building retrofitting for energy efficiency for 1000 existing buildings having covered area of 10,000 ft^2 (Dhamija 2009).

The utilization of solar technology can save 20–30% of energy through energy efficiency measures in commercial and domestic sectors (TERI 2006). The climate change agenda for Delhi made solar water heating mandatory in all industries, hostels, hotels and residential buildings on plots above 500 m^2 and Government of Delhi is also promoting subsidies on the application of solar water heating systems. Subsidy schemes of Rs 6000/for 100 l per Day (LPD) for domestic users and up to Rs. 60,000 for 1000 l per Day (LPD) for non commercial users (Dhamija 2009).

The policy instruments for energy conservation in the industrial sector are mainly regulatory instruments. The excessive increase in number of small-scale industries in Delhi is responsible for increased emissions in Delhi. So, Supreme Court of India in 1998 orders to close down the 1328 'H' category units and closure of polluting industrial units in non confirming industrial areas under the supervision of Ministry of Urban Development, Government of India as a Nodal Agency. Apart from above, Delhi Pollution Control Committee directed several industrial units to install pollution control devices, which were found polluting during the course of action implementation of Air (Prevention & Control of Pollution) Act, 1981. Other potential incentive based instruments for energy conservation by large energy intensive industries are fiscal incentives like lower duties on imports of climate-friendly technologies, accelerated depreciation allowances for the resulting investment and lower excise duties on the products.

Three thermal power plants in National Capital Territory of Delhi contributed to about 16% of emissions in the year 1991. All the three thermal power plants have installed electrostatic precipitators in all their units to control emissions. Besides this, the power plants are using beneficiated coal (ash content less than 34%) since 1999 as against the coal used earlier (ash content above 40%) to reduce emissions. The replacement of coal-driven steam turbine with gas with same efficiency reduces carbon dioxide emission (Central Pollution Control Board 2003).

7.6 Conclusion

The transport, residential, commercial, industrial and power sector are the principal sources of CO_2 emission in Delhi. All four sectors contribute approximately 16 MMT of CO_2 in the atmosphere. Transport sector contributes highest 46% to CO_2 emissions due to growth in vehicles followed by domestic Sector. The vibrant socioeconomic life creates a huge demand for mobility for work, recreation, education, and shopping. This creates immense pressure on the transport infrastructure. The reason being increased electricity consumption by the households and a large amount of biomass fuel used in slums and unauthorized settlements and general waste collection are further worsening the scenario. The commercial sector is next largest emitter followed by industrial sector. These facts are just a glimpse of the devastating impact that these CO_2 emissions will have on the ecosystems and humans over the next century in Delhi. The usage of efficient electrical appliances, switching over to sustainable environment friendly technology, i.e., renewable energy is some of the ways that can contribute toward low carbon economy in Delhi. Among all the non-convectional sources, solar energy is most feasible for Delhi which can be collected through solar thermal and photovoltaic system. The use of solar energy can supplement the demand for electricity consumption among various sectors. South, West and North Delhi have highest solar use feasibility owing to large commercial areas, huge electricity consumption, high income level and low RSPM level. The use of solar water heaters by HIG, MIG, hotels and garment industry can save 1.09, 0.4, 38.87 and 2.49 tons of GHG emissions, respectively.

The introduction of government policies resulted in some control over status of emission over Delhi. Along with this switching to cleaner fuels, phasing out old vehicles and maintenance of in-use vehicles, integrated land use and transport planning, Using information technology to reduce demand for physical transportation, plantation activities, and closing or relocating emission industries, improved equipments in the entire sector played an important role in controlling emissions in Delhi and can save mother earth from further degradation. The issue concerns each individual on planet earth and if corrective measures are not taken in right away, sustainable future earth may be on stake. People awareness and participation can lead to low carbon society in Delhi. Each one of us can make our individual contribution in combating this environmental threat in Delhi and can save earth. In order to achieve the sustainability and survivability, there should be solution oriented interdisciplinary research together with capacity building programs. Hence, it is high time to make Delhi as low carbon resilient mega city for sustainable future earth.

This chapter is the revised edition of SLUAS Science Report Vol. 5 pp. 161–176.

References

Bandyopadhyay RK (2010) Reconciling economic growth with low carbon mobility in India: addressing the challenges. Indian Infrastructure Report—Infrastructure Development in a low carbon economy. Oxford University Press, New Delhi, pp 237–257

Central Pollution Control Board (2003) Annual Report. Ministry of Environment and Forest, New Delhi

Central Pollution Control Board (2008) Annual report. Ministry of Environment and Forest, New Delhi

Department of Environment and Forest (2010) State of Environment Report of Delhi. New Delhi

Dhamija P (2010) Green productivity for sustainable energy and environment. Energy Efficiency and Renewable Energy Management Centre, Department of Environment, Government of Delhi
Dhamija P (2009–2012) Climate change agenda 2009–2012. Energy Efficiency & Renewable Energy Management Centre, Department of Environment, Government of Delhi
Government of National Capital Territory of Delhi (GNCTD) (2009) Delhi statistical hand book. Directorate of Economics and Statistics, Delhi
Hindustan Times (2009) The carbon map shows Delhi emission high. New Delhi
India Meteorological Department (IMD) (2008) Solar radiation handbook. India Meteorological Department, New Delhi
Mashelekar RA (2002) Report of the Expert Committee on auto fuel policy. Ministry of Petroleum and Natural Gas, New Delhi
Narain U, Krupnick A (2007) The impact of Delhi's CNG program on air quality. Resources for the Future, Washington
Rajghat Power Station (2009) Year wise coal used in million tons and per cent of CO_2 emission. New Delhi
RET Screen International (2007) Version 3, Software of Clean Energy Decision Support Center, National Resources, Canada
Singh RB (2016) Geographic mosaic of incredible India: introducing natural and cultural heritage. In: Singh RB (ed) Progress in Indian geography: a country report 2012–2016. Indian National Science Academy, New Delhi
Shankar U (2009) Policy Instruments for Achieving Low Carbon and High Economic Growth in India. National Institute of Public Finance and Policy, Chennai
Sharma R (2012) Growing vehicles to further choke Delhi. Zee News, New Delhi, 31 Aug 2012
The Energy and Research Institute (TERI) (2006) Urban transport energy and environment—a case study of Delhi. New Delhi

Internet sites

http://www.censusindia.gov.in/2001census/maps/atlas/NTC%20of%20Delhi.html. Accessed on 10 June 2009
www.cea.nic.in.Accessed 13 Dec 2010
www.dpcc.delhigovt.nic.in/pdf/cleanerair.pdf. Accessed 23 Aug 2010
http://www.cseindia.org/node/66. Accessed 2 Sept 2010
http://sallan.org/pdf-docs/co2_trend.pdf. Accessed 20 Oct 2010
http://www.ibtimes.co.uk/india-transport-domestic-sectors-ahead-greenhouse-gas-emissions-1492465. Accessed 19 Apr 2015
http://blogs.wsj.com/indiarealtime/2015/02/25/who-really-holds-power-in-delhi/. Accessed 10 May 2015
http://www.rediff.com/business/special/special-delhis-transport-sector-3-times-more-polluting-than-mumbais/20151221.htm. Accessed 10 May 2015

Chapter 8
Dynamics of Land Use and Climate Change in Subhumid Region of Rajasthan, India

R.B. Singh and Ajay Kumar

Abstract The impact of land use land cover on climate variability and water resources has been studied at the subhumid region (Kalisindh–Parwan subbasin of Hadoti Region) of Rajasthan in India. This subbasin in Rajasthan has undergone a tremendous change in land use land cover over the last two decades from 1988 to 2008 thereby bringing changes in energy flux. Along with land use land cover, water through hydrological cycle also helps in circulation of energy in the earth's atmosphere. To establish these relationships we have investigated the importance of climate, rainfall, streamflow, and land surface temperature interactions in the study region. The lithosphere–hydrosphere–atmosphere interaction determines the level of carbon in atmosphere and also the availability of surface and ground water energy resources. In the context of climate variability, management of land use land cover and water balance play a vital role in mitigating the adverse phenomena. Therefore, land use planning in the Kalisindh–Parwan subbasin of Hadoti Region of Rajasthan is a key factor in reducing present and future vulnerability to climate variability.

Keywords Land use land cover · Land surface temperature
Kalisindh–Parwan basin · Climate variability · Energy fluxes

R.B. Singh (✉)
Department of Geography, Delhi School of Economics,
University of Delhi, Delhi 110007, India
e-mail: rbsgeo@hotmail.com

A. Kumar
Department of Geography, SBS College, University of Delhi,
Delhi 110017, India
e-mail: mail_kumarajay@yahoo.com

Y. Himiyama (ed.), *Exploring Sustainable Land Use in Monsoon Asia*,
Springer Geography, https://doi.org/10.1007/978-981-10-5927-8_8

8.1 Introduction

The land use land cover has a significant impact on the energy balance of the earth surface through variation in albedo, thus becoming a major determinant factor in climate studies with special reference to climate variability (Boisier et al. 2013; Zhou and Wang 2010; Oke 1982; Deng et al. 2013). Along with land use land cover, water through hydrological cycle also helps in circulation of energy in the earth's atmosphere (Chahine 1992; Rind et al. 1992). Therefore in context of global climate, management of land use land cover and water balance plays a vital role in mitigating the phenomena. In developing countries like India, the population is growing at a very rapid pace which in turn puts immense pressure on limited land resources, and since green revolution time, with advent of technology the agriculture land has expanded multifold thereby contributing towards change in energy balance due to LULCC. The lithosphere–hydrosphere–atmosphere interaction determines the level of carbon in atmosphere and also the availability of surface and ground water resources. Extension of agriculture through excessive tilling not only releases more carbon in atmosphere but also alters the surface water flow (Kueppers et al. 2008; Lal and Bruce 1999; Pielke et al. 2002; Watson et al. 2000). Replacing natural vegetation with bare surfaces and fallow agriculture land often decreases the surface albedo and alters the local surface energy balance, increasing sensible heat flux and decreasing latent heat flux (Kueppers et al. 2008; Eliasson and Svensson 2003; Oke 1981). The regions with scarcity of water are among the most vulnerable to climatic variability as all the available resources are in stress condition and a miniscule change can play havoc for the inhabitants of such regions. Responsible planning of land use, and its responsive development is a key to mitigate the negative impacts of climate variability (Emergency Management Australia 2002; Gurran et al. 2008). Therefore, in this context the paper attempts to investigate the impact of changing land use land cover on climate and water resources and vice versa in dry subhumid region of India.

8.2 Study Area

This study is carried out at subbasin level and most of policies in India are being implemented at administrative level, therefore in order to assist decision support system the subbasin is further clipped by the administrative boundary of Rajasthan state. The study area lies between 24°N and 25°30′N latitudes and 75°50′E and 77°E longitudes (Fig. 8.1). Rivers Kalisindh–Parwan are tributaries of Chambal river which finally drain into river Yamuna. The catchment area of Kalisindh subbasin taken for the study is 687.8 km^2. The Kalisindh subbasin is bounded on the South by the great Vindhyan ranges. The study area lies at the edge of Malwa plateau, an area of low hills and shallow plains. The climate of the area in Kalisindh subbasins is semiarid to subhumid, the weather being mostly dry except in the monsoon

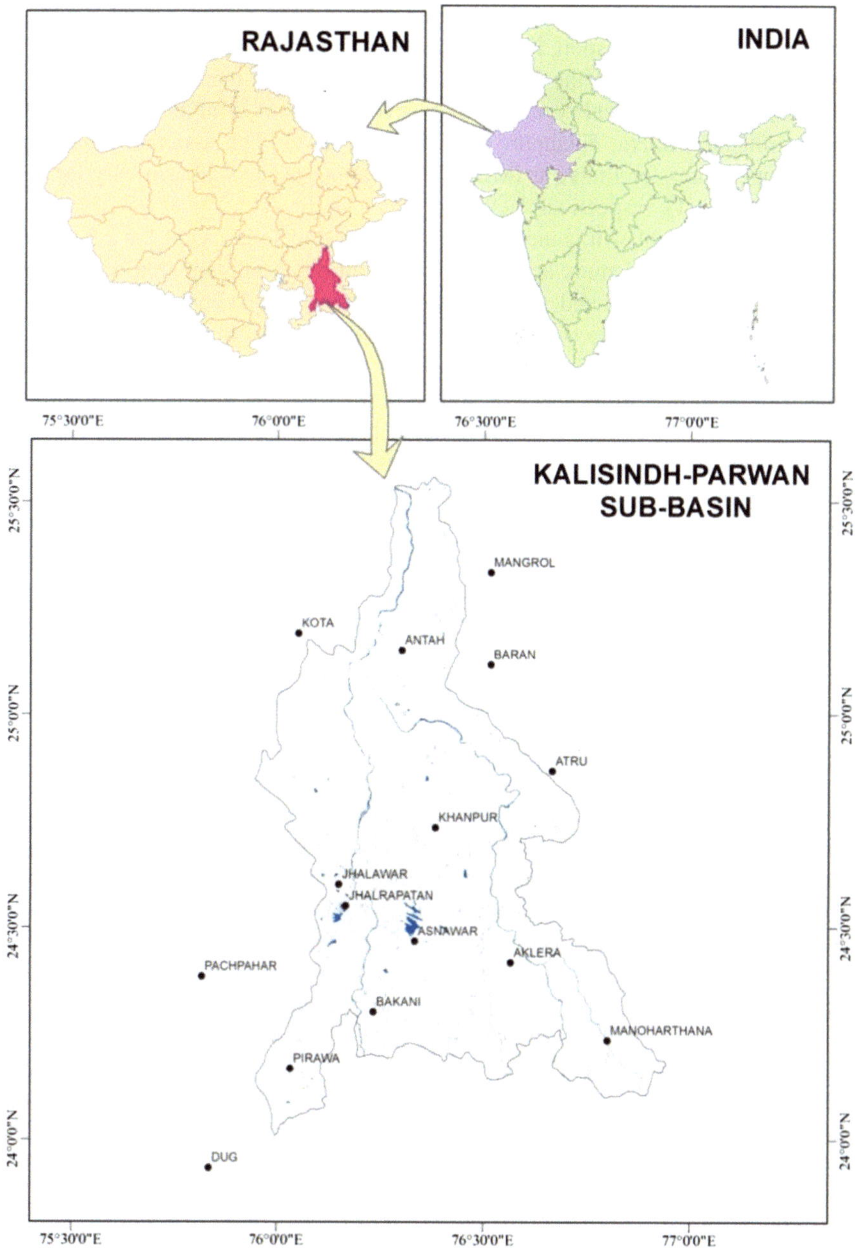

Fig. 8.1 Location of the study area

season. Summer is hot and winter is mild. About 90% of the annual rainfall is received during the monsoon, i.e., from June to October (Sharma et al. 2002). The normal annual rainfall of the Kalisindh subbasin is 1006 mm. The drought is in general of mild or normal type, however severe types of drought are recorded at Manoharthanar, Aklera, Khanpur, Bakani, and Pirawa. The maximum and minimum temperatures for Jhalawar IMD stations falling in the subbasin vary between 41.50 and 7.90 °C, respectively.

Almost entire subbasin is underlain by black cotton soil, except for a few small pockets in the north of study area where recent alluvium in plain area is found. An alluvial aquifer with limited thickness occurs along river courses. The principal means of irrigation in the study area are wells/tube wells, though some areas are irrigated by canals, tanks, etc. Groundwater is the main source of irrigation and is utilized through dug wells and tube wells. Tanks form the second most important source of irrigation in the study area. Canal irrigates only a small area. Stage of ground water development in the district is 105.15%. Seasonal fluctuation in water level based on Pre- and Post-monsoon 2008 indicate that there has been rise in water level in major part of the district. Perusal of the fluctuation data indicates that major part of the district has recorded rise in water level of more than four meter (Central Ground Water Board 2010).

8.3 Materials and Methods

The analysis in the study has been done at a subbasin level. The database has three components pertaining to land use land cover, climate, and surface water resources. The data for land use land cover has been generated using Landsat satellite images obtained from Global Land Cover Facility (GLCF) for the year 1978 and BHUVAN, National Remote Sensing Agency (NRSA), India for 2008. The satellite data after preprocessing has been classified using Erdas Imagine 9.2 software. The supervised classification has been used for both the images and a change matrix has been obtained (Singh et al. 2001; Singh and Kumar 2012). The data regarding rainfall for 1976–2010 has been obtained from India Meteorological Department (IMD) for 15 rain gauge stations viz. Jhalawar, Dug, Pirawa, Bakani, Aklera, Khanpur, Manoharthana, Jhalarapatan, Pachpahar, Asnawar, Mangrol, Antah, Atru, Kota, and Baran located in and around the study area. The month-wise rainfall data has been grouped into three categories viz. pre-monsoon, monsoon, and post-monsoon. The point data has been interpolated using kriging tool in ArcGIS 10.0. The land surface temperature data has been extracted using the thermal bands of Landsat TM satellite image of 1978 and Landsat ETM+ image of 2008. Both the satellite images used for the study are of November month. Agriculture as being influenced by the climatic condition especially rainfall becomes an important factor in determining the impact of climate variability on land use pattern, therefore data regarding agriculture especially kharif crops (Monsoon crops) for the period of 1978–2008 for Jhalawar district have been obtained from the Department of

Agriculture, Rajasthan. The condition of surface water resources is analyzed through streamflow obtained from the Water Resource Department, Rajasthan from 1977 to 2004. Multi-Linear Regression is used to find the relationship between rainfall and streamflow.

8.4 Results and Discussion

8.4.1 Land Use Land Cover

The land use land cover of a region is significantly influenced by the microclimate of the region and vice versa. Both the factors have combined effect on the water balance of the region, which again becomes an important driving factor for land use land cover change in the region. The land use land cover in Kalisindh–Parwan subbasin in Rajasthan has undergone a tremendous change over last two decades. The land use land cover of Kalisindh–Parwan subbasin in the year 1988 was dominated by bare surfaces which have been converted into agriculture land in 2008 (Figs. 8.2 and 8.3). There has been a substantial decrease in the scrub forest thereby reducing the carbon storage capacity. The scrub vegetation being a major constraint in extension of agriculture and increased mechanization of agriculture has also contributed in decrease of scrub forest. The agriculture in the region has been intensified as the total cropped area has increased from 40% of total area to 66% during 1988–2008. In the process of agriculture intensification, farmers have started frequent tilling of land which leads to the release of stored carbon. The forest area in the subbasin has also reduced from 21% of total geographical area to 10% due to encroachment by agriculture land use. The water bodies have reduced in size and numbers. As visible from the classified satellite images the flow of water in rivers has also reduced (Fig. 8.2b), which could be attributed to modification of landscape due to increased agriculture and also increased withdrawal of water for irrigation purposes (Fig. 8.4a). The easy availability of water resource from the river has also led to encroachment of river sides (Fig. 8.4b). The decrease in natural green cover has substantial impact on the microclimate of the region as vegetation has low albedos, typically ranging from 0.05 to 0.25, with forests absorbing more solar radiation than grasslands or croplands. Higher leaf area index generally increases evapotranspiration over vegetated regions in summer, provided, there is sufficient soil water, thereby surface temperature cools and precipitation may increase. Lower leaf area has the opposite effect (Bounoua et al. 2000; Buermann et al. 2001; Eastman et al. 2001). Clearing of forest in tropics for agriculture generally results in a warmer and drier climate owing to higher albedo, decreased surface roughness, and reduced evapotranspiration (Dickinson et al. 1989; Lean and Warrilow 1989; Sampaio et al. 2007; Costa and Foley 2000). Replacement of scrub vegetation with bare surfaces and fallow land also reduces rainfall by increasing albedo and reducing evapotranspiration (Dirmeyer and Shukla 1996). The influence

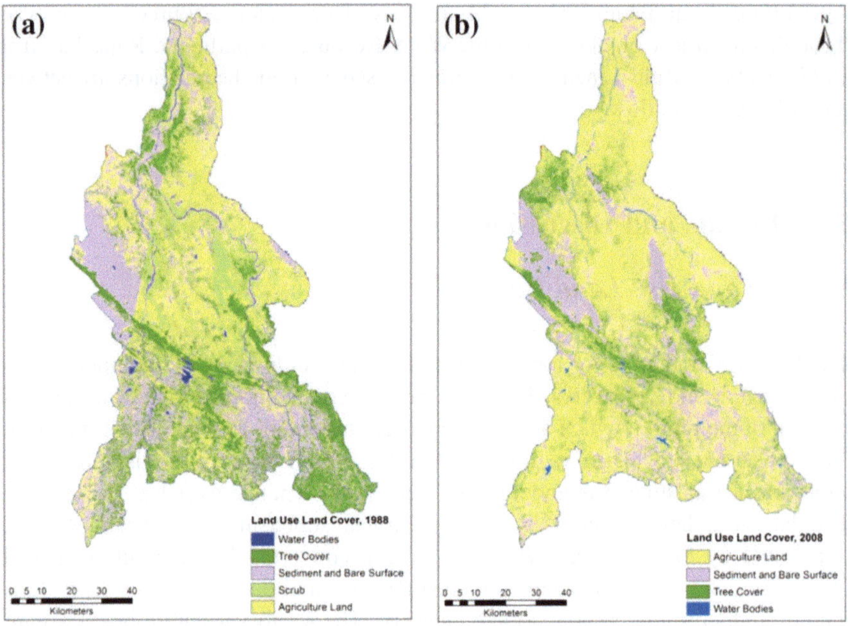

Fig. 8.2 Land use land cover map of Kalisindh–Parwan subbasin **a** 1988, **b** 2008

of surface heterogeneity arises as a result of differential surface energy fluxes and atmospheric heating. Local differences in albedo can also generate mesoscale circulations (Pielke et al. 1992). The impact microclimate of region on land use is visible through dependence of kharif crops on monsoon rainfall. The correlation coefficient between total sown area of kharif cereals and monsoon rainfall calculated for the period of 1978–2008 is 0.39 at 0.01 significance level (Fig. 8.5).

8.4.2 Climate

Land and atmosphere is a coupled system, where atmospheric radiation and climatic parameters depend on surface fluxes of energy and moisture and in turn, these fluxes depend on climatic parameters, thereby coupling the atmospheric state and surface conditions (Roy et al. 2007; Douglas et al. 2006; Douglas et al. 2008; Mahmood et al. 2013; Niyogi et al. 2007; Pielke et al. 2011). Changes in albedo, roughness, leaf area index, soil texture, and soil water due to changing land use land cover alter surface fluxes and the hydrologic cycle along with the microclimate of the region. The significant impact of land cover on microclimate arises due to changes in soil water, evapotranspiration, and the partitioning of net radiation into sensible and latent heat (Bastiaanssen 2000). The changes in microclimate also have

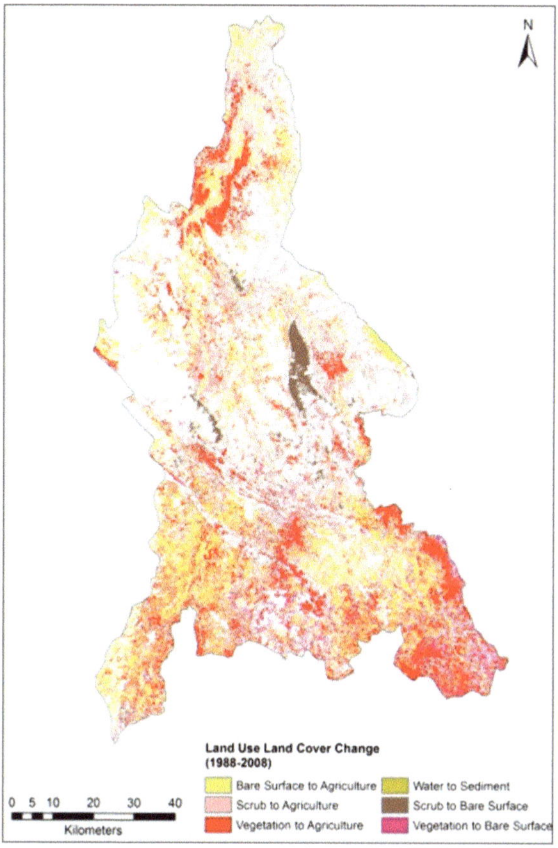

Fig. 8.3 Land use land cover change during 1988–2008

Fig. 8.4 **a** Withdrawal of river water for agriculture and **b** encroachment of river side for horticulture

an impact on streamflow thereby altering the water balance of a region through land–atmosphere interactions (Dobos 2003; Seth 1996; Sahu et al. 2012).

8.4.2.1 Rainfall

The month-wise analysis of rainfall over the study area depicts changing trends during 1978–2008 (the value of 1978 depicts average of 1976–1980 and 2008 depicts average of 2006–2010). The analysis is done for pre-monsoon (May and June) (Fig. 8.6), Monsoon (July, August, and September) (Figs. 8.7, 8.8 and 8.9) and post-monsoon (October and November) (Fig. 8.10) seasons. It has been

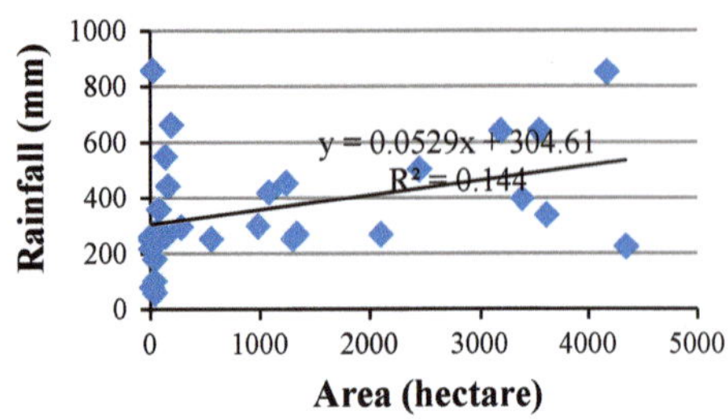

Fig. 8.5 Relationship between monsoon rainfall and Kharif cereal sown area

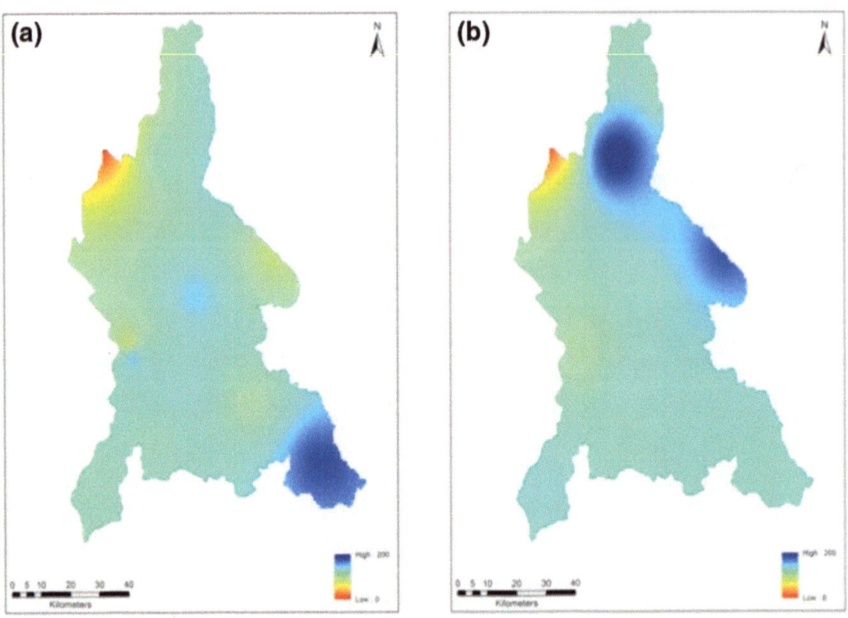

Fig. 8.6 Rainfall during pre-monsoon season **a** 1978 (averaged for 1976–1980) and **b** 2008 (averaged for 2006–2010)

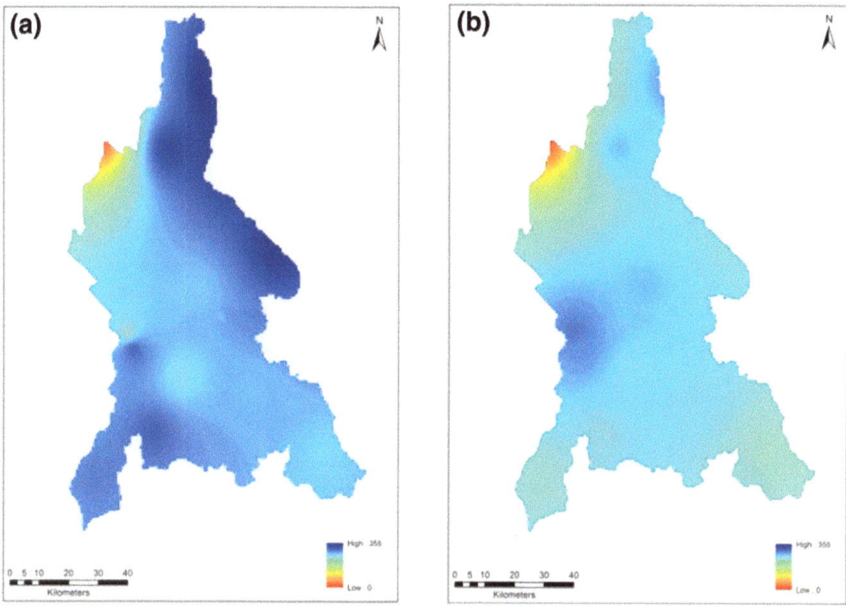

Fig. 8.7 Rainfall during month of July **a** 1978 (averaged for 1976–1980) and **b** 2008 (averaged for 2006–2010)

observed that the rainfall in pre-monsoon season has increased and in post-monsoon season it has decreased, depicting a seasonal shift in monsoon pattern. The decrease in rainfall during post-monsoon season is observed to be 21% which is very significant. The post-monsoon rainfall is essential for rabi season as it provides sufficient moisture availability for rabi crops to grow. The decrease in post-monsoon precipitation forces farmers to depend more on the irrigation facilities and withdraw a large amount of water from groundwater and river. Irrigation further leads to increase the portion of the incident solar radiation absorbed by the soil system but irrigation beyond the field capacity leads to the formation of shiny layer on the surface thereby increasing the albedo (Sahu et al. 2013). The rainfall in southeastern part of the region, where maximum amount of deforestation has occurred, has reduced significantly in almost all the months. One peculiar feature visible in all the rainfall images is that of scanty rainfall in northwestern part of the region, which is due to proximity of Kota urban area. The high built-up land in the urban area leads to increased surface temperature thereby leading to decrease in condensation and precipitation.

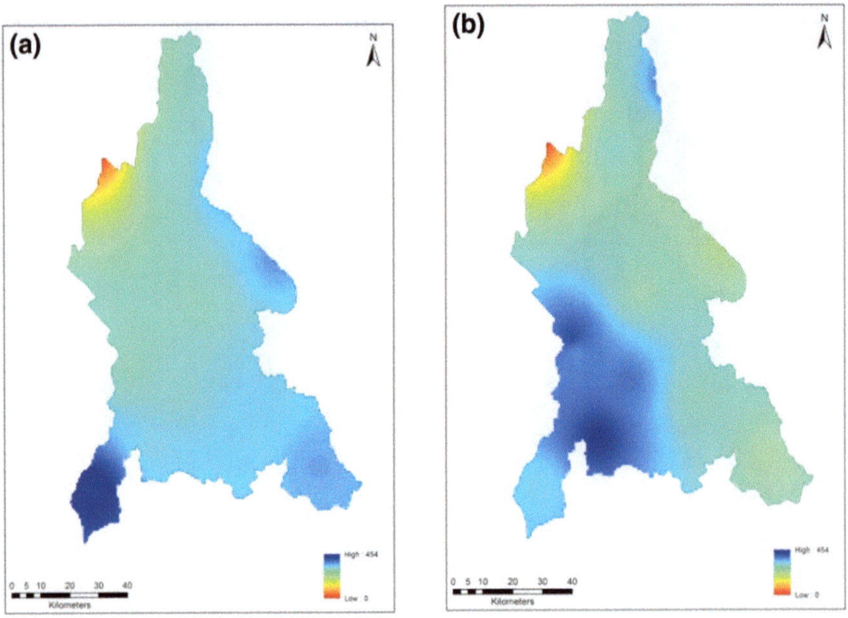

Fig. 8.8 Rainfall during month of August **a** 1978 (averaged for 1976–1980) and **b** 2008 (averaged for 2006–2010)

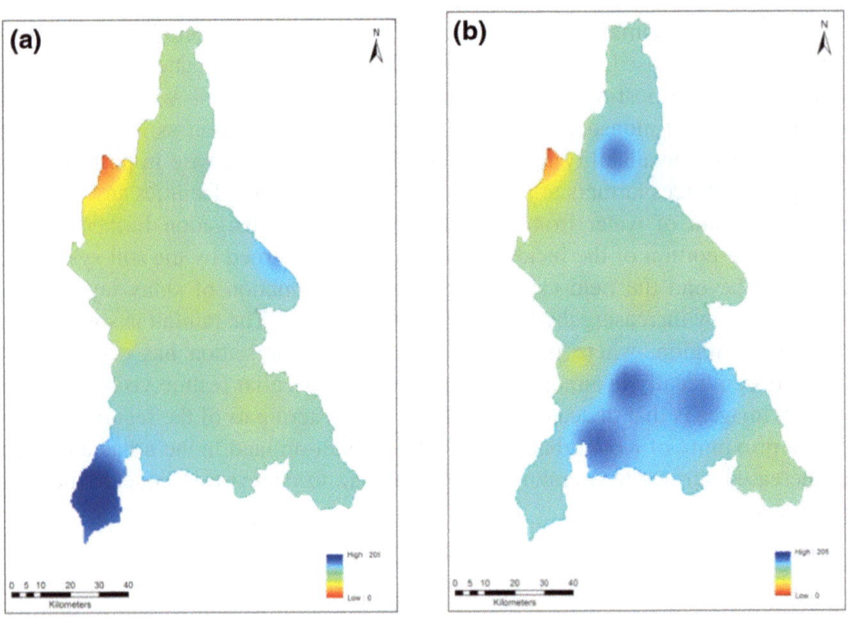

Fig. 8.9 Rainfall during month of September **a** 1978 (averaged for 1976–1980) and **b** 2008 (averaged for 2006–2010)

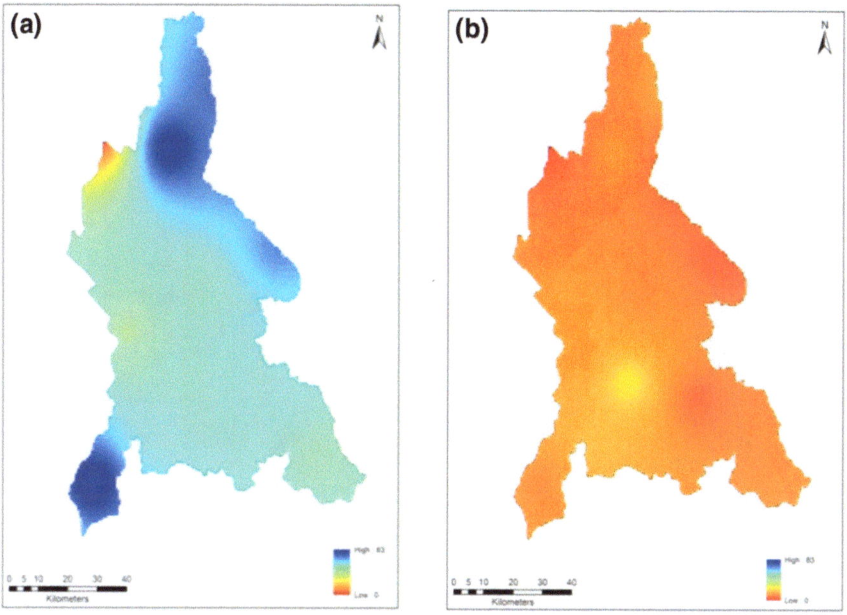

Fig. 8.10 Rainfall during post-monsoon season **a** 1978 (averaged for 1976–1980) and **b** 2008 (averaged for 2006–2010)

8.4.2.2 Land Surface Temperature (LST)

The primary driving factor for the atmospheric circulations is the land use land cover (Seth 1996; Sahu et al. 2012; Sahu et al. 2013). Land use land cover changes cause changes in LST because each land cover type possesses unique qualities in terms of energy radiation and absorption (Singh and Kumar 2012; Otieno and Anyah 2012; Ahmed et al. 2013). LST is considered to be an important variable in land–atmosphere interactions (Qu et al. 2013; Sen Roy et al. 2011; Findell et al. 2007). The analysis of land surface temperature for 1988 and 2008 depicts an increase in the LST values all over the study area. The LST for 1988 mainly lies between 25 and 32°C (Fig. 8.11a), whereas majority of region in 2008 is having LST values between 33 and 37°C (Figs. 8.11b and 8.12). The areas with intact vegetation cover show a marginal increase, whereas agriculture-intensive areas show highest increase in LST values. The water bodies are only areas where the LST has decreased owing to increased pollution from various anthropogenic sources (Fig. 8.13), which increases the albedo and decreased absorption of energy.

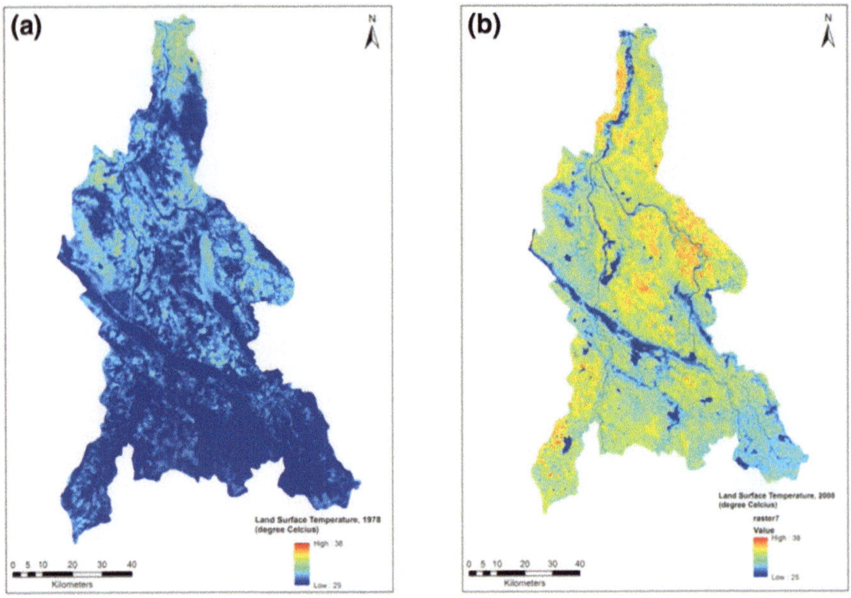

Fig. 8.11 Land surface temperature extracted from **a** Landsat TM, 1988 and **b** Landsat ETM+, 2008 for the month of November

8.4.3 Streamflow

The seasonal streamflow has been analyzed at Salavad stream gauge station that falls within the Kalisindh–Parwan Basin from 1977 to 2004. The aim is to link the hydrological cycle of precipitation and streamflow. To establish the relationship we focused the peak monsoon season of July–August–September (JAS) for the study. The cross validated Pearson's correlation measures have been (Fig. 8.14) to find the degree of linearity between the mean of the independent variable of rainfall and the dependent variable of streamflow of the relevant observations. As expected, we found low correlation of 0.42 between rainfall and streamflow observations that further signifies the impact of land use land cover as a driving force for the surface water resources.

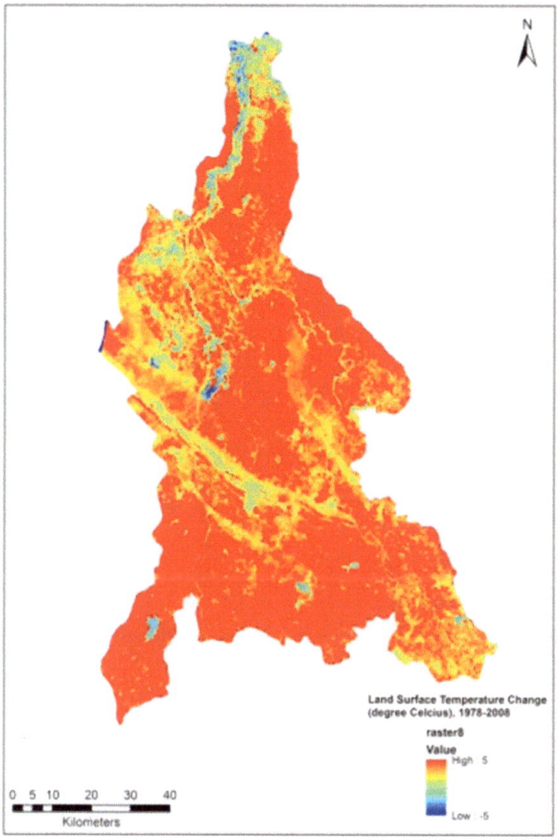

Fig. 8.12 Change in land surface temperature of November during 1988–2008

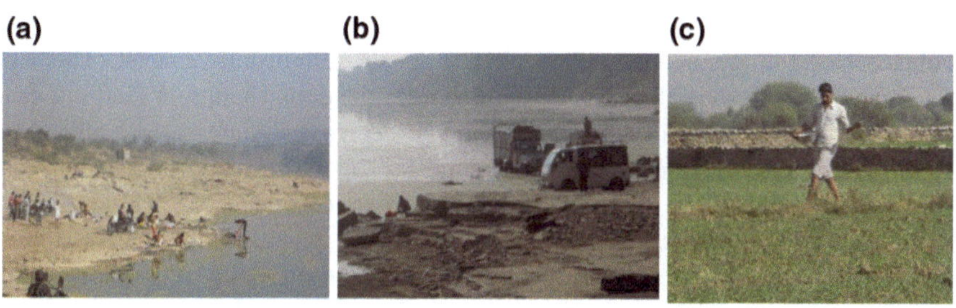

Fig. 8.13 Sources of anthropogenic pollution in Kalisindh river **a** religious bathing, **b** washing of vehicles and **c** use of fertilizers in agriculture which are washed down into the river water

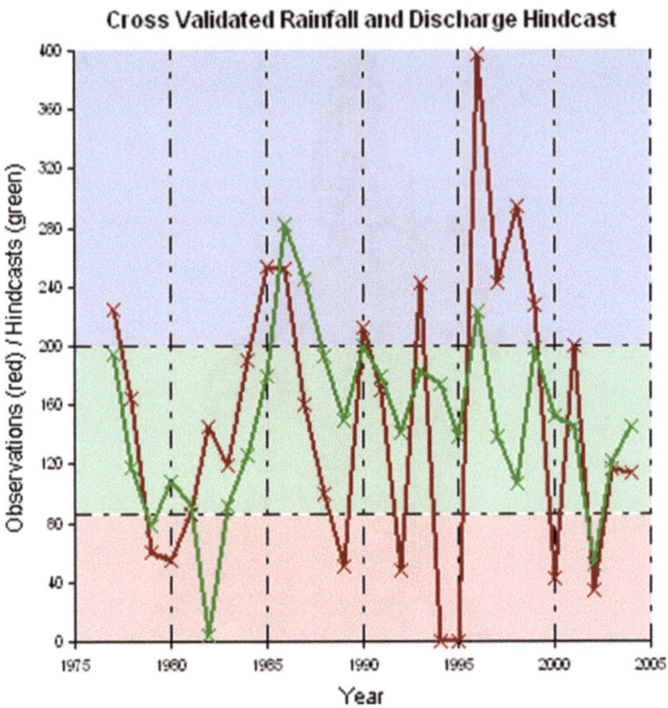

Fig. 8.14 Streamflow of JAS hindcasts (*green*) (m^3/s) based on rainfall hindcasts (*red*). The Pearson's correlation between hindcasts and observed is 0.42

8.5 Conclusion

This study has been carried out at Kalisindh–Parwan subbasin of Hadoti Region of Rajasthan in India. The subbasin clipped by the administrative boundary of Rajasthan state plays significant rule for decision support system, as most of policies in India are being implemented at lower administrative level. The Kali Sindh subbasin in Rajasthan has undergone a tremendous change in land use land cover over the last two decades in the form of decrease in bare surface and vegetation cover, and increase in agriculture land use owing to increase in demand of land and mechanization. The land use land cover of the region has significant impact on the microclimate of the region, where the overall rainfall has decreased and the land surface temperature has increased. The multi-linear regression results also show that the relationship between rainfall and streamflow is not significant thereby showing that the impact of land use land cover change is more dominant factor in the region. The increase in land surface temperature is more prominent over the agriculture land. The area with vegetation cover shows minimal increase in land surface temperature. The combined effect of these phenomena is climate variability leading to increased vulnerability of agriculture sector. Therefore, land use planning is a key factor in reducing the present and future vulnerability to climate, especially by augmenting prevention and preparedness in risk management process and by enabling response and recovery at community level. The understanding about the ecological functions that modify the microclimates, the land use land cover that alters these functions and the spatial variability in climate can be used to create more climate efficient and water conserving landscapes.

Acknowledgements We are very thankful to the India Meteorological Department for providing rainfall data. We are also thankful to the Department of Agriculture and the Water Resource Department of Rajasthan, India for providing crop and streamflow data for the research.

References

Ahmed B, Kamruzzaman M, Zhu X, Rahman MS, Choi K (2013) Simulating land cover changes and their impacts on land surface temperature in Dhaka, Bangladesh. Remote Sens 5(11):5969–5998

Bastiaanssen WGM (2000) SEBAL-based sensible and latent heat fluxes in the irrigated Gediz Basin, Turkey. J Hydrol 229(1):87–100

Boisier JP, De Noblet-Ducoudré N, Ciais P (2013) Inferring past land use-induced changes in surface albedo from satellite observations: a useful tool to evaluate model simulations. Biogeosciences 10:1501–1516

Bounoua L, Collatz GJ, Los SO, Sellers PJ, Dazlich DA, Tucker CJ, Randall DA (2000) Sensitivity of climate to changes in NDVI. J Clim 13(13):2277–2292

Buermann W, Dong J, Zeng X, Myneni RB, Dickinson RE (2001) Evaluation of the utility of satellite-based vegetation leaf area index data for climate simulations. J Clim 14(17):3536–3550

Central Groundwater Board (2010) Groundwater scenario of Jhalawar district, Rajasthan. District groundwater broacher. Ministry of Water Resource, Western Region, Jaipur

Chahine MT (1992) The hydrological cycle and its influence on climate. Nature 359(6394): 373–380

Costa MH, Foley JA (2000) Combined effects of deforestation and doubled atmospheric CO2 concentrations on the climate of Amazonia. J Clim 13(1):18–34

Deng XZ, Zhao CH, Yan HM (2013) Systematic modeling of impacts of land use and land cover changes on regional climate: a review. Adv Meteorol 2013:1–11

Dickinson RE, Jarvis PG, Monteith JL, Clark JA, Black TA, Warrilow DA (1989) Implications of tropical deforestation for climate: a comparison of model and observational descriptions of surface energy and hydrological balance [and discussion]. Philos Trans R Soc Lond B Biol Sci 324(1223):423–431

Dirmeyer PA, Shukla J (1996) The effect on regional and global climate of expansion of the world's deserts. Q J R Meteorol Soc 122(530):451–482

Dobos E (2003) Albedo. Encyclopedia of soil science. doi:10.1081/E-ESS120014334

Douglas E, Beltrán-Przekurat A, Niyogi D, Pielke RA Sr, Vörösmarty CJ (2009) The impact of agricultural intensification and irrigation on land-atmosphere interactions and Indian monsoon precipitation—a mesoscale modeling perspective. Glob Planet Change 67:117–128. doi:10.1016/j.gloplacha.2008.12.007

Douglas EM, Niyogi D, Frolking S, Yeluripati JB, Pielke RA Sr, Niyogi N, Vörösmarty CJ, Mohanty UC (2006) Changes in moisture and energy fluxes due to agricultural land use and irrigation in the Indian Monsoon Belt. Geophys Res Letts. doi:10.1029/2006GL026550

Eastman JL, Coughenour MB, Pielke RA Sr (2001) The regional effects of CO_2 and landscape change using a coupled plant and meteorological model. Glob Change Biol 7:797–815

Eliasson I, Svensson MK (2003) Spatial air temperature variations and urban land use—a statistical approach. Meteorol Appl 10:135–149

EMA (Emergency Management Australia) (2002) Planning safer communities—land use planning for natural hazards. Australian Emergency Manual Series—Manual, 7

Findell KL, Shevliakova E, Milly PCD, Stouffer RJ (2007) Modeled impact of anthropogenic land cover change on climate. J Clim 20(14):3621–3634

Gurran N, Hamin E, Norman B (2008) Planning for climate change: leading practice principles and models for sea change communities in Coastal Australia. Report no 3 for the National Sea Change Taskforce July 2008

Kueppers LM, Snyder MA, Sloan LC, Cayan D, Jin J, Kanamaru H, Kanamitsu M, Miller NL, Tyree M, Du H, Weare B (2008) Seasonal temperature responses to land-use change in the western United States. Glob Planet Change 60:250–264

Lal R, Bruce JP (1999) The potential of world cropland soils to sequester C and mitigate the greenhouse effect. Environ Sci Policy 2:177–185

Lean J, Warrilow DA (1989) Simulation of the regional climatic impact of Amazon deforestation. Nature 342(6248):411–413

Mahmood R, Pielke RA Sr, Hubbard K, Niyogi D, Dirmeyer P, McAlpine C, Carleton A, Hale R, Gameda S, Beltrán-Przekurat A, Baker B, McNider R, Legates D, Shepherd J, Du J, Blanken P, Frauenfeld O, Nair U, Fall S (2013) Land cover changes and their biogeophysical effects on climate. Int J Climatol. doi:10.1002/joc.3736

Niyogi D, Chang H-I, Chen F, Gu L, Kumar A, Menon S, Pielke RA Sr (2007) Potential impacts of aerosol-land-atmosphere interactions on the Indian monsoonal rainfall characteristics. Nat Hazards. doi:10.1007/s11069-006-9085-y

Oke TR (1981) Canyon geometry and the nocturnal urban heat island: comparison of scale model and field observations. J Climatol 1:237–254

Oke TR (1982) The energetic basis of the urban heat island. Q J R Meteorol Soc 108:1–24

Otieno VO, Anyah RO (2012) Effects of land use changes on climate in the Greater Horn of Africa. Clim Res 52:77–95

Pielke RA Sr, Pitman A, Niyogi D, Mahmood R, McAlpine C, Hossain F, Goldewijk K, Nair U, Betts R, Fall S, Reichstein M, Kabat P, de Noblet-Ducoudré N (2011) Land use/land cover changes and climate: modeling analysis and observational evidence. WIREs Clim Change 2011(2):828–850. doi:10.1002/wcc.144

Pielke RA, Cotton WR, Walko REA, Tremback CJ, Lyons WA, Grasso LD, Copeland JH (1992) A comprehensive meteorological modeling system—RAMS. Meteorol Atmos Phys 49(1–4): 69–91

Pielke RAS et al (2002) The influence of land-use change and landscape dynamics on the climate system: relevance to climate change policy beyond the radiative effect of greenhouse gases. Philos Trans R Soc Lond Ser A Math Phys Sci 360:1705–1719

Qu Y, Wu F, Yan HM, Shu BR, Deng XZ (2013) Possible influence of the cultivated land reclamation on surface climate in India: a WRF model based simulation. Adv Meteorol 2013:9

Rind D, Rosenzweig C, Goldberg R (1992) Modelling the hydrological cycle in assessments of climate change. Nature 358:119–122. doi:10.1038/358119a0

Roy SS, Mahmood R, Niyogi D, Lei M, Foster SA, Hubbard KG, Douglas E, Pielke RA Sr (2007) Impacts of the agricultural Green Revolution—induced land use changes on air temperatures in India. J Geophys Res 112:D21108. doi:10.1029/2007JD008834

Sahu N, Behera SK, Ratnam JV, Silva RV, Parhi P, Duan W, Takara K, Singh RB, El Yamagata T (2013) Nino Modoki connection to extremely-low streamflow of the Paranaiba River in Brazil. Clim Dyn. doi:10.1007/s00382-013-2006-3

Sahu N, Behera SK, Yamashiki Y, Takara K, Yamagata T (2012) IOD and ENSO impacts on the extreme stream-flows of Citarum river in Indonesia. Clim Dyn 39(7–8):1673–1680. doi:10.1007/s00382-011-1158-2

Sampaio G, Nobre C, Costa MH, Satyamurty P, Soares-Filho BS, Cardoso M (2007) Regional climate change over eastern Amazonia caused by pasture and soybean cropland expansion. Geophys Res Lett 34(17):1–7

Sen Roy S, Singh RB, Kumar M (2011) An analysis of local spatial temperature patterns in the Delhi Metropolitan Area. Phys Geogr 32:114–138

Seth A, Giorgi F (1996) Three-dimensional model study of organized mesoscale circulations induced by vegetation. J Geophys Res 101:7371–7391

Sharma HS, Sharma ML, Bhalla LR (2002) Geography of Rajasthan. Kuldeep Publications, Jaipur

Singh RB, Fox J, Himiyama Y (2001) Land use and cover change. Science Publication Inc., Enfield (NH), USA and Oxford & IBH Publication, New Delhi, pp 299

Singh RB, Kumar A (2012) Climate change, water resources and land use change in drylands of Rajasthan, India. In: Himiyama Y, Bicik I (eds) Land use/cover changes in selected regions of the world. International Geographical Union Commission on Land Use/Cover Change, vol VI, pp 13–22

Singh RB, Kumar D (2012) Remote sensing and GIS for land use/cover mapping and integrated land management: case from the middle Ganga plain. Front Earth Sci 6(2):167–176

Watson RT et al (ed) (2000) Land use, land-use change, and forestry. Cambridge University Press, Cambridge

Zhou X, Wang Y-C (2010) Dynamics of land surface temperature in response to land-use/cover change. Geograph Res 49(1):23–36

Chapter 9
Urban Land Use Land Cover Change

A Geomorphological Synoptic View of Mysore City, India

Arun Das, Koichi Kimoto, K. Jabir, Dhritiraj Sengupta, B.S. Shriharsha and M. Ravikumar

Abstract Indian cities are on the cross roads of rapid urban growth, which calls for meticulous and everlasting urban spatial planning. Cities live for longer and it should be compatible for the future growth. The previous mistakes committed in the land use planning have lead to many serious issues like floods (e.g., Chennai 2015 and Mumbai 2005 Floods) fire mishaps, traffic congestion, etc. Planning a city over a plain land is easier compared with the undulating terrain. Mysore is one such city located on an undulating terrain. This city has been declared as the cleanest city of India in 2015. Despite this, there are a few planning flaws exposed at certain point of time in the city. While planning for undulating terrain, two issues are confronted. First the need to classify the land into geomorphological regions is essential and second, if an old city is existing, the planner needs to be much more careful in choosing land use. Linking the old plan with the new plan certainly comprises seriousness of land use. The land classified on the geomorphological grounds can only give a better utilization of land suitable with the urban activities. This study provides an overview of the existing geomorphological land use land cover planning at micro-level water basins.

Keywords Urban land use land cover · Mysore city · Geomorphology
Uban planning · Micro-water basins

A. Das (✉) · K. Jabir · D. Sengupta · B.S. Shriharsha
Department of Geography, University of Mysore, Mysore, India
e-mail: arundasmysore@gmail.com

K. Kimoto
Kwansei Gakuin University, Nishinomiya, Japan
e-mail: k.kimoto@kwansei.ac.jp

M. Ravikumar
GIS and Health Sciences, JSS University, Mysore, India

Y. Himiyama (ed.), *Exploring Sustainable Land Use in Monsoon Asia*,
Springer Geography, https://doi.org/10.1007/978-981-10-5927-8_9

9.1 Introduction

Geomorphology-based land use is essential for a healthy urban planning. Urban geomorphology is the study of humans as a physical process of change whereby we metamorphoses a more natural terrain to an anthropogenic landscape/cityscape (Kuldeep Pareta 2012). Suppose if we glance at the past life of human, a close relationship exists between human and environment, especially, with respect to usage of land. The traditional way of land use planning is a good example of how human has given importance to geomorphological sites. For example, construction of forts in India, two aspects were taken into consideration and they included location of site along with availability of materials (Harish 2007). The most lonely and calm sites were chosen to construct temples, which are almost located on the top of hills, ridges, and river beds. The second-order priority was given for residential uses. Most of the villages were established along the ridge land and ridge land slopes. The stream path and river valley were chosen for wet land agricultural crops, the ridge land slopes were chosen for dry agricultural crops. But in the recent times due to over population and pressure on land resources, the selection of a landform as per the purpose of activity is not being used (Sundar 2009). Continued population growth (adding more than 70 million people to the world every year), changing dietary preferences (including more meat and dairy consumption), rising energy prices, and increasing needs for bioenergy sources are putting tremendous pressure on the worlds land resources (Foley 2009). In this context, this paper reveals the abuse and failure of land utilization which is being exceeding in the recent years. A point-based settlement like the city comprising of huge population enclosed by various forms of activity is a matter of great concern. Although the urban share of global land cover is negligible, urban land use at the local scale shows trends of declining densities and outward expansion (Harish 2007). Since within a limited land resource, the land should be utilized for various activities and it should cater the need of different activities and these must find a place of existence. As such, city land use planning has not been done on the lines of eco-friendly nor environment friendly. Urbanization in India is characterized by unplanned and uncontrolled growth leading to urban sprawl (Suneel Pandey 2015). Most of the city planning in India is short-sighted and do not have long-term spatial planning which can sustain the population and the environmental conditions for the decades to come. In case of Mysore city, the town planners (civil engineers) have given less attention to underlying geological structure and the resulting landform (Ghori 1980).

9.1.1 Geomorphological Overview of Mysore City

A city like Mysore located on the ridges of two river basins possess a huge upper ridge land area of 105 square kilometers followed by a ridge land slopes of 33.71 square kilometers and a combined valley land of mid- and upper mid-valley which converges towards the eastern direction covers 64.13 square kilometers. Mysore city being located on an undulating topographical landscape needs a judicious and meticulous use of land for different human activity (Arun Das and Koichi Kimoto 2009). Contrary to this, the urban planning authority has not given due importance to the activity related to geomorphological-based land use. It is also true that, activities of human have become more complex and it has guided by intangible elements such as land value, political motives, and lack of environmental concern. Evidently the urban planning has derailed, with respect to environmental, health, and recreational issues. Design elements in the built environment, such as street layout, land use, the transport system and the location of recreation facilities, parks and public buildings, are components of a community that can either encourage or discourage active living (Tsouros 2008) In such cases, it is the time for urban planners to consider, what kind of an urban land use mechanism can reduce the above-stated problems. In this direction, it is very much essential for us to select a proper utilization of land. To transform this into reality, the existing land condition in relation to the present activities, a detailed geomorphological oriented spatial planning has to be chalked out. Unless by implementing a proper land use model, just drawing plan on a plain sheet of paper and transforming on ground never fetches a good result.

9.1.2 Review of Literature

Prakasam C and Biplab Biswas (2011) evaluated the geomorphological resource and its importance for regional and micro-level planning and final part will look studies for water-stressed area identification and its importance on LULC of the region. F.C. Daia, C.F. Leeb, X.H. Zhangc made an attempt to evaluate the geo-environmental condition for urban land use planning based on geographical information system. Generating the result, we performed multi-criteria analysis and developed the suitability of the geo-environment for each category of buildings according to appropriately measured and weighted factors. A suitability map for each category is developed using an algorithm that combines factors in weighted linear combinations. It is demonstrated that the GIS methodology has high functionality for geo-environmental assessment. Lucian Dragut, Clemens Eisank (2011) have introduced an object-based method to automatically classify topography from SRTM data. The new method relies on the concept of decomposing land–surface complexity into more homogeneous domains. An elevation layer is automatically segmented and classified at three-scale levels that represent domains of complexity by using self-adaptive, data-driven techniques into more homogeneous domains in another context about the landform classification; and mapping has been developed as one of the most active areas of geomorphology. However, translation from continuous models of elevation and its products (slope, aspect, and curvatures) to landform divisions (landforms and landform elements) is filtered by two important concepts: scale and object ontology. Abdolmajid Ahmadi (2013) has studied about the role of geomorphological information in urban planning for sustainable development. He has concluded that establishment and development of city depends on environmental condition because the natural phenomenon has a crucial effect on the selection of the site, distribution, and urban physical development.

Geomorphological and progressive urban maps have instant applications in monitoring of urban sprawl/urban expansion and predictive modeling techniques to better forecast future areas of urban growth (Kuldeep Pareta 2012).

9.2 Background

9.2.1 The General Land Use of Mysore City

The general land use of Mysore City reflects a typical residential city (2012). The congenial climatic condition added with clean and green environment, attracts the people to live in Mysore. It is for this reason; out of the total geographical land area of Mysore city, 39.9% has been utilized for residential purpose. To cater transportation network 16.1% of land has been utilized. The industrial activity could not flourish in Mysore city as it was anticipated because of non-availability of skilled labors and higher transportation cost to import raw materials. Despite these problems, industrial zones were established in Mysore comprising of small- and medium-scale industries. They are located over the ridge land location of the city. Totally, industrial land use in this basin consumes 13.48% of geographical area. The public utility buildings and offices which are located in central part of the city accounts to 8.96% followed by open spaces and parks 13.74%. The existing land use pattern is a clear reflection of judicious planning compared to rest of the cities in India. It does not mean that, the entire planning is in terms of geomorphological landscape-based. The stream path of the city is still covered under green vegetative cover either by coconut plantation or by orchards. It is because of an elongated stream path location that these orchards and plantations stretch from the ridge land to the valley basin. Nearly 2.02% of geographical area has been occupied by water tanks and other micro-water bodies (Fig. 9.1). Varying from the classification done by MUDA (mentioned above), the general land use of Mysore city has been comprehended to six major categories such as commercial, public utilities, residential, open space, industries, and water bodies. Further for easier understanding of land use, the commercial, recreational, and residential areas are merged into one as (1) builtup land. The rest of the major land uses are the (2) agriculture (coconut plantation along stream path vegetation) which has now notified as green lands, (3) open spaces, (4) industrial layouts and (5) water bodies. The land use classification done by MUDA has not included road as one of the land use categories. But road networks are the important means of inter connecting the activities of different land. As such, in this analysis (6) road has been considered as one of the most important land use element. The spatial distribution of this land uses are described and analyzed according to the basins and further analyzed according to the micro-level geomorphological landscape that exists within the basins (Fig. 9.2; Table 9.1).

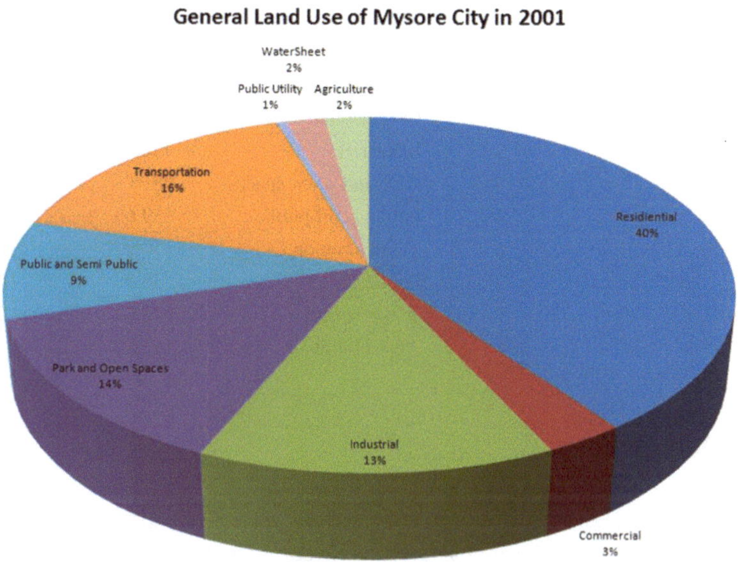

Fig. 9.1 General land use of Mysore City in 2001. *Source* Mysore City Development Plan (MUDA)

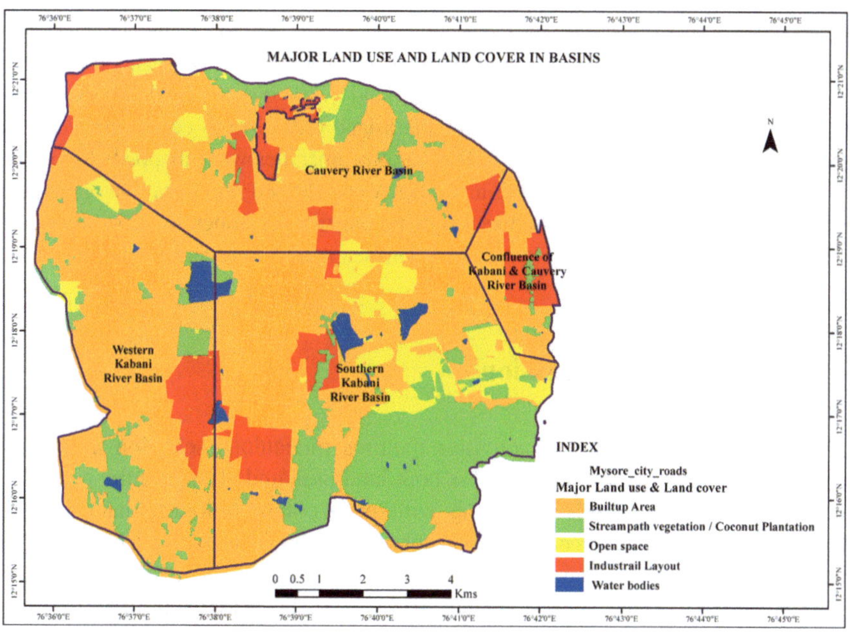

Fig. 9.2 Major land use and land cover in Mysore city (demarcation based on the Google earth image)

Table 9.1 General land use of Mysore City in 2001 (*Source* Mysore City Development Plan MUDA)

Sl. no	Land use of Mysore City	Area in Hectares	Percent
1	Residential	2849.91	39.9
2	Commercial	215.95	3.02
3	Industrial	962.61	13.48
4	Park and open spaces	981.7	13.74
5	Public semi-public	639.69	8.96
6	Transportation	1150.27	16.1
7	Public utility	36.48	0.51
8	Water sheet	143.99	2.02
9	Agricultural	162.33	2.27

9.3 Methodology

Realizing the importance of the relationship existing between land and human activity, the existing activities of Mysore has been surveyed. As an initial task of understanding the geomorphology of Mysore city, the Mysore Urban Development Authority (MUDA) boundary limit was selected as study limit. The SRTM data was downloaded and clipped to the study area boundary. Based on the Arc GIS hydrological tool, subbasins were generated. The entire city was classified into four basins such as Western Kabini River Basin, Cauvery River Basin, Southern Kabini River Basin, and the confluence of Cauvery River and Kabini river basin. Based on the elevation data, contours were generated at an interval of 20 m. Three types of geomorphological classification have been done such as ridge land (760–780 m elevation), mid-Land Plateau (740–760 m elevation), and Low plain land (720–740 m elevation). With the help of Global Positioning System (GPS) various amenities were mapped. As per the geomorphological classification the amenities were clipped and data was generated. With the help of generated data, the following existing amenities location were classified on the grounds of geomorphological landscape and were analyzed.

9.4 Results and Discussion

Choosing the land for urban activities requires meticulous selection as per the nature of the land. Mysore city possess both land uses underlying on the principles of geomorphological landscape along with non-geomorphological land use. To understand the positive aspect of geomorphological-based planning along with the negative impact of its unavailability in the planning process, two levels of discussion has been contemplated (Dutta 2011). First, the Ridgeland-based existing geomorphological land use of Mysore city and second, the micro-basins based existing land use, and further sub-geomorphological land use existing within the four basins.

9.4.1 Ridgeland-Based Geomorphological Land Use Analysis of Mysore City

The landscape of Mysore city reflects a true well-drained dendritic drainage formation. The geological controlled structures have a greater impact on the terrain (Rey 2005). A west-to-east elongated ridge land divides the city into two broad slopes, i.e., the Northern slope and Southern slope (Fig. 9.2). The triangular ridge junction is located in the west of Mysore city, exactly over CFTRI palace. The Northwest ridge stretches through Padavara Halli, Jayalakshmipuram, and Vijayanagar third stage through Hinkal and gets extended further out of the ring road. The north–to-south ridge begins from CFTRI palace stretches along District Collector Office (DC), Crawford Hall, Law Court, Chamarajapuram, Jayanagar, Ashokapuram, and Sri Ramapura layout and terminates. The north-to-south ridge which passes through Sri Ramapura and ring road junction also forms dividing ridge between Western Kabini River Basin. The west-to-east ridge passes through CFTRI palace, Mysore city railway station, Tilak Nagar, Mandi Mohalla, and Police Parade Ground up to Gousiya Nagar. At Gousiya Nagar, this ridge spilt into two, one ridge stretching towards north eastern direction and another branch stretches towards southeast direction. The northeast ridge from Gousiya nagar stretches through Kalyanagiri Nagar, Shanti Nagar up to ring road. Another branch starting from Gousiya Nagar stretches towards southeast direction passing through Sidarth Nagar, Lalitha Mahal Palace. Behind Lalitha Mahal Palace, the ridge turns further eastward and stretches through Varna Village up to ring road (Fig. 9.3). Based on the water divide developed by the ridge, the entire Mysore city has been classified into four basins. The western part of Mysore city drains towards western Kabini basin, similarly the southern part of the city drains towards southern Kabini Basin. Whereas the Northern part of the city drains towards the Cauvery Basin and eastern part of the city drains towards the confluence of Kabini and Cauvery River Basin.

9.4.2 Existing Land Cover Analysis of Mysore City Based on Micro Basins

Broadly, the geomorphology of Mysore city is classified into two major groups, Cauvery river basin and Kabini River basin geomorphology (Fig. 9.4). The Cauvery River Basin occupies the northern part of Mysore City covering 30.01% of the Mysore city area. The major portion of Mysore city drains towards Kabini River Basin which accounts to 70%. For further classification, the geomorphology of Kabini River and Cauvery River has been classified into four, based on the elevation and slope parameters. The land towards Cauvery River Basin having steeper slope, has been classified as Upper Ridge Steep Slope, Mid-Steep Slope and Valley. Since the Kabini River drainage basin covers larger area, it has been further demarcated by ridge, such as Western Kabini River Basin and Southern Kabini

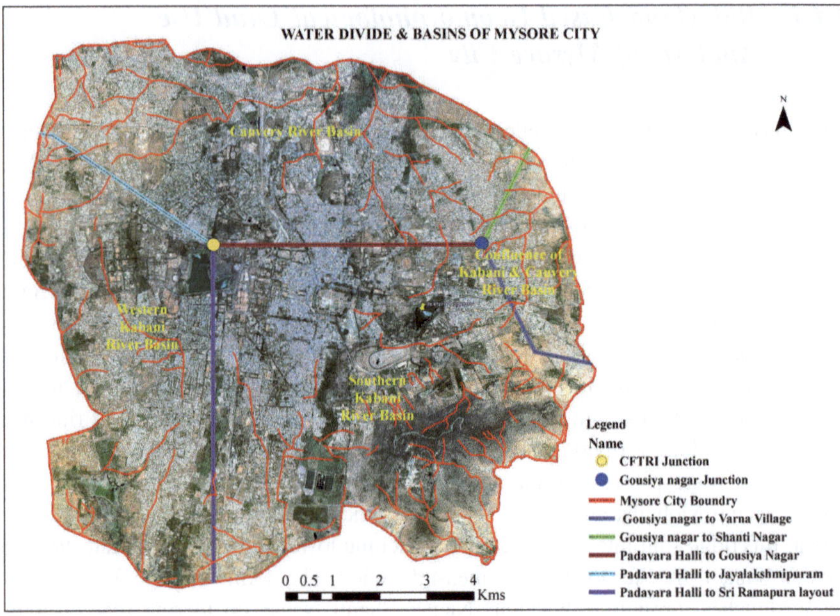

Fig. 9.3 Water divide and basins of Mysore city. *Source* Shuttle Radar Topography Mission Image

River Basin. Within these two basins, further classification has been derived based on the elevation, such as Ridge Land Zone, Mid-Land Zone, and Low Land Zone. Towards the eastern side of the Mysore city, the drainage networks are seen drained towards Cauvery and Kabini confluence section. Based on this classification, the detailed interpretation has been provided in the forthcoming paragraphs.

9.4.3 Land Cover Analysis of Cauvery River Basin

From 1897 to 1923, Mysore state witnessed the persistent of Plaque, which killed thousands of people who were residing within the fort of Mysore (Thriveni 2013). The Mysore king took initiative to restructure the city planning by entrusting this job to a British engineer. He selected Mysore city ridge land of Cauvery and Kabini river basin for first phase planning. Most elevated land was reserved for monumental buildings, posh localities, play grounds, and administrative buildings. The Cauvery basin forms the northern part of the city which possess built-up area of 58.69% in low land and 75.1% mid-land whereas 72.5% in ridge. The steep gradient of this basin has hampered the built-up land in the low land and ridge land landscapes, whereas the road network is almost parallel to one another in all the three geomorphological landscape. The glaring aspect of Cauvery basin region is stream path vegetation. Especially, the mid-land and low land possess significant

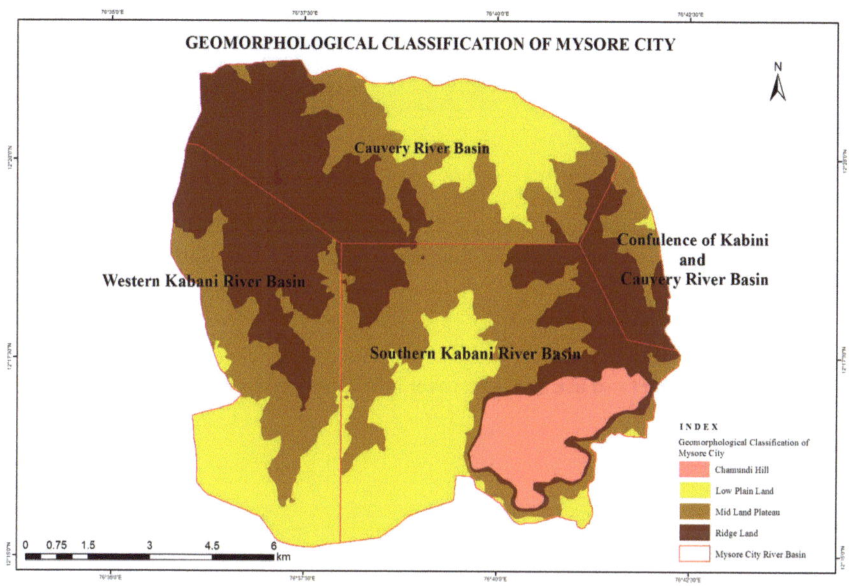

Fig. 9.4 Geomorphological classification of Mysore city. *Source* Shuttle Radar Topography Mission Image

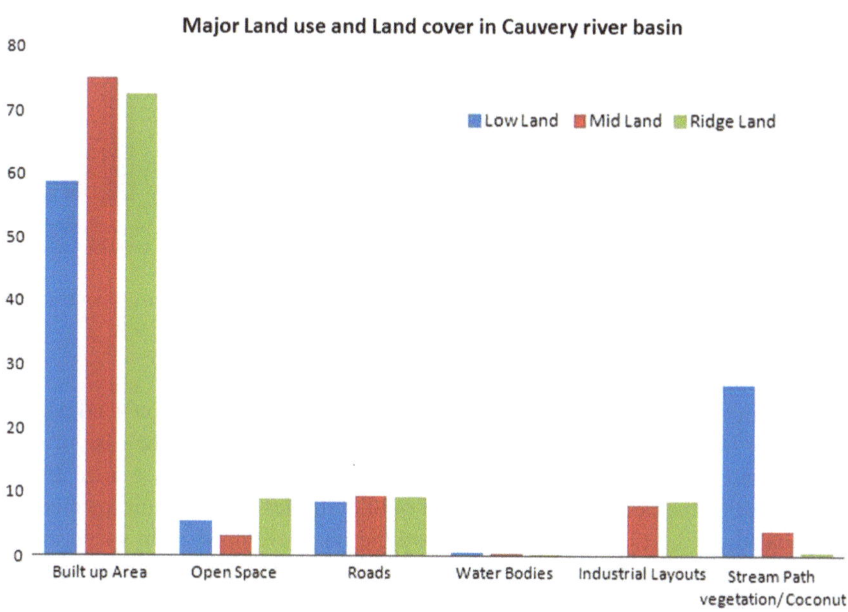

Fig. 9.5 Major land use and land cover in Cauvery river basin. *Source* Satellite Image

Table 9.2 Major land use and land cover in Cauvery River Basin

Sl. no	Major land use and land cover (in percent)	Low land	Mid land	Ridge land
1	Built-up area	58.69	75.1	72.5
2	Open space	5.38	3.19	8.95
3	Roads	8.47	9.34	9.29
4	Water bodies	0.53	0.34	0.08
5	Industrial layouts	0	8.1	8.65
6	Streampath vegetation/coconut plantation	26.92	3.93	0.53

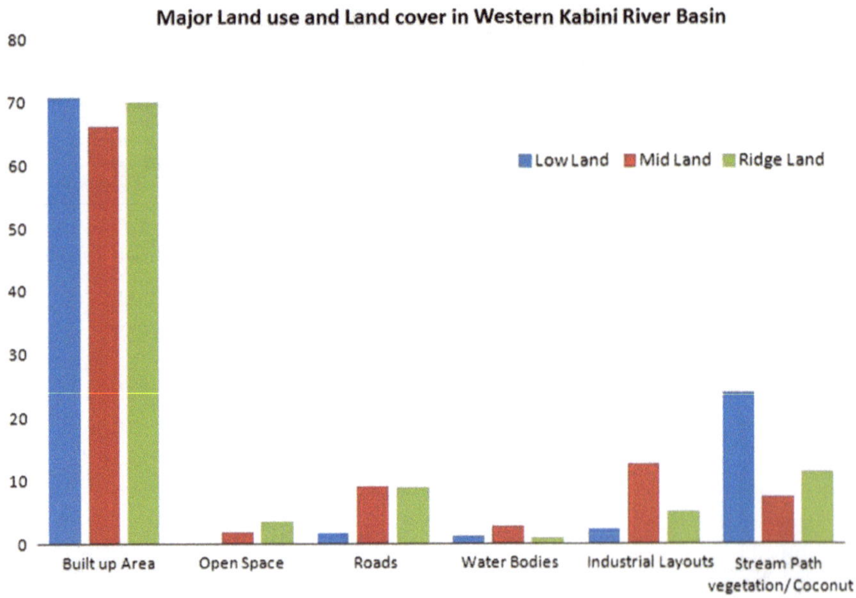

Fig. 9.6 Major land use and land cover in Western Kabini River Basin. *Source* Google earth Image

stream path vegetation (Bhibhu Prasad Nayak 2013). Whereas, the low land possess 26.92% of land under stream path. Also, it is quiet vice-versa in the case of open space, where the ridge land possesses higher percent of open space area. In this basin, most of the heritage structures and public utilities like stadium, race course are situated on the ridge land, followed by the non-developed open space in low land moderately developed with more number of parks and open spaces in mid land (Fig. 9.5; Table 9.2). This is a clear reflection of judicious use of land according to the functional characteristics of the activities.

9.4.4 Land Cover Analysis of Western Kabini River Basin

The micro-classification of Kabini River Basin has also been categorized similar to Cauvery basin. In this basin, the major land use is residential, and it is predominant under all the three types of geomorphological landscapes. The total area under open space in the ridge land of the Western Kabini river basin occupies 3.56%. The mid-land region possesses 66.20% whereas the low land and ridge land possesses almost close to 70.79%. The next major land use in this basin has been occupied by the stream path vegetation and coconut plantation. The low land plains of Western Kabini river basin possess 23.90% followed by 11.37% by ridge land and 7.36% by mid-land. The fact is that, the stream path concentrations are more on the ridge and low land rather at the mid-land (Charsley 1997). The road network is quiet complementary to geomorphological land according to the level of development in this region. Since the low land terminates near the river channel and much of land use activities are not yet taken place, the scope for city development is quiet favorable. This is reflected from the fact that, only 1.68% of land has been used for the proposal of road network. Whereas the good concentration of the residential layouts and educational institutes in this particular basin is a good sign of road connectivity which accounts to 9.04% under mid-land and 8.88% under the ridge land region respectively (Fig. 9.6; Table 9.3). During 1898, the City Improvement Trust Board had developed plan and notified the area for residential and education purposes in western Kabini river basin such as University of Mysore and residential layouts like Swaraswathi puram, and vontikoppal. The present Kuvempu Nagar layout was under agriculture land in this region (2012).

9.4.5 Land Cover Analysis of Southern Kabini River Basin

The Southern Kabini River Basin, was the second fastest grown segment of the city, adhering to strict land use planning. This can be visualized from the graph shown where coconut plantation, open space, and streampath vegetation have a good amount of land under ridge and mid-land. The built-up land is considerably lower compared to other basins but ranks first in the use of land which accounts 60–43% under various landforms. The low-lying land possesses higher built-up land space. The lesser is found in mid-land and much lesser found in ridge land which accounts to 42.65%. This is a clear reflection of a good land space allocation for various activities (Fig. 9.7; Table 9.4).

Table 9.3 Major land use and land cover in Western Kabini River Basin

Sl. no	Land use and land cover (in percent)	Low land	Mid land	Ridge land
1	Built-up area	70.79	66.2	70.03
2	Open space	0	1.96	3.56
3	Roads	1.68	9.04	8.88
4	Water bodies	1.27	2.79	0.98
5	Industrial layouts	2.36	12.64	5.17
6	Streampath vegetation/coconut plantation	23.9	7.36	11.37

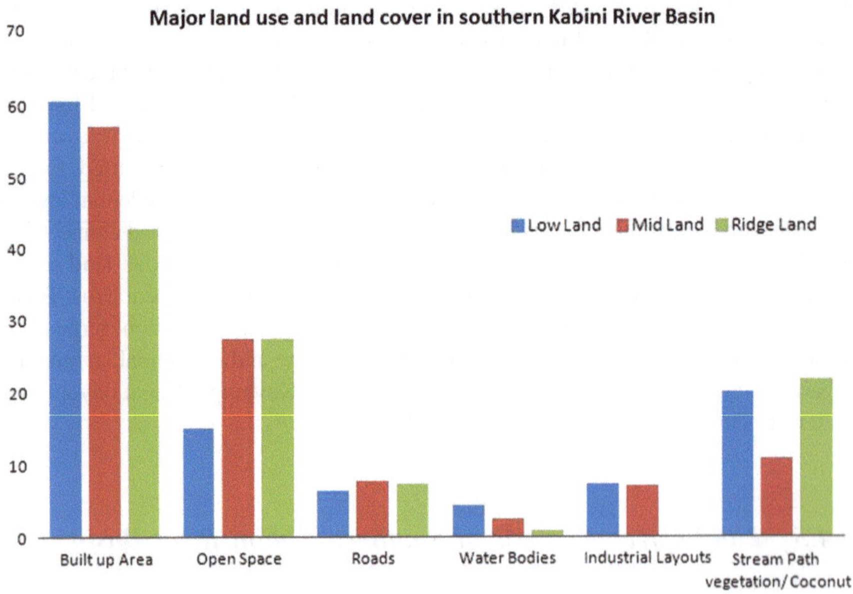

Fig. 9.7 Major land use and land cover in southern Kabini River Basin. *Source* Google earth Image

9.4.6 Existing Land Cover Analysis in the Confluence of Cauvery and Kabini Basin

Confluence of Cauvery and Kabini river basin land use depicts a different scenario. The urban land use is very much limited in this basin compared to other basins of Mysore. This can be seen from the statistics generated from the satellite image. The overall built-up area in this basin accounts to 60% on an average, whereas the coconut plantation and stream path vegetation are good in mid-land region accounting to 5.14%. Open space is almost nil in mid-land. In ridge land topography, 67.81% is occupied by built-up land and the next major share is seen by road which occupies 5.91% and remaining land use such as coconut plantation and

Table 9.4 Major land use and land cover in Southern Kabini River Basin

Sl. no	Land use and land cover (in percent)	Low land	Mid land	Ridge land
1	Built-up area	60.41	56.84	42.67
2	Open space	15.08	27.44	27.44
3	Roads	6.42	7.64	7.28
4	Water bodies	4.39	2.53	0.9
5	Industrial layouts	7.31	7.09	0
6	Streampath vegetation/coconut plantation	20.08	10.82	21.72

Table 9.5 Major land use and land cover in the confluence of Cauvery and Kabini River Basin

Sl. no	Land use and land cover (in percent)	Low land	Mid land	Ridge land
1	Built-up area	50.89	56.84	42.67
2	Open space	0	0	1.17
3	Roads	0	3.98	5.91
4	Water bodies	0	0.28	0.07
5	Industrial layouts	49.11	29.85	21.77
6	Streampath vegetation/coconut plantation	20.08	10.82	21.72

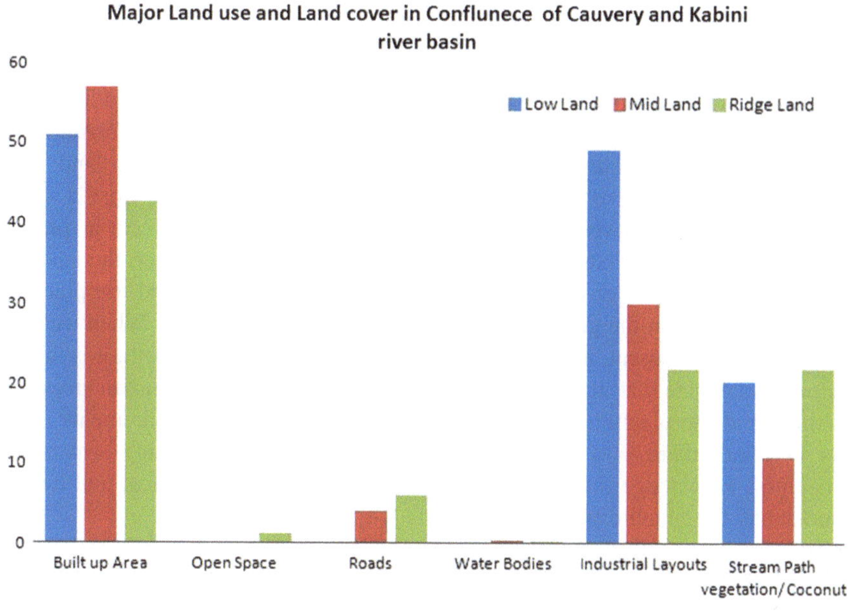

Fig. 9.8 Major land use and land cover in confluence of Cauvery and Kabini river basin. *Source* Google earth Image

stream path vegetation is 3.26%. The open space accounts to just 1.17%. Since the city in its initial stages of expansion has moved towards the northern and then towards the northwestern direction. In the later years it extended towards the southern and followed by eastern direction. The increased land value influenced the farmers to sell their land for non-agricultural activity resulting into higher percent of land coming under residential and other built-up purpose (Table 9.5; Fig. 9.8).

9.5 Conclusion

The existing inter-mix land use planning of Mysore city is due to the negligence shown towards implementing geomorphological applications in spatial (urban) planning. The expansion of all the amenities and utilization of land use in the city for various purposes has been concentrated within a radius of 5 km, which is not a good sign of sustainable planning (Kalpana Kannabiran 2012). Going through the distribution of amenities in different basin, it envisages that, planners before Independence have given due consideration towards utilizing the landscape on the scientific study of geomorphology. These land use points are now categorized as monumental structures. Even the ancient cities like the remnants of Mohenjodaro and Harappa reveal remarkable feats of uniform urban planning and carefully executed layout, water supply and drainage (India, H. o. 2011, Karen and Dhakal 2014). This clearly envisages how good the ancient planners were, without any formal education, they could able to deliver a best suitable geomorphological oriented spatial planning. What is lacking in us? Why we are not like our ancient planners? In case of Mysore city, there were many lung spaces provided inside the city region in the form of huge parks and playgrounds such as the Curzon Park, Oval, and Pavilion Grounds. Apart from the open spaces and built-up land, educational institutions like Maharanis, Maharajas, Hardwick, and Sharda Vilas Colleges are located over the ridge and lower ridge land, which are at present notified as monumental structures. Beyond 3 km from the city center, the land use planning is out of proper plan. This clearly reminds that, urban land use planning needs to be done on the grounds of geomorphological background. When we look and estimate the present, along with the future trend of the city growth, it is indeed alarming to realize that, if 70% of the population start living in the urban areas, the existing land use planning will certainly fail to cope up to the requirements. With this in mind, there is a strong need to advocate spatial planning which can encompass knowledge that can interpolate between multiple physical and human interactions for suitable sustainable cities.

Acknowledgements Urban Land Use Land Cover Change: A Geomorphological Synoptic View of Mysore City, India was an endeavor of many people who have contributed directly and indirectly. I personally wish to thank all my students for their efforts in generations of remote sensing data and also in the preparation of maps and charts. This research work would not have been carried out without the financial support from the JSPS Japan. Prof. Himiyama being the LULCA project director recognised and encouraged me for five years. The final outcome are the two papers published in this book and also many more papers. I owe my indebtedness to Prof. Koichi Kimoto who introduced me to Prof. Himiyama.

References

Arun Das S, Koichi Kimoto A (2009) Counter urbanization in Indian cities-the case of Mysore. Geographische Rundschau, International edn

(2012) Authority, Mysore City development plan: under JNNURM scheme. Mysore: Government of Karnataka

Bhibhu Prasad Nayak PK (2013) Livelihood of local communities and forest degradation in India: issues for REDD+. Delhi, India

Charsley K (1997) 'Children of the Forest' or 'Backwards or Communities? The ideology of tribal development. Edinburgh Papers In South Asian Studies Number 7

Dutta AD (2011) Encroachment creating alarming situation. Guwahati, Assam, India

Foley J (2009) The other inconvenient truth: the crisis in global land use

Ghori G (1980) Urban geography of the city of Mysore. Mysore, Karnataka, India

Harish PS (2007) Significance of Forts during the Vijayanagara Period

India, H. o. (n.d.) (2011) Ancient History Encyclopedia. Retrieved from Library of Congress: http://www.ancient.eu/article/294/

Kalpana Kannabiran EA (2012) Legal clinics and adivasi rights: a report of national workshop. Econ Polit Wkly 47:21–24

Karen CS, Dhakal S (2014) Human settlements, infrastructure and spatial planning. Climate change 2014: Mitigation of climate change. Contribution of Working Group III to the Fifth Assessment Report of the Intergovernmental Panel on Climate Change. Cambridge University Press, Cambridge

Kuldeep Pareta SP (n.d.) (2012) Geomorphic effects on urban expansion: a case study of small towns in central India

Rey PF (2005) Introduction to strctural geology. University of Sydney, Sydney

Sundar N (n.d.) (2009) Violent social Conflicts in Indias forests: society, state and the market, Delhi. www.academia.edu/8291031/ India 2009

Suneel Pandey SS (2015) Urban environment. IIT Kanpur, Kanpur

Thriveni D (n.d.) (2013) Plague Outbreak Eradication Campaign under Colonial Mysore. Int J Sci Res Publ 1–3

Tsouros PE (2008) A healthy city is an active city: a physical activity planning guide. The Regional Office for Europe of the World Health Organization, Denmark

Chapter 10
Population and Land Use in Semiarid Area—A Case of Karnataka, India

Koichi Kimoto

Abstract Changes in land use due to population pressure will be certain at the statistical and macro level. However, in the event of undesirable changes, we must direct the situation to improve by social activities of human beings. The era of simply entrusting the way to improvement with technology and policies is over. Since the 1980s, the emergence of Community-Based Natural Resource Management that has become active in various parts of the world will be an appearance. However, substantial results of CBNRM have not been seen. In this chapter, we will look back over the history of deforestation in India in a case of Western Ghats, which is the area caused by the deforestation and is an area with newly generated and unique characteristics.

Keywords Geography · Forest · Periphery · Protected areas · Western Ghats India

10.1 Introduction

There is a strong correlation between demographic dynamics and LUCC. The Population and the Land Use (LU) are related by ensuring food for nurturing a certain population, such as there is an expansion of cultivated land by population increase, and conversely an expansion of cultivation-abandoned land by population reduction. Expansion and reduction of various lands such as residential areas and factories are determined by the supply-demand relationship to the land with the land price and rent as an indicator.

K. Kimoto (✉)
Center for Hands-on Learning Programs, Kwansei Gakuin University, Nishinomiya, Japan
e-mail: k.kimoto@kwansei.ac.jp

Y. Himiyama (ed.), *Exploring Sustainable Land Use in Monsoon Asia*, Springer Geography, https://doi.org/10.1007/978-981-10-5927-8_10

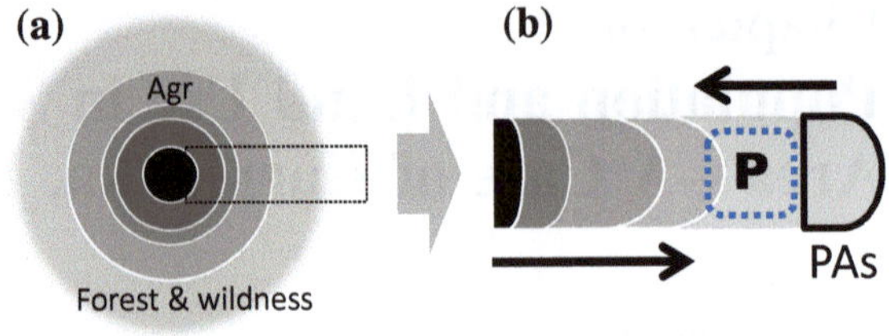

Fig. 10.1 Periphery (P) in double meaning **a** forest and wildness as a frontier, **b** making peripheral areas

The above explanation framework assumes the "frontier" of land use, which is explained by the center-peripheral structure. In the peripheral area, when the LU pressure drops, the spontaneous renewal assumes LU regeneration (Or it does not assume such a case).

However, as concerns about deforestation are worrisome on a global scale, the forest cannot simply be regarded any more as peripheral land use. Comprehensive management comes to be required for all land use including forest and wilderness unsuitable for agriculture and to be considered technical and social measures for them.

In particular, when forests and wilderness are designated as protected areas (PAs) such as national parks and wildlife sanctuaries, these periphery areas are with "periphery" in the double meaning as a periphery around the area in the center-peripheral structure, and around the PAs (Fig. 10.1). The pressure of population and land demand from the urban area was exerted, while the forest became wall rather than at the frontier of development due to the setting of PAs. A chaotic area appeared around the PAs. These peripheral areas are a crucial point of management in regional context.

10.2 Purpose and Method

In this chapter, we will examine the process of deforestation and the formation process of the peripheral area from the relation between population and land use, taking up the area around the Nagarhole National Park in the south of the Western Ghats.

We have conducted two projects, FMaRG (2010–2012) and ReGFF (2013–2016), under the frame of LUCC study (Fig. 10.2).

Historically said, Geography has been the study for a comprehensive understanding of "the relation between human and nature." But, geographers have abandoned this feature and geography has been divided into three parts, namely human geography, geography, and physical geography. In parallel, the object of each part has been divided. Human geography focuses on the human phenomena and Physical geography does on physical ones. Under this situation, small letter geography kept straying and loses its object. But, this small geography has still had a key clue for developing the future science as capital letter geography (Fig. 10.3).

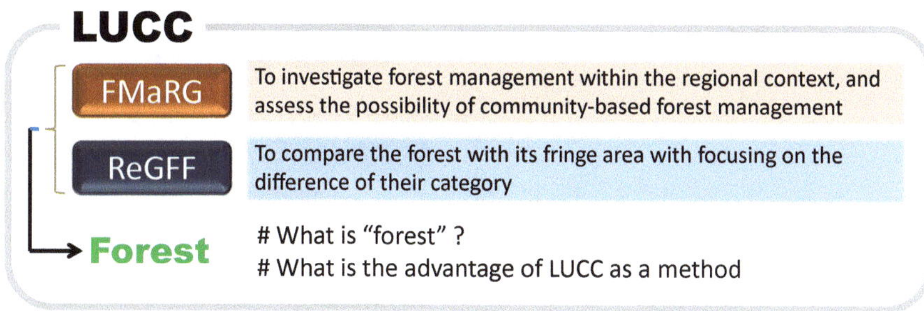

Fig. 10.2 Forest management as regional governance (FMaRG:2010–2012) and regional governance in forest and its fringe (ReGFF) 2013–2016

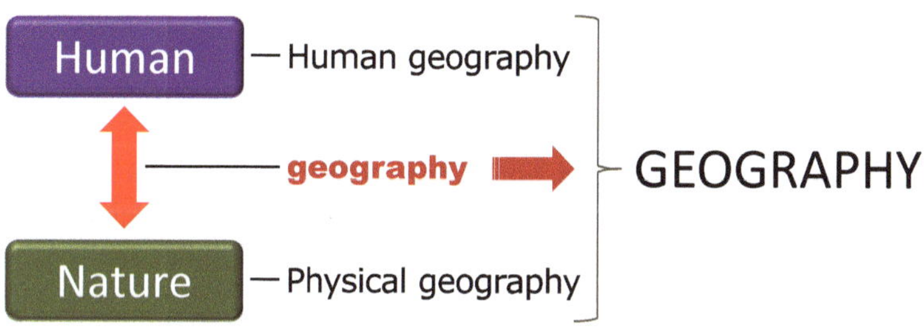

Fig. 10.3 Geography in history and in future

Even if there is room for discussion as to whether geography is science and technology that will directly contribute to solving concrete problems, geography, in situations where interdisciplinary research is required on a global scale such as Future Earth (FE), has an important role as a research framework for "comprehensive examination of the relationship between human and nature," and will be expected to refine as a methodology. Capital letter geography is a hope of future science.

By paying attention to LUCC, it becomes possible to examine the "relationship between human and nature" more concretely (Fig. 10.4). On that occasion, when considering the population as an independent variable and looking at the relationship between LU and the population, livelihood including food, clothing, and housing will come to emerge.

Land use is not merely the state of land or the fact of change, but also gives us an important clue to grasp the way of life. In the scales shown in Fig. 10.1, that is, not on individual villages but in the administrative areas established for the sake of statistical convenience, on the scale of the area related to the forest formed according to the object of forest and land use. It will be possible to consider the possibility of governance in the area as well.

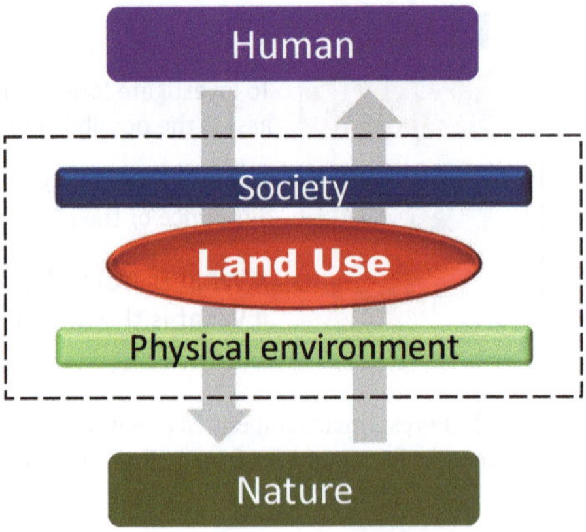

Fig. 10.4 Geography and land use

10.3 Fact: Deforested Area

The study area is Nagarhole National Park (NNP) in Western Ghats and its peripheral area. Western Ghats is a mountain range that runs parallel to the western coast of the Indian peninsula. The range runs north to south along the western edge of the Deccan Plateau and separates the plateau from a narrow coastal plain along the Arabian Sea. It is a UNESCO World Heritage Site (designated in 2012).

In the nineteenth century, the Western Ghats was managed in British India as a place to produce fine materials mainly of teak wood. After the independence of India, the area continued to decrease due to industrial pulp production and dam construction. From the colonial period to the 1980s, the Western Ghats were damaged both in quality and quantity.

NNP has been preserved for a long time as a king's hunting field in the days of the Indian State of Mysore, so a better environment has been maintained compared to other areas of the Western Ghats belonging to the territory of British India. NNP was named officially as NNP in 1983. NNP was renamed Rajiv Gandhi National Park in the 1988 and became the site of the Project Tiger in 1999. The area is 643.39 km^2 and the altitude is 700–960 m (Figs. 10.5 and 10.6).

10.4 Population Pressure

Figure 10.5 shows the situation of deforestation in the study area. One of the causes of deforestation is the character of the original vegetation in this area. In the Western Ghats mountain range, dense forest spreads in the mountainous region on the western side of the Deccan Plateau, but in the gently sloping facing east, it is vegetation centered on the semiarid climate and the central vegetation, relatively easy to cut down.

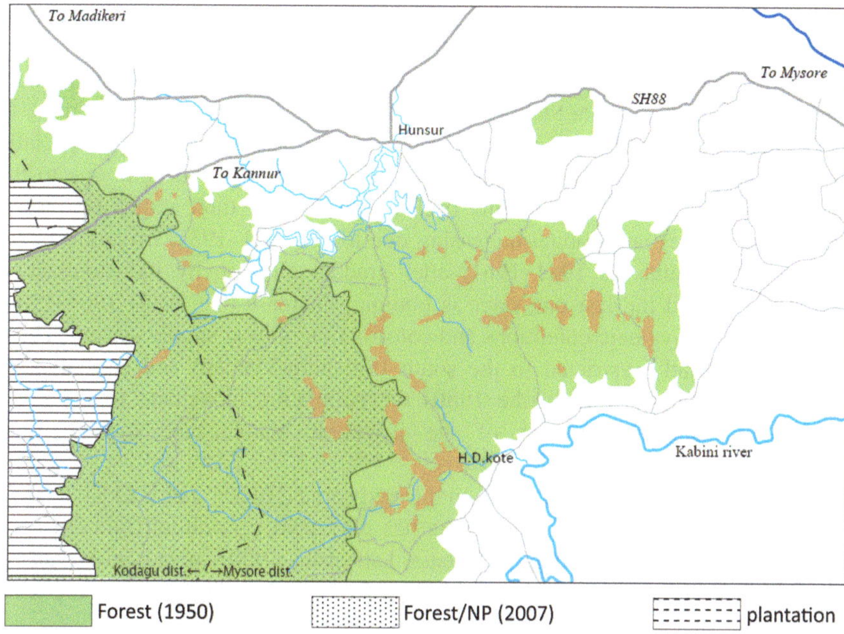

Fig. 10.5 Deforestation in Hunsur and H.D.Kote taluk, Karnataka, India

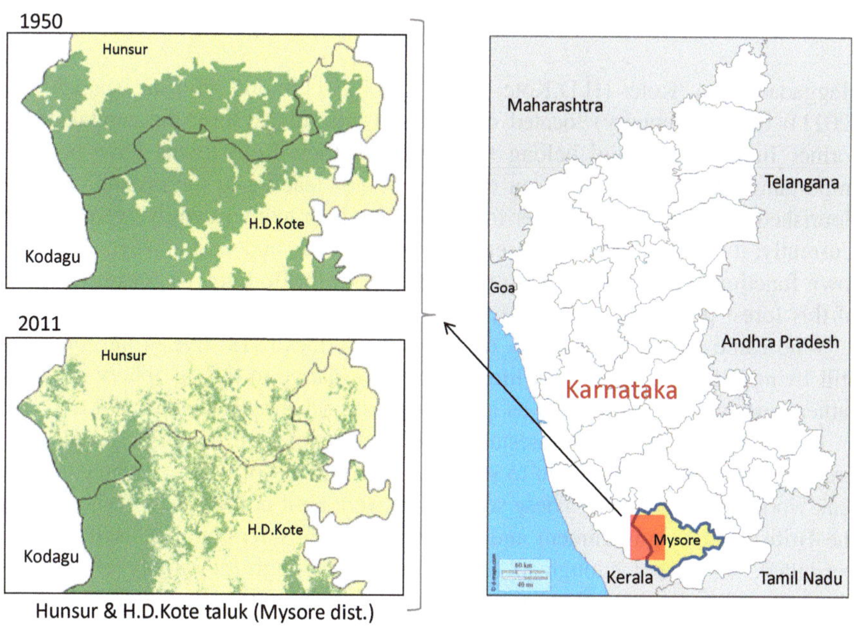

Fig. 10.6 Deforestation in Hunsur and H.D.Kote taluk (in vegetation)

Actually, the cause of deforestation is not due to the characteristics of vegetation itself. So-called anthropogenic effects should be taken into account.

Before entering concrete consideration of this area, we will review the population dynamics of Karnataka State (Table 10.1). During Mysore State era, similar to the trend of British India, the population had been gradually declining; especially the great famine at the end of the nineteenth century and the decline due to the World War I are noticeable. After the 1920s, we can divide three stages as follows; the first two stages are gradual increasing due to (1) 1921–1931: irrigation maintenance (south) and (2) 1941–1961: after gradual increase by industrialization (coastal area), and the last one stage is booming increase due to (3) 1961–1981: improvement of hygienic conditions and construction of food supply system. These trends were largely restricted to the south (Southern-Maidan) and coastal areas (Coastal and Malnad), but in the 1980s the population movement to the north started with the development of the irrigation maintenance in the north (Northern-Maidan) (Ranganath 2010).

When examining the above trends in the study area (Hunsur and H.D.Kote taluk, Karnataka), since almost the whole area has dray land agriculture (rain-fed agriculture), until the electric pump in the latter half of the 1990s spread, an expansion in an irrigation farmland has not preceded population influx. There were some other factors different from the general trend in the state.

10.5 LUCC and Land Category

Heggadadevana Kote (H.D.Kote, pop.14,313(2011)) and Hunsur (pop.50,865 (2011)) is taluk (county) located on the west side of Mysore, the capital of the former Indian State, and belong to Mysore district. Both towns have been an important transportation point on the major roads leading to the coastal area and flourished as a distribution center for timber in which many sawmills were located. Currently, it is also a point of dissemination of tobacco and various vegetables, and town function is diversifying with population growth. Both towns became a center of this forest peripheral area.

Who came to this area? Many of Scheduled Tribes (STs) live in the forest are still living a traditional way of life. STs are not native to this periphery area, but rather a newcomer in this area as a rehabilitation scheme in the late 1990s.

In the early stages of deforestation (the late nineteenth century), people who produce firewood settled down to support the livelihood of laborers who cut down teak wood (Kimoto 2014). These settlements were built in the places designated by the British Indian government and the Indian State. Since logging itself was not permitted, it was not enough to alter land use. Through the supply of firewood charcoal to the plantation that has spread rapidly during this period, and through the transaction of goods from STs who have connected to the outside of the forest through the firewood charcoal settlement and plantation, they were establishing a living base certainly, though the settlement was small in scale.

Table 10.1 Decadal variation in population growth by regions in Karnataka: 1901–2001

State/ regions	1901–1911	1911–1921	1921–1931	1931–1941	1941–1951	1951–1961	1961–1971	1971–1981	1981–1991	1991–2001
Karnataka	3.60	−1.09	9.38	11.09	19.36	21.57	24.22	26.43	21.12	17.25
Coastal and Malnad	−0.51	−1.11	5.71	7.81	17.24	32.67	24.19	24.33	13.13	11.47
Northern-Maidan	3.15	−4.98	9.96	11.05	15.60	19.27	22.53	23.60	23.00	18.76
Southern-Maidan	6.44	3.35	10.42	12.81	24.25	19.55	25.90	30.15	22.62	14.35
All India	5.75	−0.31	11.00	14.22	13.31	21.51	24.80	24.66	23.51	21.34

GOI (2007): Karnataka Development Report, p. 169

The establishment of a revenue village advanced at the same time had a decisive significance for the deforestation. The firewood charcoal residential area is like a dot, but the installation of the revenue village was coverage expansion. The revenue village was set up for tax collection from farmland. In the case of the existing village, the village is regarded as a revenue village itself. But in an area like a forest where no village exists, the territory of the revenue village had been set, with prior to the birth of the actual settlement of revenue village (Fig. 10.7). In particular, since the enactment of the Indian Forest Law (1927), many of the reserved forests were certified as revenue villages. Revenue village under the jurisdiction of the Ministry of Finance (MoF) will overlap forests under the Forest Department (FD). As a result, the FD generated revenues through timber and non-timber production, and the MoF implements tax collection after agricultural activities have actually started on the farmland of logging site. In this case, people's influx (i.e., population pressure) did not lead to deforestation, but people were flowing in due to changes in the category of land use, resulting in a reduction in vegetation forests.

After the Independence, it can be divided into two periods. In the 1970s, it is a land distribution project as a social welfare policy implemented. During this period, the land distribution policy to the landless farmer was implemented in the settlement village where the condition was bad, and many landless farmers flowed in. The inflowing people included farmers outside the state and fishermen in the coastal area, and the composition of residents in the area diversified.

In the 1980s, the increase in population and the increase in the ratio of forest cover became prominent in the form of increasing the ratio of total crop land and decreasing the ration of unused land in the crop land (Fig. 10.8).

Figure 10.9 shows an example of population increase due to an increase in the rehabilitation of STs (X) settled in the revenue village. X is a rehabilitation village located along the trunk line 5 km west of Hunsur. In the early 1970s, it was a settlement established by STs Migration policy but not revenue village belonging to revenue village (RV2). Residents in the village are not farmers but are engaged in public works such as management of nearby reserved forest (RF) and road construction. X village is included in RV2 village as a revenue village, but it does not appear on census.

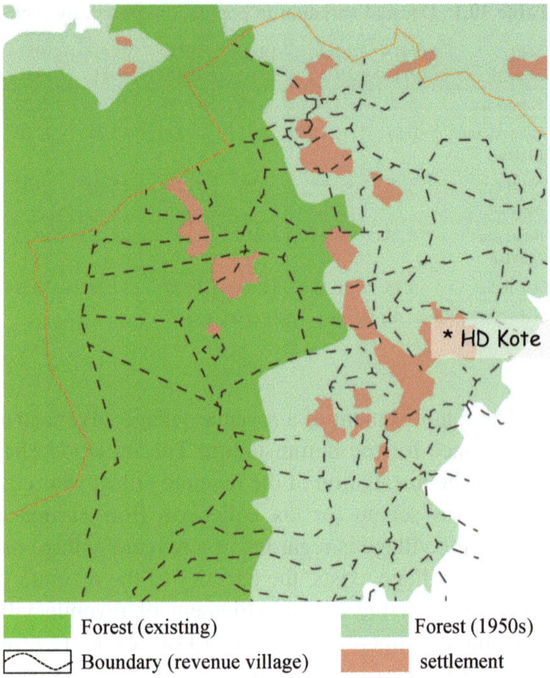

Fig. 10.7 Revenue village and forest

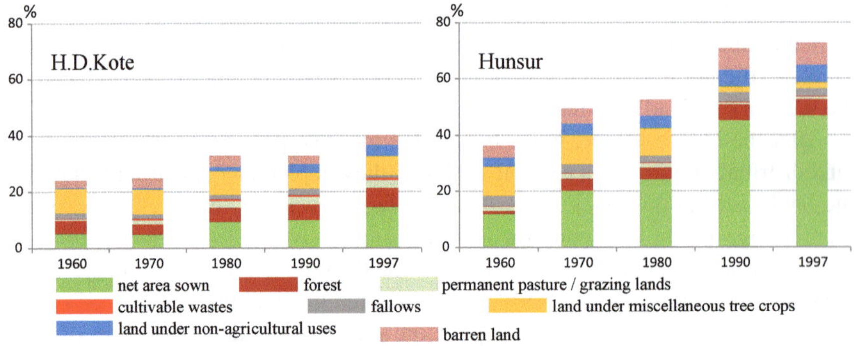

Fig. 10.8 Total cropped land (taluk, %)

The ratio of forest cover is rising statistically due to afforestation activities carried out in the revenue village. However, this afforestation activity is unrelated to existing forest regeneration like National Park, and many are carried out along the road, along the canal, on the boundary of the farmland, in the grassland (Figs. 10.10 and 10.11).

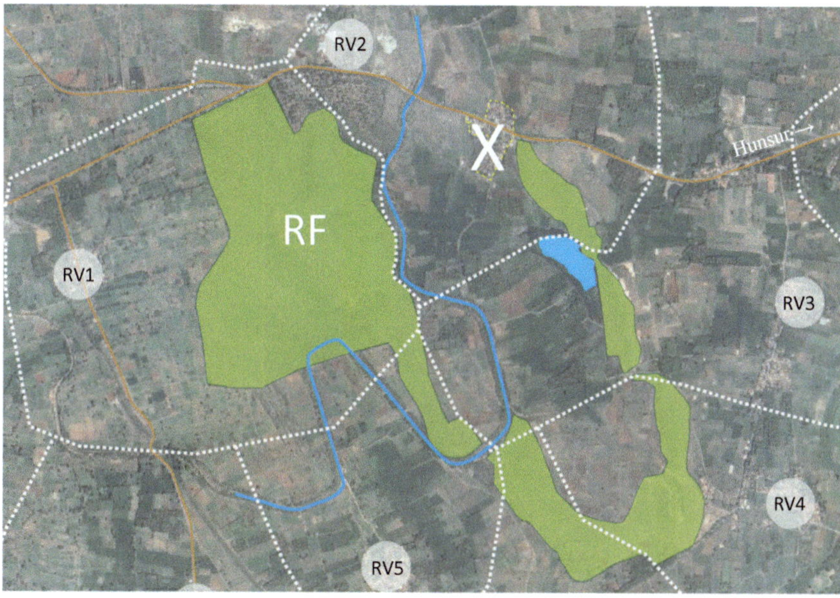

Fig. 10.9 Rehabilitation village for STs in Hunsur taluk (X)

Fig. 10.10 Afforestation on the bank of canal in X village

Fig. 10.11 Afforestation (Eucalyptus) in the village forest

10.6 Conclusive Remarks

In this chapter, we have been studying the relationship between land use change and population pressure based on intensive field studies. As a result, it became clear that the process of deforestation was not based on population pressure itself, but was policy-induced. Miscellaneous people gathered in a framework made by policies and formed a new region around "forest" where access was blocked as a PAs.

This result was a process not visible from statistical and macro research. This process is not just a transition of "change," but as a change in the structure seen in the relationship between human and nature (Fig. 10.12).

In future, we have to take care of this newly established region around the forest. The strategy might be promoted from the community-oriented policy, such as Community-Based Natural Resource Management. But, we have to consider a region according to the scale of PAs, that is, the range of consideration has not to be set by individual settlements (including revenue village), but by the community group suitable for management. Regions generated by deforestation are relatively inexperienced, and those who live in do not have a history of engagement with forests in particular. If aiming for community-based management, it is desirable not

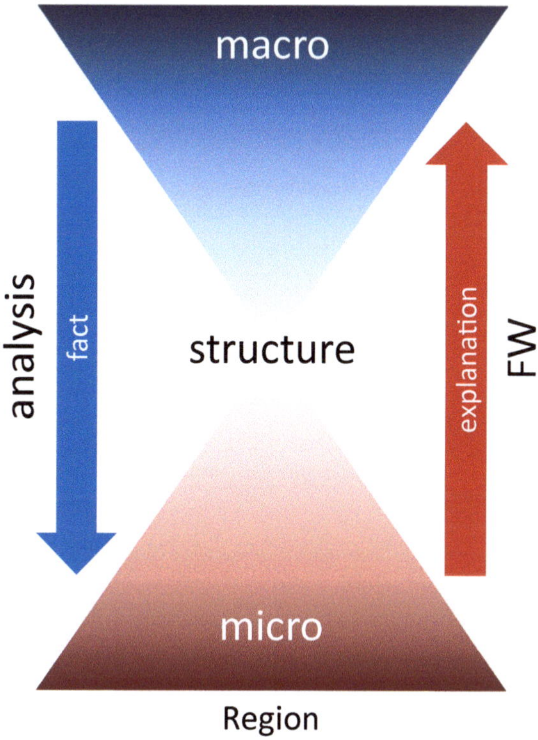

Fig. 10.12 Region in multi-scale and its structure

to create a local knowledge based on "tradition," but to invent an artificial relationship (and with their cooperation). In order to build a relationship between human and nature, we must also invent relationships within the human group.

References

Kimoto K (2014) Forest management as regional governance: an experience in India. Glob Environ Res 18(2):177–182

Ranganath (2010) Geography of Karnataka. Mysore

Part IV
Land Use Change in Japan

Chapter 11
Land Use Change in Tokyo Prefecture Viewed from the Medium Scale Topographic Maps

Yukio Himiyama and Tetsuya Fukase

Abstract Grid-type raster data files of land use in Tokyo Prefecture ca. 2005 have been compiled based on 56 sheets of 1:25,000 topographic maps of Japan, and compared with those of the earlier periods and those of the other prefectures, both of which produced by the present authors' team at Hokkaido University of Education, and were analysed spatially and numerically. The study has shown a detailed pattern of land use in Tokyo Prefecture ca. 2005. It indicates that further urban expansion is unlikely in Tokyo Prefecture because of the scarcity of flat land and that the production of specialty vegetables is becoming an important part of agriculture there. The Japanese Government is said to stop printing the 1:50,000 topographic maps, which started in 1897, and may do the same with the 1:25,000 topographic maps due to the harsh situation of the national budget and because of the advancement of "digitalization." It is therefore important to make use of the high-quality land use information on these maps in order to assist reevaluation of their value. It is hoped that this study contributes to such evaluation, and the survival of these maps, as well as to the study of land use/cover change.

Keywords Topographic map · Land use · GLP · Tokyo

11.1 Introduction

The Geography Institute of Hokkaido University of Education Asahikawa Campus is known in Japan for its study of land use/cover changes based on the topographic maps produced by Geospatial Information Authority (renamed from Geographical Survey Institute in 2010), or GSI, of Japan since 1897. At first 1:50,000 topographic maps of ca. 1900, ca. 1950, and ca. 1985 were used to reconstruct the land use of each period and to construct data bases of land use changes. Atlas—Environmental Change in Japan (Himiyama et al. 1995) shows a number of national maps of land

Y. Himiyama (✉) · T. Fukase
Hokkaido University of Education, Sapporo, Japan
e-mail: himiyama.yukio@a.hokkyodai.ac.jp

Y. Himiyama (ed.), *Exploring Sustainable Land Use in Monsoon Asia*,
Springer Geography, https://doi.org/10.1007/978-981-10-5927-8_11

use and the related tables, figures, and texts in order to show, describe, and explain land use changes, their background and their consequences, and related problems. In 2005, the Institute started to use 1:25,000 topographic maps instead of 1:50,000 maps, partly because the 1:50,000 maps were then made from 1:25,000 maps, hence the land use information on them were less updated and were almost a mere reduced-size copy of 1:25,000 maps. Incidentally, GSI stopped revision of the 1:50,000 maps in 2009. In another word, over a century-long history of 1:50,000 topographic maps ended in 2009 and it gave way to 1:25,000 topographic map which covered the entire country for the first time in 1983. The academic achievement of the study based on 1:25,000 topographic maps include the studies on land use change in Niigata Prefecture (Himiyama and Himiyama 2005), Fukuoka Prefecture (Himiyama and Nakamura 2006), Gunma Prefecture (Himiyama and Honda 2008), Kyoto Prefecture (Himiyama and Adachi 2009), Osaka Prefecture (Himiyama and Takei 2012), and Chiba Prefecture (Himiyama and Takamatsu 2013).

The present study focuses on Tokyo Prefecture as its study area (Fig. 11.1). It is simply called Tokyo unless otherwise mentioned. The main part of Tokyo is located in the southwest of Kanto District, with the Tokyo Bay to the east and south, Chiba Prefecture to the east, Saitama Prefecture to the north, Yamanashi Prefecture to the west and Kanagawa Prefecture to the south. It forms part of Tokyo Metropolitan Area, with rapid urban expansion and population increase in its western part near the Tokyo Bay. On the other hand, in more rural part of the prefecture, population decline has been eminent. On the Tokyo Bay coast is a huge stretch of reclaimed land which accommodates port facilities, various industries, which form the major

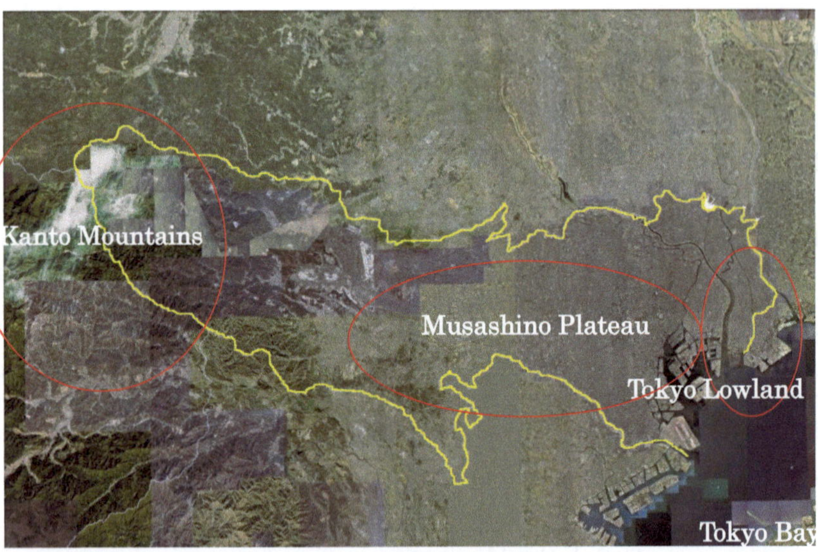

Fig. 11.1 Location of Tokyo Prefecture (after Google Earth)

part of the Tokyo-Chiba Industrial Zone. Tokyo is the capital of Japan, with 12,880 thousand (2015) residing in its administrative area of 2,191 km^2 and 35,000 thousand in its metropolitan area. Understanding in land use of this heartland of Japan is essential for its security, economic activity, and welfare. This study explores the use of the 1:25,000 topographic maps covering the administrative area of Tokyo Prefecture. The 1:25,000 topographic map is the largest scale map covering the entire territory of Japan. Institute of Geography, Hokkaido University of Education Asahikawa Campus started to use 1:50,000 topographic maps produced by Geographical Survey Institute of Japan and its predecessor as a tool for studying land use change, and its great primary achievements are demonstrated in the above Atlas" 1995, Himiyama et al.). From 2002 Himiyama and his associates started to use 1:25,000 topographic map series in addition to the 1:50,000 topographic maps in order to grasp the latest changes of land use.

The study will show how the land use of Tokyo Prefecture has been changing since ca. 1900, and what problems are arising in relation to land use changes and their background and effect. Through the study, the use of this topographic map series will be explored.

11.2 Reconstruction of Land Use

In this study, some 56 sheets of 1:25,000 topographic map were used to cover Tokyo Prefecture. The method of reading land use on the paper map, digitizing the information, and making data files for map drawing and analyses originates from the one proposed in Himiyama et al. (1990) and follows the one proposed by Himiyama and Honda (2008) and Himiyama and Adachi (2009). The land use classification used here is shown in Table 11.1. It is basically the one proposed by Himiyama and Honda (2008), but *palm* was newly added. The 1:25,000 topographic maps used in the study are those purchased in May 2008, i.e., the latest ones at that time, and represent the land use of around 2005.

11.2.1 Making Land Use Data Files

(a) **Coloring Maps by Land Use Type**

1:25,000 and 1:50,000 topographic maps of Japan have rich information of land use. Land use on the topographic maps is colored by color pencils according to its type, as shown in Table 11.1, which shows land use classification. The classification is based on Himiyama and Adachi (2009) which follows Himiyama and Nakamura (2006).

Table 11.1 Land use classification

Land use type	Code number	Color of pencil used
Paddy field	10	Yellow
Upland field/grass	11	Pale bitter orange
Mulberry garden	12	Yellowish green
Tea garden	13	Brown
Orchard	14	Bitter orange
Other tree crops	15	Bright yellow
Broad-leaved forest	20	Deep green
Needle-leaved forest	21	Green
Mixed forest	22	Blue
Bamboo	23	Ocher
Creeping pine	24	Reddish brown
Bamboo grass	25	Reddish purple
Palm	26	Emerald
Settlement	30	Red
(School)	31	Red
(Public office)	32	Red
(Temple)	33	Red
(Shrine)	34	Red
(Airport)	35	Red
(Military)	36	Red
(Vacant)	37	–
(Cemetery)	38	Fringe with gold line
(Park)	39	Ultramarine
Road	40	
Railway	41	
Golf course	1##	Fringe with light water
Ski ground	2##	Fringe with purple
Other sports/recreation	52	Dark grey
[Military exercise]	3##	Fringe with vermilion
[Natural park]	4##	Fringe with purple
Rough land	60	Light grey
Gravel/bare land	61	
Wetland	62	
River	63	
Lake	64	
Sea	99	

(1) Items in ( ) are all included in “settlement” when the largest land use type is recorded, while their code numbers are recorded at the end of the list of land use types observed in each grid cell.
(2) Items in [ ] are recorded according to the type of forest found at the location concerned with 3 or 4 affixed in front

(b) **Reading Land Use**

The topographic map has nine small dividing marks on the four edge lines so that one can divide the map into ten equally spaced strips, both horizontally and vertically. These marks are used to make more detailed $20 \times 20 = 400$ grid squares by drawing 19 dividing lines horizontally and vertically. The land use information on the map is then read for each grid square.

(c) **Inputting Data**

The land use information read as above for each grid square is then recorded by using Microsoft WORD or any other software. The first line is for data file name, survey year, and the name of map sheet and the actual data are written from the 2nd line to 401st line, corresponding to each of the 400 grid squares. A data line contains the following information: *x*-coordinate, *y*-coordinate, the code number of the land use on the top-left corner of the grid, code numbers of the largest land use, the second-largest land use, the third-largest land use, and all the land use found in the grid. The code number of each land use type is found in Table 11.1.

11.2.2 Using Data Files

The data files are then processed to make various grid maps. There are three different types of map presented here, namely:

- General land use map, which shows the largest land use type in each grid cell.
- Two-level density map of one type of land use, which differentiates grid cells with this land use as the dominant (D) land use, and those with this land use as existent (E) land use.
- Three-level density map of one type of land use, which differentiates grid cells with this land use as the dominant (D) land use, those with this land use as second dominant (S) land use, and those with this land use as existent (E) land use.

The area of each land use type in the prefecture was calculated according to the method developed by Himiyama and Ota (1993) and Himiyama and Motomatsu (1994). It is a method based on systematic point sampling. Land use type on the top-left corner of each grid square is recorded and counted by the type of land use. The simple formula below is used to reflect latitudinal effect, and adjust the total land area of the prefecture so that it equals the official figure issued by the government, i.e., 2187 km^2.

$$A = 0.323 \cos\theta((35 + 30/60 + (Y/240))3.14159265/180)\text{km}^2$$

11.3 Land Use in Tokyo in ca. 2005, General

Figure 11.2 is the general land use map of Tokyo in ca. 2005 produced in the present study. Urban land use is spread from the prefectural boarder with Chiba Prefecture in the east to 139°15′W around Ohmi City, Akiruno City, and Hachioji City. Forest is concentrated in the mountainous west. Agricultural land is almost invisible on the map. There are some islands in Tokyo Prefecture, and they are largely covered by forest and rough land, except Ohshima and Hachijo-jima where some settlements and dry fields are visible.

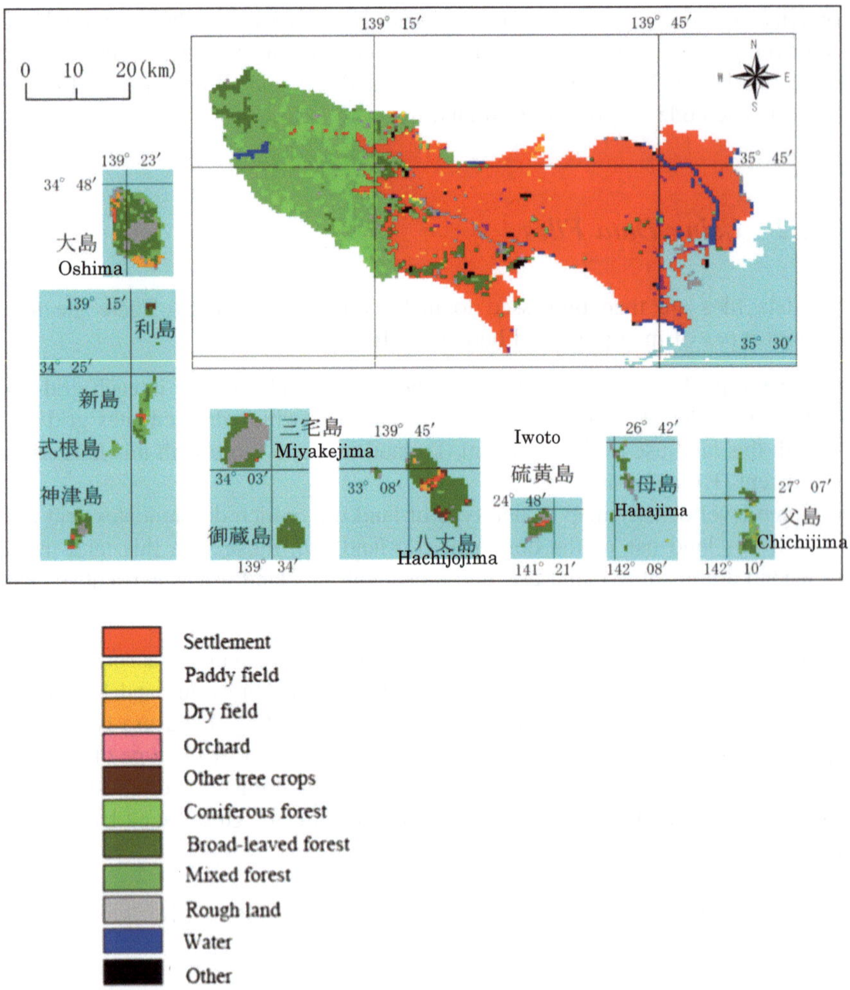

Fig. 11.2 Land use in Tokyo in ca. 2005

Table 11.2 Land use structure of Tokyo (ca. 1900–ca. 2005)

	ca. 1900		ca. 1950		ca. 1985		ca. 2005	
	km^2	%	km^2	%	km^2	%	km^2	%
Urban	327	15.90	704	33.29	964	45.02	1023	46.77
Settlement	252	12.26	553	26.14	767	35.82	888	40.61
Road	67	3.26	139	6.57	159	7.43	118	5.43
Railway	8	0.39	13	0.61	38	1.77	16	0.73
Agricultural	563	27.38	316	14.94	164	7.66	98	4.47
Paddy field	184	8.95	88	4.16	4	0.18	9	0.39
Upland field	283	13.76	191	9.03	143	6.68	68	3.11
Mulberry	96	4.67	38	1.79	0	0	1	0.02
Tea	0	0	0	0	0	0	1	0.04
Orchard/tree crops	0	0	0	0	17	0.79	20	0.91
Forest	1008	49.03	952	45.01	828	38.67	855	39.10
Broad-leaved	421	20.48	377	17.82	283	13.22	285	13.01
Needle-leaved	201	9.78	126	5.96	121	5.65	140	6.39
Mixed	381	18.53	449	21.23	423	19.76	409	18.69
Bamboo	4	0.19	0	0	0	0	14	0.66
Palm	–	–	–	–	–	–	8	0.35
Other	158	7.68	142	6.71	185	8.64	211	9.66
Rough land	141	6.86	121	5.72	143	6.68	174	7.96
Water	17	0.83	17	0.80	29	1.35	35	1.61
Sports	0	0	0	0	13	0.61	2	0.10
Other	0	0	4	0.19	0	0	0	0
Total	2056	100.00	2115	100.00	2141	100.00	2187	100.00

Table 11.2 shows land use in Tokyo ca. 1900, ca. 1950, ca. 1985, and ca. 2005. The data for ca. 1900, ca. 1950, and ca. 1985 are taken from Himiyama (1995), while the data of ca. 2005 are the result of the present study. Although there are some differences of scale of the maps and land use classification, they are considered minor. Figure 11.3 shows the trends of the four major land use categories, namely *urban*, *agricultural*, *forest*, and *other*. Rapid increase of *urban* is obvious. In ca. 2005, *urban* and *forest* together occupied over 85% of the prefectural territory, indicating the weakness of agriculture. Constant increase of *urban* and constant decrease of *agricultural* is evident.

11.4 Trend of Urban Land Use

The area of *urban* in Tokyo was 327 km^2 (ca. 1900), 704 km^2 (ca. 1950), 964 km^2 (ca. 1985), and 1023 km^2 (ca. 2005), with a linear rapid increase till ca. 1985, then a moderate increase afterward. The day time population in Tokyo is about 20%

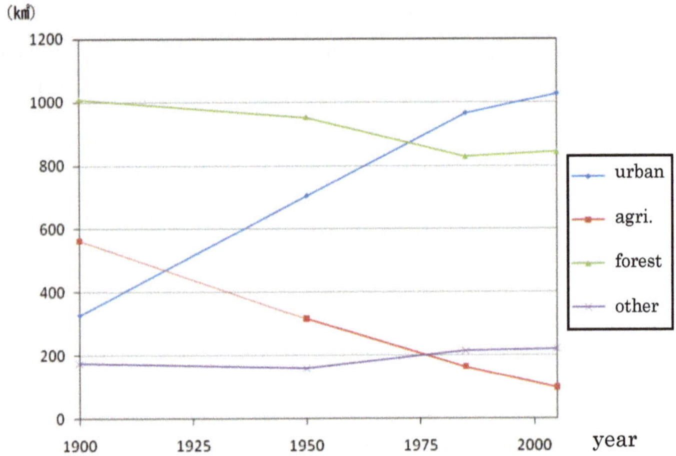

Fig. 11.3 Land use change in Tokyo ca. 1900–ca. 2005

larger than that of night time, so the actual population pressure for urban space might have been even higher. The population of the ward area peaked in 1965 at 8.84 million, and started to decrease due to suburbanization and passed the line of 8 million in the early 1990s.

11.4.1 Settlement

Settlement in Tokyo was 252 km^2 (ca. 1900), 553 km^2 (ca. 1950), 767 km^2 (ca. 1985), and 888 km^2 (ca. 2005) in respective years, showing a linear expansion till ca. 1985, then the speed eased afterward. It has been increasing continuously, particularly during ca. 1950–ca. 1985, which includes the period of the rapid economic growth of Japan. Residential, commercial, and industrial areas were developed in the former agricultural areas, forest, and on the coastal reclaimed land.

Figure 11.4 is the three-level density map of *settlement.*

Settlement dominant cells are largely found in the area east to the 139°15′ line. The west from this line is mountainous and urbanization is blocked. In the plains–mountains border area of Tama Ward, many second dominant and existent cells of the *settlement* are found, indicating the front-line nature of urbanization. In the ward area of Tokyo, some 8.8 million people, or 68% of the prefectural population, live. In contrast, the islands have been losing population, particularly after 2005. Figure 11.5 exemplifies the townscape of Central Tokyo, with many office buildings, while Fig. 11.6 shows the vast stretch of townscape dominated by low buildings. Figure 11.7 is part of the 1:25,000 topographic map showing forest around the settlement, in which Okutama Station (Fig. 11.8) is located.

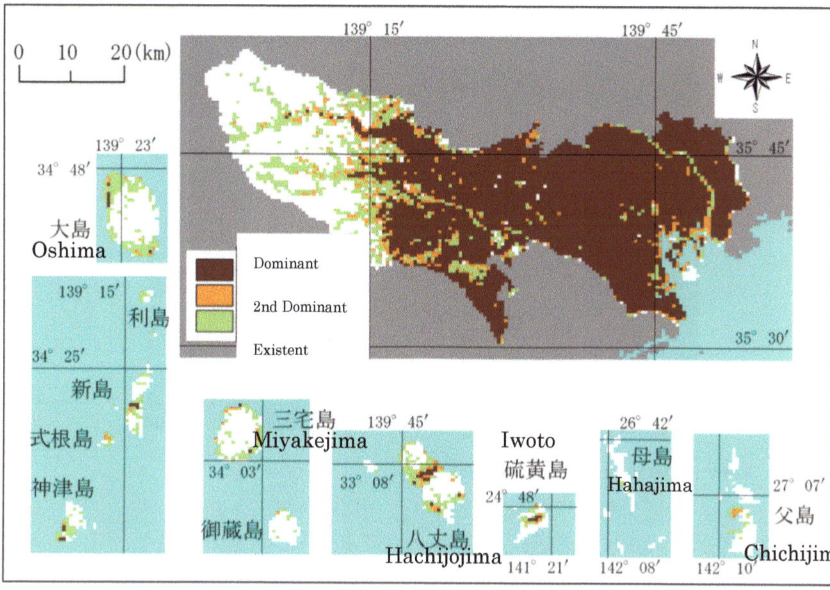

Fig. 11.4 Settlement in Tokyo in ca. 2005

Fig. 11.5 View to the east from Downtown Tokyo

11.4.2 Roads and Railways

Road and *railway* are exaggerated on the paper map, but this fact is not taken into account in order to avoid inconsistency of measuring by different people. Therefore, the area size of *road* and *railway* is larger than reality.

The area of *road* increased from 67 km^2 (ca. 1900), 139 km^2 (ca. 1950), 159 km^2 (ca. 1985), then decreased to 118 km^2 (ca. 2005) which is likely to be caused by the change of the scale of the map from 1:50,000 to 1:25,000. The case of railway is similar, as it increased from 8 km^2 (ca. 1900) to 13 km^2 (ca. 1950),

Fig. 11.6 View to the west from Downtown Tokyo

Fig. 11.7 Coloured topographic map of part of Okutama Town

38 km^2 (ca. 1985), then decreased to 16 km^2 (ca. 2005). Some of the decrease is actual, as some local lines and part of railway facilities were closed due to motorization.

Fig. 11.8 Okutama Railway Station

11.5 Trend of Agricultural Land Use

Agricultural in Tokyo was 563 km^2 (ca. 1900), 316 km^2 (ca. 1950), 164 km^2 (ca. 1985), and 98 km^2 (ca. 2005). It is a sharp and contiguous decrease from ca. 1900 to ca. 2005, i.e., quite opposite to urban land use.

11.5.1 Paddy Field

Figure 11.9 shows that the *paddy field* existent cells are found in the central area of the prefecture and in Hachijo-jima Island. *Paddy field* in Tokyo was 184 km^2 (ca. 1900), 88 km^2 (ca. 1950), 4 km^2 (ca. 1985), and 9 km^2 (ca. 2005). It may be surprising to find that there was fairly large paddy field area in Tokyo even after WWII. The paddy field in Hamura City is famous for its tulips which are planted after the harvest of rice and harvested before planting rice in spring.

Figure 11.10 is part of the 1:25,000 topographic map containing the *paddy field* closest to the center of Tokyo. Figure 11.11 shows the paddy field marked in Fig. 11.10. It is close to the Arakawa River, and there are some dry fields nearby as well.

Hachijo-jima Island is the only island of Tokyo which has paddy field. Some of the rice produced in this island is used to make sake, i.e., Japanese rice wine.

11.5.2 Upland Field and Grass

Upland field/grass in Tokyo was 283 km^2 (ca. 1900), 191 km^2 (ca. 1950), 143 km^2 (ca. 1985), and 68 km^2 (ca. 2005), showing a constant decline. It is far more widely

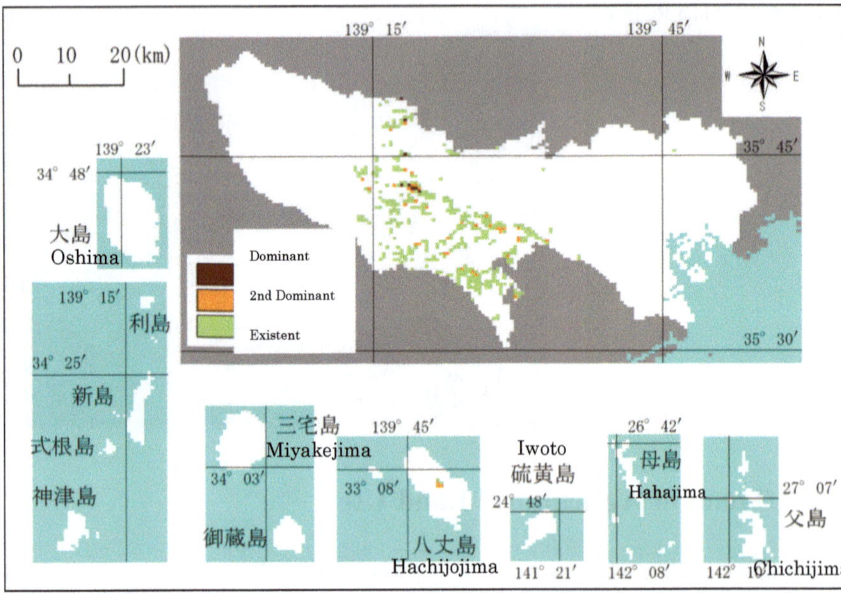

Fig. 11.9 Paddy field in Tokyo in ca. 2005

Fig. 11.10 Paddy field in central Tokyo

distributed than the paddy field, and even dominant and second dominant cells are found in many locations although most of the colonies consist of a few cells (Fig. 11.12). The largest concentration of upland field/grass is at Kiyose City at the

Fig. 11.11 Paddy field in central Tokyo (marked in Fig. 11.10)

northern edge of the prefecture (Fig. 11.13). Agricultural production is about 30 billion yen/year and half of it are vegetables, such as carrot and spinach. Tokyo supplies 10% of its demand for vegetables by itself. Some of the locally produced vegetables are sold by the farmers directly, as seen in (Fig. 11.14). Dairy farming and sweet potato production are also seen. The only *upland field/grass* in central Tokyo is the one in the main campus of the University of Tokyo, belonging to Faculty of Agriculture (Fig. 11.15).

11.5.3 Orchard/Other Tree Crops

Orchard and other tree crops in Tokyo were 0 km^2 (ca. 1900), 0 km^2 (ca. 1900), 17 km^2 (ca. 1985), and 20 km^2 (ca. 2005). Figure 11.16 shows the distribution of *orchard*. Existent cells are widely scattered in the central part of the prefecture west to 139°15′, indicating that farm households still remain in big number, but that they are mostly quite small and only part time in most cases. There is one dominant cell, which is in Inagi City in the south. The area of an *orchard* in Tokyo is about 6 km^2. The commercial production of blueberry started in Kodaira City in Tokyo, and it is increasing.

Other tree crops include the land for Paulownia, wax tree, paper mulberry and garden trees, and nurseries, but in Tokyo, only garden trees and nurseries are found. Figure 11.17 shows the fairly wide distribution of other tree crops in the central area of the prefecture, with some concentration along the northern bank of the Tamagawa River. Hachijo-jima, a Pacific island of subtropical climate, has many existent cells and even some dominant cells, indicating the importance of other tree

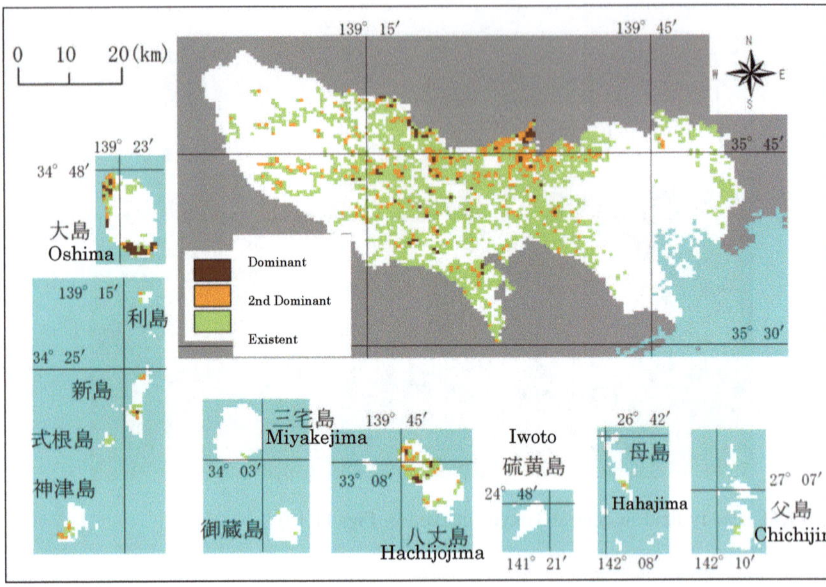

Fig. 11.12 Upland field/grass in Tokyo in ca. 2005

Fig. 11.13 Upland field/grass in Kiyose City

crops such as aloe, strelitzia, and hibiscus (Fig. 11.18). Another small island of Toshima also has some dominant cells. The island is well known for its camellia planted both for getting oil and for tourism.

Fig. 11.14 Vegetables sold on the stand

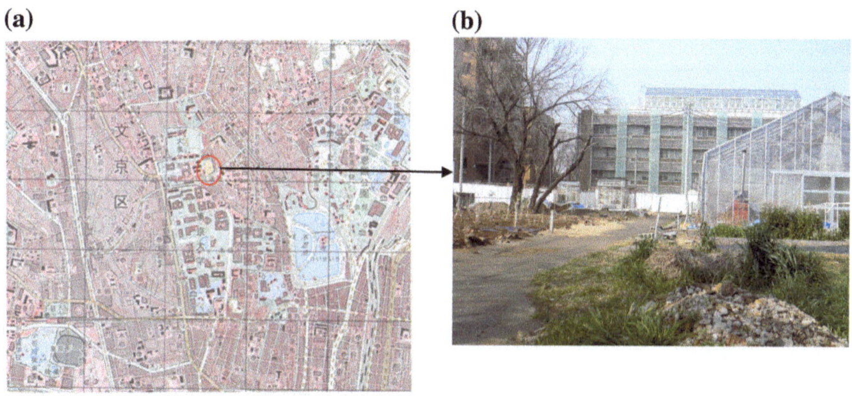

Fig. 11.15 **a** Upland field/grass in Central Tokyo in ca. 2005. **b** Upland field/grass in Tokyo University, central Tokyo

11.6 Trend of Forest

Forest in Tokyo was 1008 km^2 (ca. 1900), 952 km^2 (ca. 1950), 828 km^2 (ca. 1985), and 855 km^2 (ca. 2005). It has been a long time gradual decline, but between ca. 1985 and ca. 2005 it increased slightly. *Forest* occupies about 39% of the area of the prefecture. Tokyo has high altitude forest of 2000 m elevation at Mt. Kumotori and subtropical forest of Ogasawara Islands in the Pacific. 60% of the forest in the prefecture is artificial, which is high in comparison to the national average of 46%. The forest of 70% of the forest owners is smaller than 0.01 km^2, and 30% of forest owners live outside the municipality where the forest is located. They are unfavorable for land use planning and management, and for forestry and forest management.

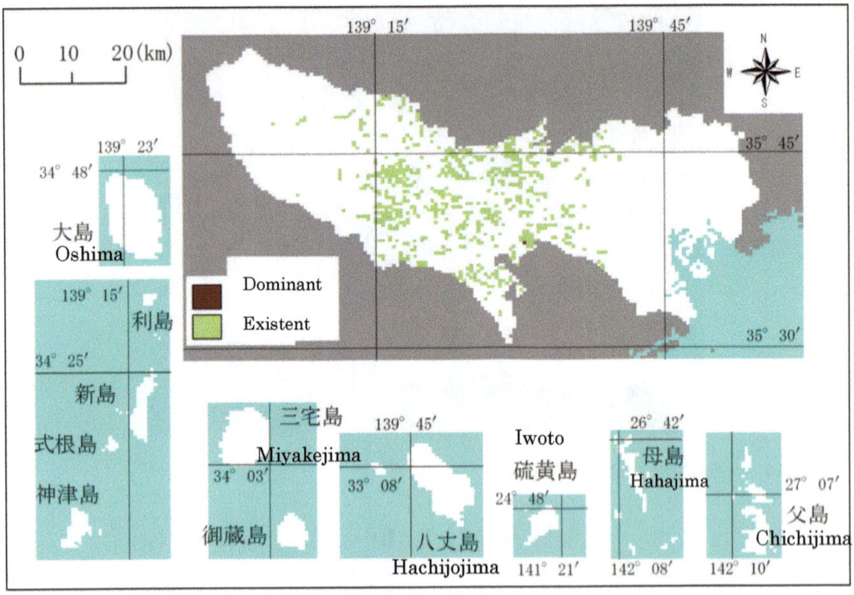

Fig. 11.16 Orchard in Tokyo (D, E) in ca. 2005

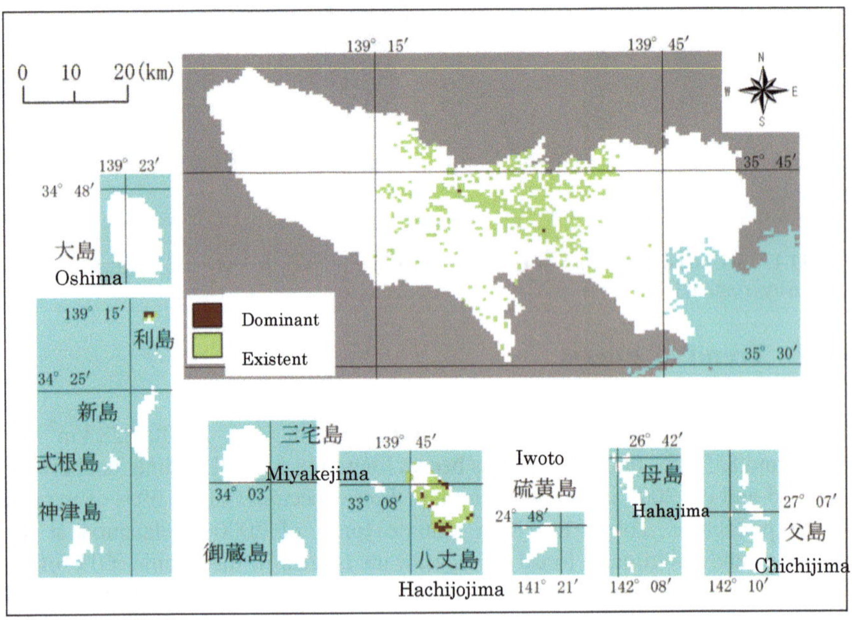

Fig. 11.17 Other tree crops in Tokyo in ca. 2005

Fig. 11.18 Strelitzia in Hachijo-jima

Most of the lost forest was "coniferous forest," and it was mainly due to urban expansion, golf course development, excavation of soil and stones, and industrial waste damping.

11.6.1 Broad-Leaved Forest

Figure 11.19 shows the distribution of *broad-leaved forest*. There is a clear tendency of gradual increase of existent cells from more urbanized east toward more rural west, but it ends roughly at around the longitudinal line of 139°15′ where the hilly area is taken over by mountain area. The area of *broad-leaved forest* was 421 km^2 (ca. 1900), 377 km^2 (ca. 1950), 283 km^2 (ca. 1985), and 285 km^2 (ca. 2005). The sharp decrease from ca. 1950 to ca. 1985 is mainly due to the development of new residential towns such as Tama Newtown in the 1960s. The development of new residential areas left some small forests here and there, so the decrease of the area of broad-leaved forest is seen in Fig. 11.19 as the many numbers of existent cells and a limited number of dominant cells. Figure 11.20 is the broad-leaved forest in Hachijo-jima.

11.6.2 Needle-Leaved Forest

Figure 11.21 shows the distribution of *needle-leaved forest*. The dominant cells and existent cells are mostly seen to the west of the longitudinal line of 139°15′. The area of *needle-leaved forest* was 201 km^2 (ca. 1900), 126 km^2 (ca. 1950), 121 km^2 (ca. 1985), and 140 km^2 (ca. 2005). Most of the *needle-leaved forest* in Tokyo are artificial, and the trees are planted and cut according to demand. The Japanese

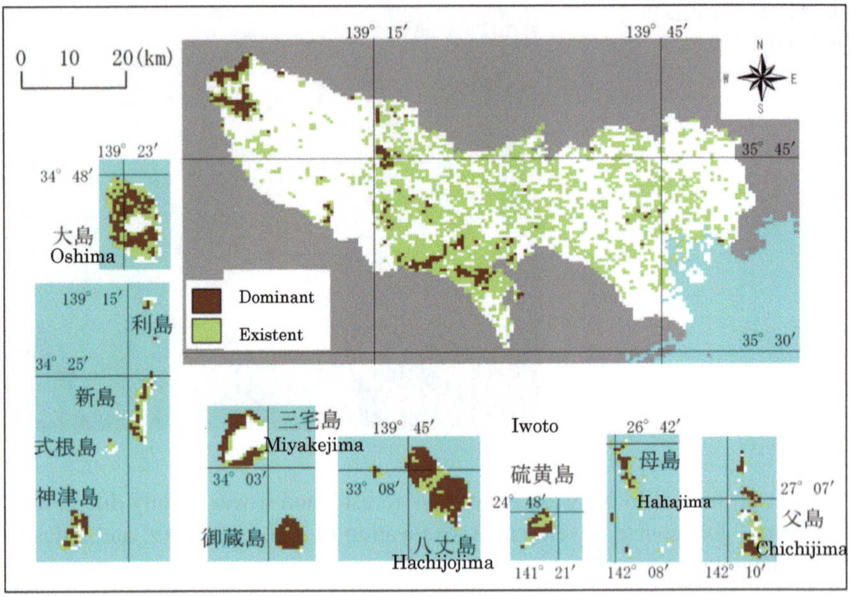

Fig. 11.19 Broad-leaved forest in Tokyo in ca. 2005

Fig. 11.20 Broad-leaved forest in Hachijo-jima

Government eased the condition of importing timber from abroad in the late 1970s and the early 1980s, and this change caused a sharp decline in the price of the domestic timber, which in turn deteriorated forestry and forest management in Japan. The increase of needle-leaved forest is not only a result of such unfavorable circumstances but also the cause of forest abandonment or ill management which may result in disasters.

11.6.3 Mixed Forest

Mixed forest in the present study is defined as the area with the symbols of "coniferous forest" and *broad-leaved forest* existing side-by-side on the topographic map. *Mixed forest* in Tokyo Prefecture was: 381 km^2 (ca. 1900), 449 km^2 (ca. 1950), 423 km^2 (ca. 1985), 409 km^2 (ca. 2005). It has been increasing steadily due to the increasing planting of coniferous trees in the areas with broad-leaved trees. Figure 11.22 is the two-level density map of *mixed forest*. Dominant cells prevail around the longitudinal line of 139°15′ and in the area west to it. It is rare in the central and eastern part of the prefecture, and the dominant cells are found only in such special place as Meiji Shrine.

11.7 Trend of Other Land Uses

11.7.1 Rough Land

Figure 11.23 is the two-level density map of the *rough land*. A linear distribution of dominant cells is seen along the Tama River at the center of the map. *Rough land* in Tokyo Prefecture was 141 km^2 (ca. 1900), 121 km^2 (ca. 1950), 143 km^2 (ca. 1985), and 174 km^2 (ca. 2005). It includes the sandy/gravel land besides the river

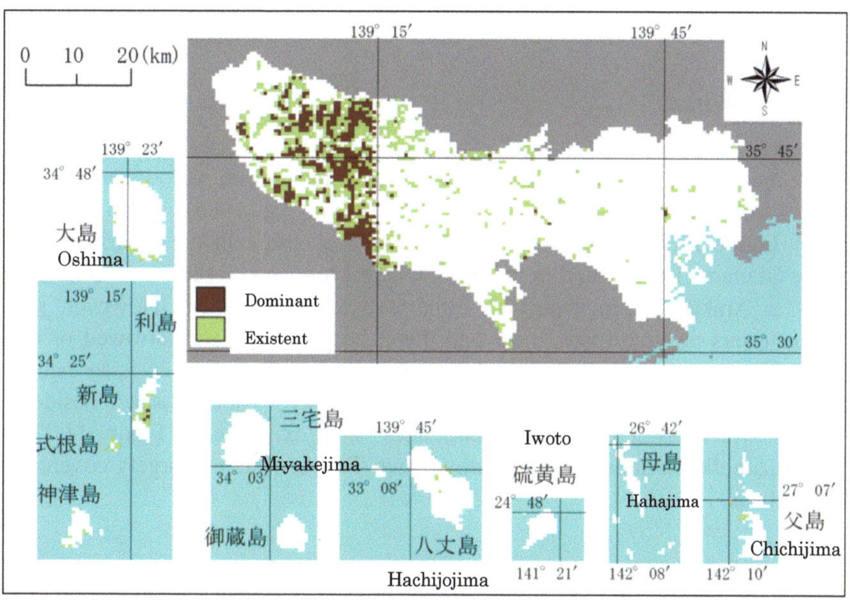

Fig. 11.21 Coniferous forest in Tokyo in ca. 2005

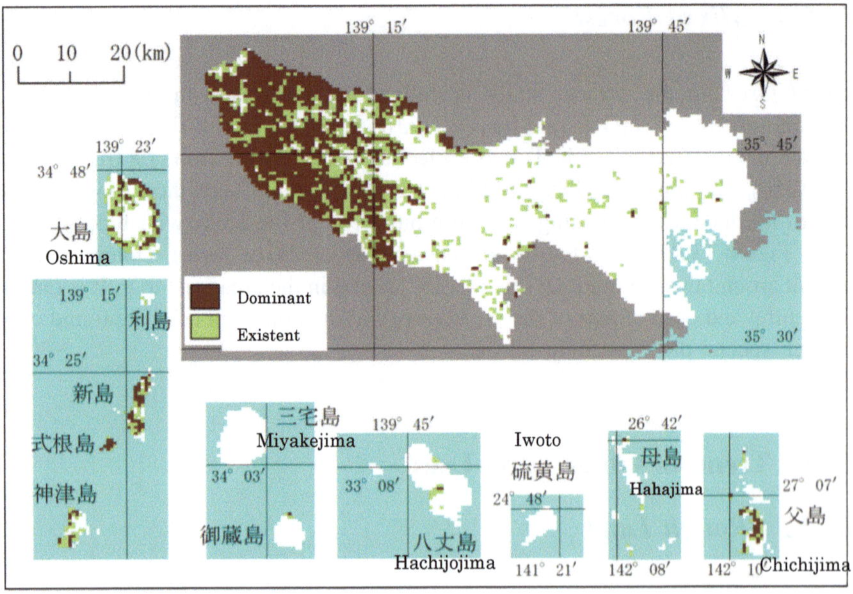

Fig. 11.22 Mixed forest in Tokyo in ca. 2005

flow (Fig. 11.24). Oshima and Miyakejima have large colonies of dominant cells, which correspond to volcanoes. The colony of dominant cells on the coast is the land filling site which will be used for various uses, eventually.

11.7.2 Water

Water means *lake*, *river,* and *wetland*, and does not include sea. *Water* in Tokyo was 17 km^2 (a. 1900), 17 km^2 (ca. 1950), 29 km^2 (ca. 1985), and 35 km^2 (ca. 2005). Figure 11.25 shows that the water dominant grid cells are seen in the west where Okutama Lake is and in the north where Murayama reservoir is. Such large rivers as Arakawa, Edogawa, and Tamagawa are also evident. However, some smaller rivers are interrupted here and there, because they are narrowed or even covered by concrete panels due to dense urbanization. Figure 11.26 is an example of a smaller river in the densely built-up area in Tokyo. An open space beside the stream, which is commonly seen, does not exist here. Figure 11.27 is Shinobazunoike, a famous pond in downtown Tokyo, which is largely covered by water plants.

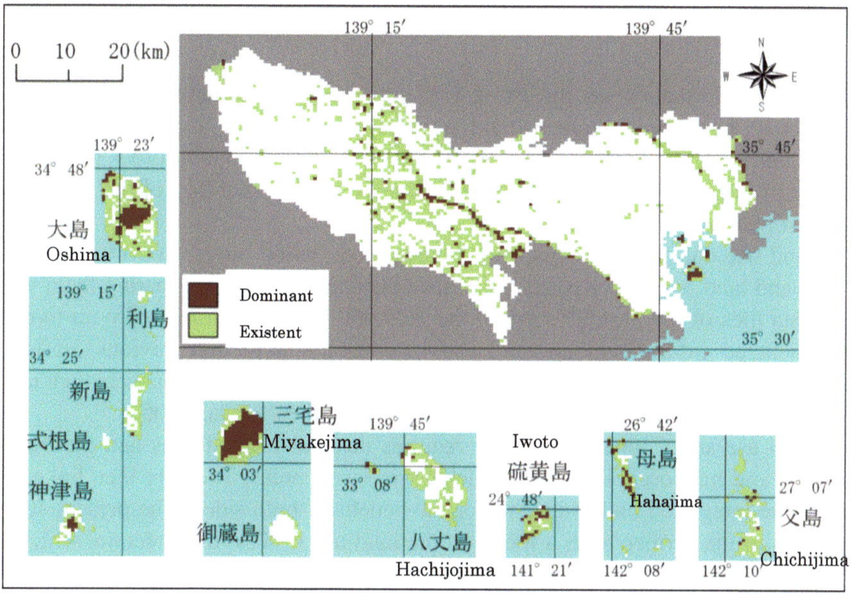

Fig. 11.23 Rough land in Tokyo in ca. 2005

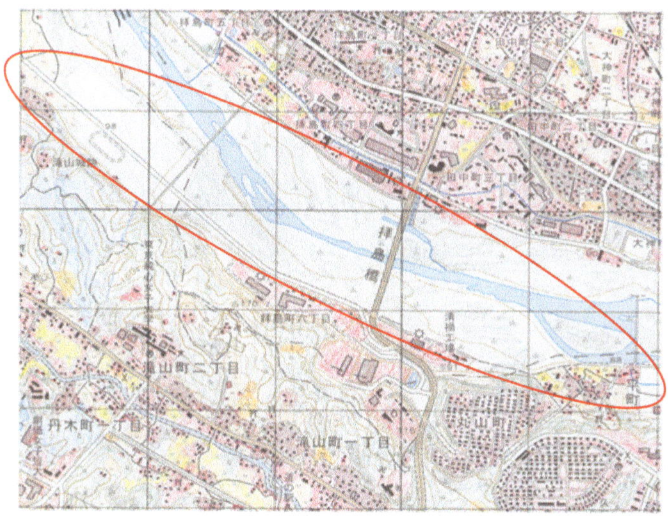

Fig. 11.24 Rough land in Tokyo

11.8 Conclusions

Land use information on the latest 1:25,000 topographic maps covering Tokyo Prefecture have been digitized and used to make various maps and statistics, compared with the land use data of the different time periods in order to find the trends of different land uses and used to explore the vulnerability of the prefecture against natural hazards.

Land use data files and maps have been made based on 1:25,000 topographic maps and land use in Tokyo ca. 2005 have been analysed by comparing them with the data files of the past. The use of the 1:25,000 maps meant the more up-to-date and detailed information on land use in comparison to the previous 1:50,000 topographic maps. The topographic maps colored with color pencils were then used to reconstruct land use and to make land use data files for ca. 2005.

Land use study for sustainability requires detailed local studies and more systematic and macroscopic studies which cover larger areas. They are then linked to get more integrated understanding for the benefit of both sides. The present study belongs to the former, with a focus on the land use of Tokyo Prefecture, which is 2187 km^2 in area. Similar studies have been carried out in several other prefectures so that in future the entire country of Japan is grasped both at national and prefectural levels.

This study has been carried out as part of SLUAS (Sustainable Land Use for Asia, JSPS Science Fund Basic Research (S)). The authors are most grateful to

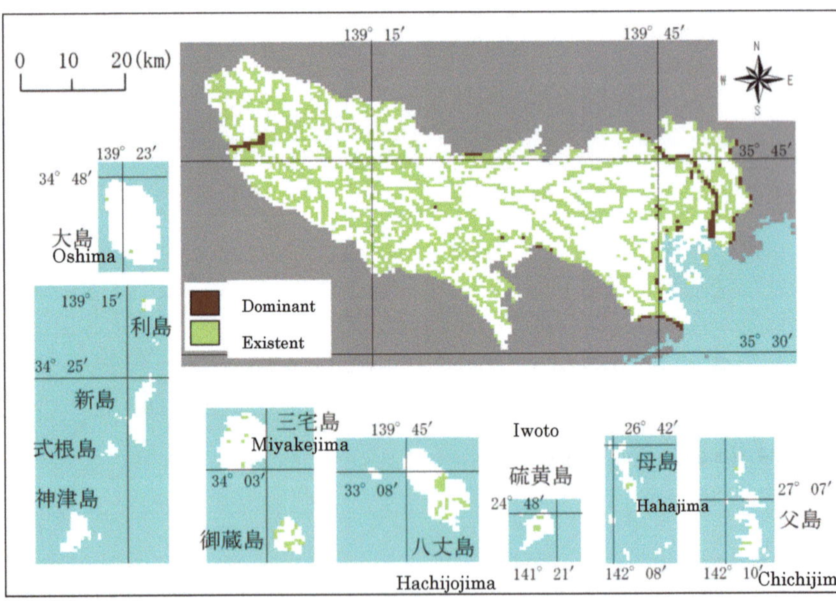

Fig. 11.25 Water in Tokyo in ca. 2005

Fig. 11.26 River at Suidobashi, Downtown Tokyo

Fig. 11.27 Pond at Ueno, Downtown Tokyo

those who kindly helped us. Our thanks are also extended to the other members of SLUAS for their support and advice.

Acknowledgements This chapter is the revised edition of SLUAS Science Report Vol. 1 pp. 103–121 (in Japanese).

References

Himiyama Y et al (eds) (1995) Atlas—environmental change in modern Japan. Asakura Shoten, Tokyo

Himiyama Y, Hisahara K, Iwaki H (1990) Construction of data files of land use in Japan ca. 1900. Geographical Reports of the Institute of Geography at Hokkaido University of Education Asahikawa Campus, vol 6, pp 1–5

Himiyama Y, Ohta N (1993) Land use change of Hokkaido since ca. 1920. Rep Taisetsuzan Inst Sci 28:1–13

Himiyama Y, Motomatsu H (1994) Land use change in the Tohoku District since ca. 1910. Rep Taisetsuzan Inst Sci 29:1–16

Himiyama Y, Himiyama S (2005) Change of agricultural land use in Niigata Prefecture since 1970. Rep Taisetsuzan Inst Sci 39:57–68

Himiyama Y, Nakamura N (2006) Land use in Fukuoka Prefecture ca. 2000 viewed from the 1:25,000 topographic maps. Rep Taisetsuzan Inst Sci 40:87–100

Himiyama Y, Honda T (2008) Land use in Gunma Prefecture ca. 2000 viewed from the 1:25,000 Topographic Maps. Rep Taisetsuzan Inst Sci 42:35–48

Himiyama Y, Adachi K (2009) Land use in Kyoto Prefecture ca. 2000 viewed from the 1:25,000 topographic maps. Rep Taisetsuzan Inst Sci 43:55–68

Himiyama Y, Takei T (2012) Land use in Osaka Prefecture, its change and vulnerability to natural hazards viewed from the topographic maps. Rep Taisetsuzan Inst Sci 46:53–73

Himiyama Y, Takamatsu K (2013) Land use in Chiba Prefecture, its change and vulnerability to natural hazards viewed from the topographic maps. Rep Taisetsuzan Inst Sci 47:13–32

Chapter 12
Flood Risk and Mitigation Under Changing Land Use

Lessons from the Kumozu Fluvial Plain in Japan

Shigeko Haruyama and Atsuko Suzuki

Abstract In this study, we describe how flood hazard maps of the Kumozu fluvial plain were developed based on geomorphologic land classification maps and discuss the status of comprehensive disaster management with respect to flood mitigation in the lower reaches of the Kumozu River Basin in Mie Prefecture, Japan. Open levee-retarding basin systems are one of the traditional flood-mitigation strategies employed along the lower and middle reaches of river basins in Japan. Rapid changes in land cover and land use in the Kumozu River Basin have increased flood risk, and rapid urban expansion in the river basin has led to broad-scale encroachment of residential areas into agriculture land, reducing the area of flood buffer zones available for use as retarding basins. By assessing flood damage along the lower reaches of the Kumozu River Basin while taking into the consideration the role of traditional open levee-retarding basin systems, we clarify the utility of such systems in terms of flood mitigation and the importance of land use planning based on geomorphological land classification maps. Taken together, these results demonstrate the importance of incorporating nonstructural measures into river basin management.

Keywords Geomorphologic land classification map · Land use change
The Kumozu River Basin

12.1 Introduction

Despite the implementation of flood control measures including the construction of hydrologic infrastructure such as embankments, floodways, and seawalls in river basins, river management in the lower and middle reaches of rivers, and dam and reservoir construction in the upper reaches of rivers, floods due to cyclical torrential

S. Haruyama (✉) · A. Suzuki
Faculty of Bioresources, Mie University, 1577 Kurimachiyacho,
Tsu, Mie, Japan
e-mail: haruyama@bio.mie-u.ac.jp

Y. Himiyama (ed.), *Exploring Sustainable Land Use in Monsoon Asia*,
Springer Geography, https://doi.org/10.1007/978-981-10-5927-8_12

rainfall events and changes in river basin land cover continue to cause severe damage and economic loss in Japan (Kondo 2005). Flood damage has not decreased for a broad portion of Monsoon Asia, especially in the lowlands of Southeast Asia (Haruyama 2015), due to an increase in irregular rainfall events associated with global warming and rapid land use change that has altered hydrologic conditions over entire river basins. Urban areas have become increasingly vulnerable to flooding due to the construction of underground malls and land reclamation efforts (Oya and Haruyama 1987, Hara et al. 2007, Murooka et al. 2008, Murooka and Haruyama 2005, Haruyama 2012, Haruyama et al. 2011).

For Japan, 2004 was a particularly drastic year in terms of flooding caused by torrential rainfall events. River basins throughout Japan, including the Kumozu River Basin (Mie Prefecture), Shinano River Basin (Niigata Prefecture), Maruyama River Basin (Hyogo Prefecture), and Asuwa River Basin (Fukui Prefecture), suffered severe flood damage, resulting, in some cases, in injury or loss of human life. Specifically, severe and intensive rainfall caused flood damage to the fluvial and coastal plains in lower reaches, landslides on sloped land, and mud flows in steep valleys of the abovementioned river basins. Rapid land use change due to urban expansion has been occurring in flood-prone areas of Japan since the rapid economic growth period of the 1970s. Haruyama and Tsujimura (2009) pointed out that the severe flood damage in both Toyooka city (in the lower Maruyama River Basin) and Fukui city (in the lower Asuwa River Basin) occurred at geomorphologically vulnerable points and that the severity of flooding in these areas was influenced by changes in land cover and land use. In addition, they pointed out that the evacuation and rehabilitation processes in these regions differed depending on the flood-mitigation strategies adopted by local communities. River improvement through the construction of so-called "hard infrastructure" along rivers causes hydrologic changes to river basin systems that create social problems related to the frequent occurrence of flood damage to newly developed residential areas on the periphery of urban areas.

Haruyama and Murai (2004) argued that the open levee-retarding basin systems used in the middle and lower reaches of the Kumozu River Basin not only represent traditional practice but also serve a useful flood mitigation/prevention function based on local knowledge regarding flood reduction. Open levee-retarding basin systems are a common, traditional flood-mitigation strategy employed in Japan. The Shingen-Tsutsumi embankment, which is part of the Kofu basin river management system, is a famous example of such an open levee-retarding zone system.

Matsumoto (2009) attempted to clarify the formation of micro-landforms and combinations of micro-landform elements in the lower Kumozu River Basin and developed a geomorphologic land classification map of flood-affected areas. Using this map, Matsumoto analyzed the region's flood-carrying capacity based on fluvial morphology, comparing this to maximum flood levels recorded at several hydrologic observation points along the river basin during severe historical floods. Kon (2005) evaluated flood control and prevention in the Tsurumi River Basin based on an urban retarding basin in the Yokohama metropolitan area.

Open levee-retarding basin systems that have served as traditional local flood-mitigation systems should be considered social capital and analyzed for their potential as present-day flood mitigation and reduction methods. In this study, we discuss land use change and the efficacy of flood control based on "river improvement" in the Kumozu River Basin as well as the potential role of open levee-retarding basin systems in future land use policy. We analyze land use patterns in 1920, 1959, 1985, and 2008 in terms of seven main land use categories (residential, paddy field, upland crop field, orchard, industrial, public space, and golf course) that involve the establishment of artificial surfaces or the removal of the original land cover (i.e., forests). Next, we analyze the fluvial landscape of the Kumozu River Basin in terms of micro-landforms and conduct an economic assessment of flood risk and damage to demonstrate the importance of continuing to use open levee-retarding basin systems in regional flood control planning.

12.2 Geographical Setting of the Kumozu River Basin

The Kumozu River originates on Mt. Mitsumine and joins other main branch rivers including the Temata, Segawa, and Nakamura Rivers before dividing into the Kumozu and Old Kumozu Rivers near Ise Bay. The river is only 55 km long and drains a watershed of 550 km^2. The population of the river basin was approximately 90,000 as of 2010. Average annual rainfall exceeds 2200 mm in the upper reaches of the river basin, with heavy rainfall due to typhoons typically occurring in September. The gradient of the river is steep in the upper reaches (1/300), becoming gentler in the lower reaches (1/1450) and nearly flat in the coastal plains (1/10,000). There are two knickpoints along the river, at 16 and 22 km from the river mouth.

The Kumozu River Basin has experienced severe flooding events in the past (Table 12.1). The Ise Bay Typhoon in September of 1959 caused severe flood damage in central Japan that also impacted the lower reaches of the Kumozu River. The Ise Bay Typhoon inundated 2531 ha in the Kumozu River basin and also caused a sea wall to collapse. Current river management and discharge levels are based on peak discharges recorded at observation stations during the Ise Bay Typhoon. Subsequent to the Ise Bay Typhoon, the Kumozu River Basin has been struck by several typhoons accompanied by high tides and storm surges. A flood in 1974 inundated 2589 ha spanning both lower and middle reaches of the river and damaged agricultural areas. A flood in 2004 inundated a 786 ha along the middle reaches of the main Kumozu River and caused landslide-related damage in the upper reaches. During a flood in 1985 caused by torrential rainfall, peak discharge recorded at the Ide-bridge observation station was 5400 m^3/s. Since the Ise Bay Typhoon, various river improvements, including the construction of embankments and seawalls, river bed dredging, and embankment reinforcement have been carried out along the lower and middle reaches of the Kumozu River (Tokai Noseikyoku, Tsu Statistical Information Center, 1999, 2000, 2001, 2002, 2003, 2004 Periods).

Table 12.1 Historical record of floods in the Kumozu River Basin (data collection from National Mie River and Road Office)

Flood occurrence year	Continuous 24 h rainfall (mm)	Discharge at Kumozu bridge (m^3/s)	Number of immersion above floor level	Number of inundated houses below floor level	Flooding area (ha)
1959 August	223	2600			
1959 September	261	4400	943	1581	2531
1961 June	234	2700			
1961 October	268	3000			
1965 September	193	3200	23	160	795
1971 August	233	2600	30	753	1656
1971 September	189	2900	196	3562	1121
1974 July	303	3900	48	561	2589
1976 September	261	2100	1	102	355
1985 July	257	5400	405	928	977
1990 September	239	3700	9	43	132
1994 September	166	2600	38	199	272
1995 September	244	3500		4	9
2004 September	238	4800	28	92	786

In addition, channel widening and river bed dredging along the middle reaches and construction of coastal infrastructure to protect against high tides and storm surges have also been undertaken over the last 60 years.

In addition, river management in terms of discharge control was adjusted to match the smaller peak discharge observed during the 1974 flood. A new Kumozu River management plan aimed at flood reduction was adopted in 1984; this plan called for reduction of peak discharge by 1900 m^3/s in the upper reaches of the river through storage volume in the Kimigano Dam and Reservoir and for reduction of peak discharge in the middle reaches through the use of retarding basins behind open levees. It was determined that peak discharge in the lower reaches during flood events should be no more than 6100 m^3/s, with the Old Kumozu River carrying a peak discharge of 2500 m^3/s and the main Kumozu River carrying a reduced peak discharge of 3600 m^3/s at its lowest point at the Karasu observation station.

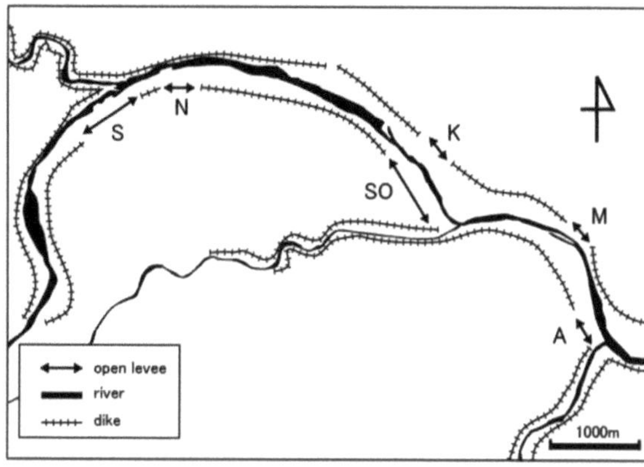

Fig. 12.1 Open levee and retarding locations of the Kumozu River Basin

The current river management plan comprises the following elements: storm surge prevention measures; channel bed cross section widening; upstream discharge reduction from the Kimigano Dam, which was completed in 1972; and a land use management plan incorporating open levee-retarding basin systems aimed at flood mitigation in the middle reaches of the river. At present, there are six zones with open levee-retarding basin systems (Fig. 12.1).

Changes in both land use and land cover have impacted the occurrence of severe natural disasters and the vulnerability of fluvial and coastal plains by altering the vegetation and creating man-made landforms that, in turn, have altered surface flow and hydrologic features of a given river basin. In the study area, land use patterns have changed rapidly over the last 100 years (Iwata 2010). The years 1921, 1959, 1985, and 2008 mark boundaries between different economic eras (i.e., different stages of urban and industrial development) that have resulted in rapid land use change and are important time points for understanding the evolution of land use in the region. Thus, we chose to analyze changes in land use patterns and surface alteration in the lower and middle reaches of the Kumozu River Basin at these four-time points.

In 1921, the predominant land uses were paddy fields and upland crop fields, with scattered rural residential areas located on the fluvial and coastal plain. Accordingly, the primary industry in the region was agriculture. The rural areas typically comprised upland crop fields near sparse villages. Land use units were generally located on natural levees, and residents' lifestyles and land use patterns were typically oriented toward the maintenance of surface soil and landform conditions suitable for agricultural production. Paddy fields were typically located in back swamps and the delta plain, because the poor drainage and continuous presence of subsurface water in these micro-landforms rendered them unsuitable for the production of crops other than wetland rice prior to the introduction of irrigation and drainage systems (Fig. 12.2).

Orchards and mulberry farms were generally located on natural levees along the river due to the good drainage and sandy texture of the soil in these locations. Some

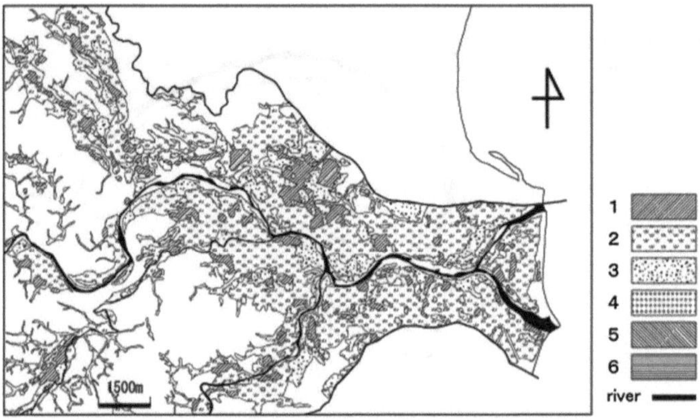

Fig. 12.2 Land use map of the Kumozu River Basin in 1921 (Explanatory note; *1* residential zone, *2* paddy field, *3* upland crop field, *4* orchards, *5* industrial zones, *6* official places)

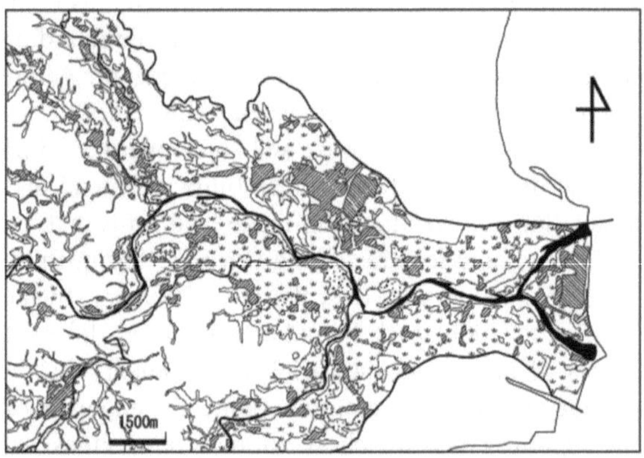

Fig. 12.3 Land use map of the Kumozu River Basin in 1959 (explanatory note is same as Fig. 12.2)

fruit trees were grown on hill slopes in the northern Kobeki region along the Kumozu River. In the middle reaches of the river, two retarding basins behind open levees were located in the south of the river. These retarding basins were used as paddy fields because paddy fields would be able to tolerate inundation for periods on the order of one week. Some upland crop fields were located on the natural levees near the river banks. Rural villages were also located on natural levees, due to their higher elevation, which was considered advantageous by residents. Land near the river mouth was historically used as paddy fields. However, due to their slightly higher elevation and good drainage relative to the surrounding flood plain, the sand dunes and sand ridges that had formed over time were utilized as upland crop fields or as sites for rural fishing villages. Such sandy areas have been used as residential areas and as religious sites since medieval times (Fig. 12.3).

Land use patterns in Japan changed dramatically after the Second World War, led by the development of agricultural fields to establish a sustainable food supply

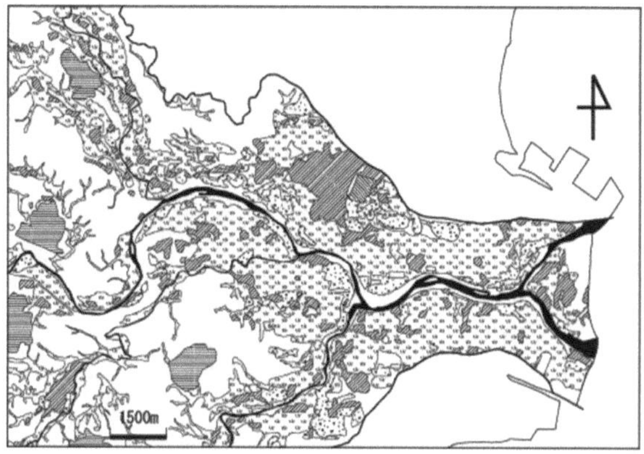

Fig. 12.4 Land use map of the Kumozu River Basin in 1985 (explanatory note is same as Fig. 12.2)

and the expansion of industrial zones. During this period, the area of paddy fields increased throughout Japan, whereas that of upland crop fields did not. Paddy fields were developed in the middle reaches of the Kumozu River as part of the effort to establish food security. Upland crop fields in this region decreased in area and were replaced by Japanese pear and other orchards. Agricultural areas in coastal areas were converted to industrial zones. A railroad network spanning the entire region was also established during this period.

In 1985, a new land use category, golf courses, was introduced on hill sides and hilly regions near the river, leading to further deforestation of the catchment area. The introduction of golf as a weekend leisure activity was a sign of the improved economic status of working individuals. Rapid land use changes associated with the urban expansion of Tsu city into the middle reaches and deforested hill slope areas surrounding the Kumozu River was also dramatic. Residential areas expanded into the lower fluvial and coastal plain and hilly fluvial terraces. The flood plain, and specifically the area near the railway station, which comprises a natural levee with a back swamp, shifted from agriculture to residential and industrial use. This rapid change in land use reduced the area of flood buffer zones (Fig. 12.4).

Starting in 2008, land use patterns have undergone dramatic new changes. The area of golf courses has expanded further into the gentle hill slopes and vanishing forested areas. Meanwhile, the area of forests and paddy fields in the lower fluvial plain has been decreasing due to the expansion of residential and industrial areas. These land use changes have reduced the area of flood buffer zones in the lower reaches of the Kumozu River Basin. Heavy industry and light industry zones have expanded into coastal areas in the lower reaches of the Kumozu River, and terraces have been targeted for future development. This expansion of industrial and residential zones has increased the severity of flooding and other natural disasters in the region by lowering the permeability of the surface soil. The main land use in the coastal area along the Ise Bay is becoming residential (Fig. 12.5).

Comparing land use patterns along the Kumozu River at the four-time points above reveals the dramatic changes in land use that have occurred over the last 100 years. The area of paddy fields expanded up to 1970 but has subsequently

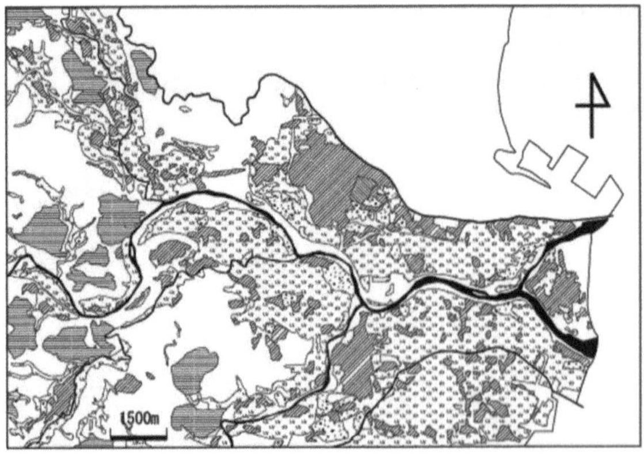

Fig. 12.5 Land use map of the Kumozu River Basin in 2008 (explanatory note is same as Fig. 12.2)

decreased. The area of golf courses began increasing after 1970 and continues to increase today. In the 50 years since the Ise Bay Typhoon in 1959, the frequency and severity of flood events have increased as a result of rapid changes in land use and land cover that have increased risk and vulnerability to flooding. In general, the development of sloped land, which involves slope cutting, deforestation, and disturbance of the surface soil, results in higher peak discharges by reducing water retention by surface soil and encourages rapid flows that carry debris. The reduction of paddy field area shortens the arrival time and increases the magnitude of floods by decreasing the area available for flood retardation. Similarly, the expansion of residential and industrial zones also increases the intensity of floods by reducing the area available for flood retardation. The dramatic changes in land use in the middle reaches of the Kumozu River have resulted in floods of greater intensity and duration (Figs. 12.2, 12.3, 12.4 and 12.5).

12.3 Flood Mitigation and River Improvement

We used a geomorphologic landform classification map of the lower Kumozu River Basin to analyze the relationship between micro-landforms and flood damage in the lower reaches of the Kumozu River at three flood stages in the retarding basins of open levee-retarding basin systems: 0.5, 1.0, and 1.5 m flood depths. The lowest reaches of the Kumozu River, with an elevation between 0 and 2 m, are extremely vulnerable to storm surges, tsunamis, and flooding due to torrential rainfall. The highest points (2.0 m elevation) in the lower fluvial plain of the study area are the tops of sand dunes and natural levees consisting of deposited sand. The most severe historical flood ever recorded inundated 258,000 m^2 of the A zone, 1,665,000 m^2 of the M and K zones (Fig. 12.1), 1,375,000 m^2 of the S and N zones (Fig. 12.1), and 255,000 m^2 of the S zone (Fig. 12.1). We used the storage capacity of individual retarding basins based on micro-landform analysis to estimate the total area of retarding basins: 5,875,000 m^2.

Retarding basins in the middle reaches of the Kumozu River include back swamps, natural levees, and old river channels. The respective areas (and relative proportions) of these landforms are 4,026,318.8 m^2 (68.5%), 1,790,912.2 m^2 (30.5%), and 57,769.0 m^2 (1%). The storage volume of the retarding basins of each landform type was estimated for three flood stages (0.5, 1.0, and 1.5 m) (Table 12.2). Briefly, the estimates were generated as follows. During peak inundation of 0.5 or 1.0 m, it was assumed that back swamps and old river channels would be inundated but natural levees would not. Thus, the total storage volume of retarding river basins including back swamps and old river channels, when peak inundation is 0.5 m, is estimated to be 2,042,043.9 m^3. Similarly, during peak inundation of 1.0 m, the total storage volume of retarding ponds is estimated 4,084,087.8 t. We further assumed that, during peak inundation of 1.5 m, back swamps, old river channels, and natural levees would be inundated to a depth of 1.5 m. Thus, the total storage volume of retarding basins when peak inundation is 1.5 m is estimated to be 7,021,587.8 m^3 (Table 12.2).

If all of the open levees feeding into retarding basins were to be closed, flood water would rapidly flow to the lower reaches of the river, causing severe flooding in Karasu Ward. The magnitude of damage is expected to vary as a function of peak inundation depth (0.5, 1.0, and 1.5 m in the scenarios analyzed in this study). The patterns and risk of flooding are also expected to vary for different micro-landforms, which include sand dunes, sand ridges, natural levees, deltas, and old river channels, highlighting the importance of landforms as mechanisms for flood mitigation. The relative distributions of different landform types in terms of total area in Karasu Ward are as follows: sand dunes (14.3%), sand ridges (6.2%), natural levees (16.2%), back swamps (33.9%), old river channels (4.2%), and other floodplain areas (25.2%). The relative proportions of these different landform units determine the level of flood risk and vulnerability. For peak inundation of the retarding basins behind open levees of 0.5 m, the flood water can be discharged to Karasu Ward through the old river channels; the back swamps, floodplains, delta, and old river courses will experience stagnant flooding but natural levees and sand dunes will not. For peak inundation of 1.0 m, the flood water discharged to Karasu Ward through the old river channels and back swamps is expected to reach 1.84 and 1.34 m, respectively, while the sand dunes and sand ridges will not be flooded (Tables 12.3, 12.4, 12.5 and 12.6). For peak inundation of 1.5 m, the flood water discharged to Karasu Ward through the old river channels and back swamps are expected to reach 2.69 and 2.19 m, respectively, while the sand dunes and sand ridges will not be flooded (Tables 12.3 and 12.4).

Table 12.2 Estimated inundation storage volume in the retarding ponds (unit: t)

	0.5 m in depth	1.0 m in depth	1.5 m in depth
Back swamp and delta	2,013,159.4	4,026,318.8	6,039,478.2
Natural levee	0.0	0.0	895,456.1
Former river course	28,884.5	57,769	86,653.5
The total storage volume	2,042,043.9	4,084,087.8	7,021,587.7

Table 12.3 Estimated flooding retarding volume at Karasu Wards in each 0.5, 1.0, and 1.5 m inundation in retarding pond

Land form	Relative height (m)	Area(m^2)	Inundation depth (m) in max. 0.5 m at retarding pond	Flood volume at Karasu(t) in 0.5 m at retarding pond	Inundation depth (m) in max.1 m at retarding pond	Flood volume at Karasu(t) in 1 m at retarding pond	Inundation depth (m) in Max.1.5 m at retarding pond	Flooding volume at Karasu (t) in 1.5 m at retarding pond
Sand dune	2.5	556,771	0	0	0	0	0.19	108,320
Sand ridge	2	242,867	0	0	0	0	0.69	168,582
Natural levee	1.5	632,272	0	0	0	0	1.19	755,281
Back swamp and delta	0	1,322,191	0.6	0	0.6	0	2.19	2,901,616
Former river course	0	1,322,191	1.1	78,7014.8	1.1	78,7014.8	2.69	439,952
The other flood plain	0	982,624	1.1	1,076,204	1.1	1,076,204	2.69	1,647,731
The total		3,900,000		2,042,043		2,042,043		7,021,587

Table 12.4 The inundation depth in each landform at Karasu Ward (unit: m)

Landform unit	Relative height(m)	The maximum of the inundation depth of retarding pond		
		0.5 m	1.0 m	1.5 m
Sand dune	2.5	0	0	0.19
Sand ridge	2	0	0	0.69
Natural levee	1.5	0	0.34	1.19
Back swamp and delta	0.5	0.6	1.34	2.19
Former river course	0	1.1	1.84	2.69
The other flood plain	0	1.1	1.84	2.69

Table 12.5 The annual agriculture production income in the ordinal year and flooding year

	Crop	Production/ha (kg)	Yield (kg)	Income/year (JPY)
The ordinary year	Spring cabbage	4120	82,400	8,240,000
	Winter cabbage	3826	76,520	6,121,600
	Soy bean	1724	34,480	4,310,000
	wheat	2780	27,800	7,180,740
	Second soy bean	1724	17,240	2,155,000
	The total income			28,007,340
The flood year	Spring cabbage	5590	111,800	11,180,000
	Winter cabbage	4080	81,600	6,528,000
	Soy bean	840	16,800	2100,000
	wheat	2570	25,700	6,638,310
	Second soy bean	840	8400	1,050,000
	The total income			27,496,310

Table 12.6 The ratio of damage per unit in each inundation depth

The inundation depth (m)	Damage ratio (%)
0–0.5	30
0.5–1.0	50
1.0–1.5	70
1.5–2.0	100
2.0–2.5	100
2.5–3.0	100

Inundation depths for each landform type were calculated and mapped for 0.5- to 1.5-m flood stages based on the total storage volume of retarding basins. As can be seen from the inundation maps, the industrial zone in Karasu Ward is predicted to experience extensive and deep inundation under all evaluated scenarios (Figs. 12.6, 12.7 and 12.8).

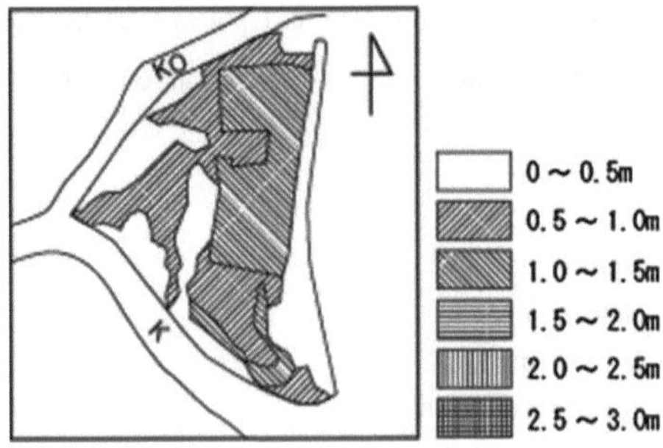

Fig. 12.6 Flood distribution map in 0.5 m inundation in retarding pond (explanatory note showing the depth of inundation)

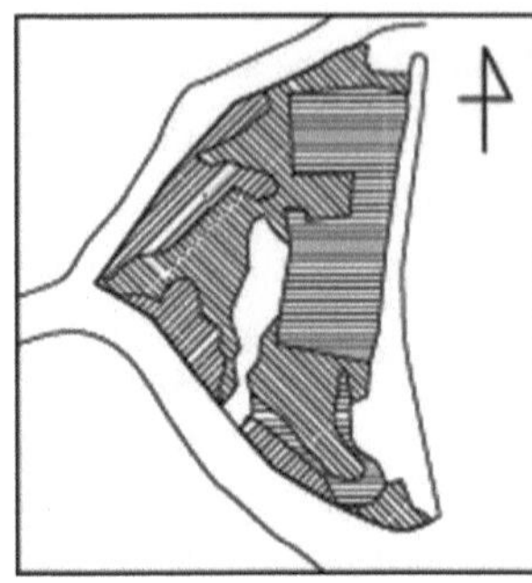

Fig. 12.7 Flood distribution map in 1.0 m inundation in retarding pond

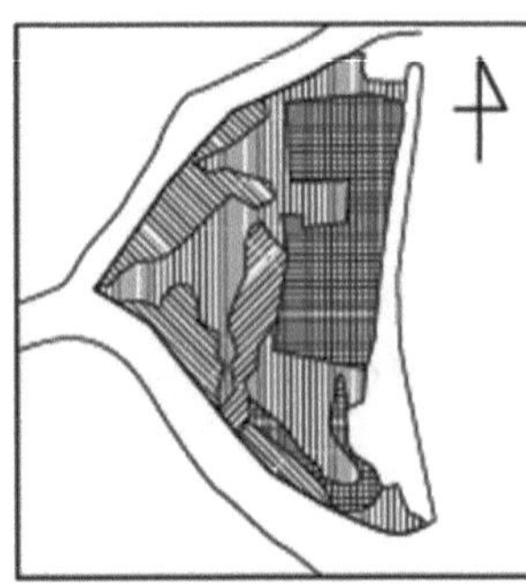

Fig. 12.8 Flood distribution map in 1.5 m inundation in retarding pond

12.4 Land Use Patterns and Vulnerability to Flooding

We compared estimated flood damage at three flood stages to survey data regarding farming household income from agricultural production in relevant flood-prone areas. Estimates of total annual income for farming households was calculated based on data from statistical year books published by Mie Prefecture. We also examined the results of a questionnaire survey conducted by the Mie office of the Japan Agriculture (JA) Group related to recent land use patterns and vulnerability to flooding. In addition, we investigated severe flood damage between 2008 and 2014 in terms of agricultural income and production in the retarding basins behind open levees.

To assess flood impact, we compared economic losses due to flooding to income from agriculture production for each landform type. Our analysis revealed that the magnitude of flood damage in the study area is related to crop status and that the cropping calendar, which determines the timing of agricultural production, is also an important factor to take into consideration when assessing flood risk.

In the study area, spring cabbage is grown from February to June, followed by soy bean from June to September and winter cabbage from September to January. Flooding due to typhoons and torrential rainfalls usually occurs in the summer during the soy bean growing season. Using the cropping calendar for the study area, we calculated annual gross income from agricultural production of each crop (10 kg units) in both ordinary and flood years. For a typical 30-ha upland crop field, annual gross income from agricultural production was 28,007,340 JPY in an ordinary year and 27,396,310 JPY in a flood year (2014), representing an income loss of 611,030 JPY or 2% (Table 12.5).

The gross production of paddy fields was calculated based on the statistical information data by Mie Prefecture (2000–2009). The annual output of a typical paddy field in the study area was 87,633,800 JPY in a typical ordinary year and 842,883,333 JPY in a flood year (2014). Damage to residences was estimated based on general housing prices from the housing statistical yearbook published by Mie Prefecture. The mean price of residential buildings was 178,220 JPY/m^2, and the mean price of residential land in the study area was 5740 JPY/m^2. To estimate damages, the land price was weighted by the damage ratio associated with each flood level (Tables 12.6 and 12.7).

Of the total flood-prone area (4,084,088 m^2) in the study area, 96.1% (3,924,957 m^2) is used for paddy fields, 1.9% (78,536 m^2) is used for upland crop fields, and 1.4% (58,819 m^2) is residential area. The area of natural levees is 1,790,912 m^2, of which 68.1% is used for upland crop fields in retarding basins. Flood damage to paddy and upland fields is estimated to be 7,771,375 JPY when peak inundation is 0.5 or 1.0 m and to be 6,809,803 JPY when peak inundation is 1.5 m.

In contrast, flood damage to residential areas at the three flood stages is estimated to be 101,286,318 JPY (0.5 m peak inundation), 168,810,530 JPY (1.0 peak inundation), and 823,753,714 JPY (peak inundation). Flood damage in the industrial zone in Karasu Ward for peak inundation of 1, 2, and 3 m is estimated to be 3358 JPY/m^2, 10,071 JPY/m^2, and 20,142 JPY/m^2, respectively. Of the landforms in Karasu Ward, the most vulnerable to flooding, old river channels, comprises 596,523 m^2 of residential area, 283,991 m^2 of industrial area, and 59,280 m^2 of upland crop fields, paddy fields, and orchards.

Table 12.7 Flood inundation and landform role in each land use unit

Flood depth (m)	Inflow (t)	Landform	Relative Height (m)	Area (m^2)	Flood depth (m)	Land use area(m^2)			Damage (JPY)
						Residence	Industry	Upland crop field	
0.5	2,042,044	Former river channel	0	1,145,899	1.1	596,523	283,991	59,280	6,635,756,133
		Flood plain	0.5	1,322,191	0.6	425,211	13,804	115,576	
		Natural levee	1.5	632,272	0	194,363	0	97,975	
		Sand dune	2	242,869	0	88,741	0	0	
		Sand ridge	2.5	556,771	0	523,880	0	0	
1.0	4,084,088	Former river channel	0	1,145,899	1.8	596,523	283,991	59,280	8,797,153,901
		Flood plain	0.5	1,322,191	1.3	425,211	13,804	115,576	
		Natural levee	1.5	632,272	0.3	194,363	0	97,975	
		Sand dune	2	242,869	0	88,741	0	0	
		Sand ridge	2.5	556,771	0	523,880	0	0	
1.5	7,021,588	Former river channel	0	1,145,899	2.7	596,523	283,991	59,280	14316894866
		Flood plain	0.5	1,322,191	2.2	425,211	13,804	115,576	
		Natural levee	1.5	632,272	1.2	194,363	0	97,975	
		Sand dune	2	242,869	0.7	88,741	0	0	
		Sand ridge	2.5	556,771	0.2	523,880	0	0	

12.5 Conclusions

Open levee-retarding basin systems serve as a key flood-mitigation mechanism for fluvial lowlands of the Kumozu River Basin. If levees along the middle reaches of the Kumozu River were to be completely closed, lower reaches of the Kumozu River would experience higher peak inundation and greater flood damage as a result of typhoons and torrential rainfall. Flood damage is expected to differ for land under different uses and to vary with inundation depth in the middle reaches of the river. Changes in land use and land cover over the past 100 years have increased the risk of flooding in the future. The results of this study demonstrate that (i) open levee-retarding basins are not only a traditional river management system, but are a rational system for lower river basin management that should be incorporated into future flood control planning and (ii) land use policy has important implications for flood mitigation.

References

Hara Y, Haruyama S, Okhubo S, Takeuchi K (2007) Landfill development causes changes in distribution of flood-prone areas: case study in Laguna Plain, Metro Manila The Philippines. In: Hurumai H et al (eds) Southern Asian water environment. IWA Publishing, London, pp 11–18

Haruyama S (2012) Climate change impacts on the Mekong River Delta. In: Anbumozhi ADB (ed) Climate change in Asia and the Pacific. SAGE Publication, Thousand Oaks, pp 130–137

Haruyama S (2015) Climate Change and Natural Disaster in Asia. In: Conference proceedings of the three day international conference on Climate change—inconvenient truths. St. Priminous University Press, pp 1–12

Haruyama S, Murai A (2004) Geomorphology and flood plain management for flooding—the lesson from the Kumozu River. J Suiri Kagaku 278:21–32 (in Japanese)

Haruyama S, Tsujimura A (2009) Community-based activities for flood mitigation by the smallest reginal community, E-Journal GEO 4(1):1–20 (in Japanese)

Haruyama S et al (2011) Flood mitigation and land use. Kokonshoin Publisher, Tokyo (In Japanese), p 223

Iwata S (2010) Tsu city—the local city construction history. NourinTokei Kyokai Publisher, Tokyo (in Japanese)

Kon T (2005) The effect of the multipurpose retarding area of the Tsurumi River Basin. J River 61 (6):71–75 (in Japanese)

Kondo T (2005) The development of counter measures for heavy rainfall. River 61(6):3–5

Matsumoto M (2009) The research of the geomorphology of the middle and lower basins of the Kumozu river. Graduate thesis of Mie University. (**in Japanese**)

Mie Prefecture (2000–2009) Statistical information of Mie prefecture. Data Box home page http://www.pref.mie.jp/databox/

Murooka M, Haruyama S (2005) Coastal erosion risk map of the southern Red River Delta, Vietnam. J Rural Plan 7:109–114

Murooka M, Haruyama S, Masuda Y (2008) Land Cover Detected by Satellite Data in the Agricultural Development Area of the Sanjiang Plain, China. J Rural Plan 26:197–202

Oya M, Haruyama S (1987) Flooding and urbanization in the Lowlands of Tokyo and vicinity. J Nat Disaster Sci 9(2):1–12

Tokai Noseikyoku, Tsu Statistical Information Center (1999, 2000, 2001, 2002, 2003, 2004 periods) The statistical year book of Mie prefecture—agriculture, forestry and fishery section. Mie Prefecture Office Publisher (**in Japanese**)

Part V
Land Use Change in Other Countries/Regions in Asia

Chapter 13
Human Impacts on the Landcover Change of the Inle Watershed in Myanmar

Kay Thwe Hlaing, Shigeko Haruyama and Saw Yu May

Abstract The land cover of this basin showed the changes clearly. In 1990, closed forests were found in northern edge, eastern and southwestern part of the watershed area, near Pinlaung range, Kyauktalone range and Pindaya-Ywangan area, however, in 2000, these closed forests were changed to open forest, scrub grassland, and agriculture. In some part of the watershed area, near Taunggyi, scrub land was changed into open forest because of the reforestation program. Agriculture extension occurred in the lake watershed area especially in western and northwestern part of the lake, near Kalaw-Aung Ban, Heho Valley, and Thamakhan Plain. One of the unique characteristic of Inle Lake is hydroponic cultivation. Besides, tomato cultivation is an important economy of Inle Lake. This type of cultivation is practiced on the naturally floating island and it is effected the water surface area of the Inle Lake in the following ways such as after using the old floating garden islands, they are decomposed to the lake bottom and the lake will be shallower, and the extension of floating garden cultivation causes more shrinkage to the water surface area in the study area. The floating garden area was increased from 1990 to 2010 years. On the other side, water surface area is decreased from 1990 to 2010 year because it is influenced by climatic condition, extension of floating garden, and population growth of the study area.

Keywords Land cover change · Inle watershed · Remote sensing
Geographical information systems

K. T. Hlaing (✉)
Department of Geography, Yangon University, Pyay Road,
Yangon City, Myanmar
e-mail: kthlaing@gmail.com

S. Haruyama
Mie University, Tsu, Japan

S. Y. May
University of Taunggyi, Taunggyi, Myanmar

Y. Himiyama (ed.), *Exploring Sustainable Land Use in Monsoon Asia*,
Springer Geography, https://doi.org/10.1007/978-981-10-5927-8_13

13.1 Introduction

The Inle Lake, the second largest freshwater lake in Myanmar located in Nyaung Shwe Township, Taunggyi Township, Yatsauk Township, Ygwagan Township, Kalaw Township, Pindaya Township, and Pinglaung Township, Southern Shan State, is one of the destinations for domestic and international tourism. Recently, the water quality degradation, land cover changes, and water surface area changes due to the natural as well as anthropogenesis activities are dominant in the Inle Lake and its watershed area. Therefore, we need to report for the land cover condition, settlement pattern, and water quality condition for the Inle Lake. Characteristic elements of land cover and land use change in the Inle Watershed area were extracted from near infrared, infrared, and red data on multi-temporal Landsat Thematic Mapper (TM) images and Landsat Electromagnetic Thematic Mapper (ETM) images. Maximum likelihood supervised classification and post-classification change detection techniques were applied to Landsat images acquired in 1990 and 2010, respectively, to map land cover changes in the Inle Watershed area.

13.2 Previous Investigations and Purpose of Our Study

Butkus and Myint (2001) pointed out that the Inle Lake is impacted by many factors including pesticides from agriculture, chemical dyes from textile processing, excess siltation from watershed erosion, and the dumping of wastes and garbage. As a result, the fish (Inle Carp) serving as the main public food supply is becoming scarce. In addition, public disease rates have dramatically increased. These problems are thought to primarily be caused by excessive use of pesticides. BANCA (2006) reported that Inle Lake in Myanmar is now facing climate change challenges. The two key issues that need to be addressed are depletion of water area and deterioration of water quality. Due to climate change conditions, the evidence of water reduction in the lake is getting more visible in April 2010. These issues are emergent threats to the local Inle people of the lake.

Asian Development Bank (2006) reported that the population in and around Inle Lake has been growing steadily over the past 25 years. In the 1450 km^2 area constituting the wetlands and upland village tracts around the lake as well as villages in the lake itself, the total population estimate for 2005 was 143,793; a large proportion (90%) of these households are considered rural. Volk et al. (1996), Ngwe Sint and Catalan(2000) and Su and Jassby (2000) examined overview studies note the dramatic infilling of the lake with sediment, a process that threatens the lake ecosystem, as well as, the local economy and the factors contributing to the loss of Inle Lake, the most have speculated that forest conversion, logging, and shifting cultivation are the primary sources of increased sediment load sand sedimentation in this lake.

The length of the lake has reportedly declined from roughly 58–18 km and its maximum width has decreased from 13 to 6.5 km during the past 100–200 years, although the earlier dimensions are believed to be overestimated. The drainage area and storage capacity of the lake have been estimated at 5612 km^2 and 3.5×107 m^3, respectively, while annual inflow and water residence time are estimated as 1.1×108 m^3/year and 0.32 year. Several reports and overview studies note the dramatic infilling of the lake with sediment, declining due to decreased water quality clarity levels associated with suspended sediment and eutrophication, with natural environmental changes of width and length of the lake (Volk et al. 1996; Su and Jassby 2000; Win 1996; Swe 2011; Than 2007). Some of these recent investigations and interests appear to be associated with the increasing tourism in the area. However, no studies have identified such sediment linkages, nor have these land use and land cover terms been consistently used in reports from the region. Sidle et al. (2006) explained that "degraded" forest conditions, where larger trees were removed and current regrowth consists of scrub forest, are not considered conversion, because much of the inherent rooting strength within the soil mantle is retained and the scrub forest has the potential to evolve with time.

The purpose of this paper is to verify the changes of environmental degradation surrounding Inle Watershed Area and dimensions for the contemporary period, describe various observed processes contributing to present day water quality, and discuss issues related to the sustainable development of the landscape around the lake and within the lake. Our research is supported by field observations obtained during several visits to the areas surrounding the lake, discussions with local and regional experts, and analysis of Landsat TM and Landsat ETM imagery compared with land cover maps of the watershed area using Remote Sensing and Geographic Information Systems.

13.3 Geographical Setting of the Inle Watershed

13.3.1 Physical Environment

Inle Lake is situated in Nyaungshwe Township, southern Shan State in Myanmar, between Latitude 19 58′ 0″ and 20 43′ 05″ North, Longitude 97 46′ 30″ and 97 55′ 30″ East (Fig. 13.1). It is the second largest lake in Myanmar at about 22.53 km long and 11.27 km wide. The Inle Lake region is characterized by a large, flat valley running north to south which is surrounded by mountain ranges and its catchment area is 4110.69 km^2 comprising land systems, ridges, mountains, plains, and basins and altitudes ranging from 870 to 2045 m above sea level (Fig. 13.2).

Over 200 villages surround the lake and inhabit the immediate watershed. Inle Lake has been classified as a shallow freshwater lake because of its depth ranging from 2 to 6 m during summer and rainy season. The lake is situated at 883.9 m above mean sea level, occupies the central part of a trough between two mountain

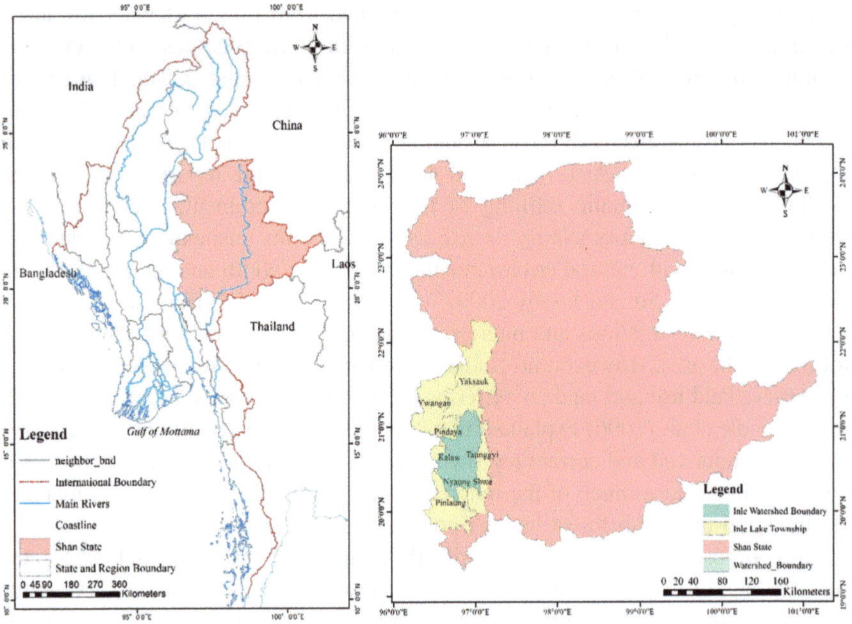

Fig. 13.1 Location of the Inle Watershed, in lower Myanmar

ranges, which runs almost from north to south. This area of the Shan Plateau is formed with mountain ridges which split up and was parallel, formed the flat-bottomed basins or valleys namely, the Thamakhan Basin, the Heho Basin and the Nyaung Shwe Valley (Yaungshwe Valley), which are separated from each other by a mountain. It is evident that drainage must be taken from these three basins and directly drained into the Inle Lake. Generally, the watershed can be divided into three parts; the eastern hilly part, the northern part, and the western hilly part, and several streams and rivers flow through these hilly areas and enter into the lake. The following table shows that the watershed area of Inle Lake which was characterized by 11 distinct land systems with an area of some 2169 square miles designated as follows (Table 13.1).

The whole of Shan States including Inle region is formed with limestone of early Paleozoic to the Mesozoic Era. The rocks around the lake are mainly limestone, dolomites, and marls of Ordovician and Permo-Carboniferous age formation. Structurally, the lake is bounded by a fault running east to west and can be called as a graben. Along the western shores of the lake, the composition of silt and sandstone was deposited. Moreover, the rocks present in the western ridge are of lower resistance as compared to those in the eastern ridge, which rises more steeply above sea level of the basin.

The drainage area and storage capacity of the lake have been estimated at 5612 km^2 and 3.5×10^7 m^3, respectively, while annual inflow and water residence time are estimated as 1.1×10^8 m^3 $year^{-1}$ and 0.32 year (Volk et al. 1996; Ngwe Sint

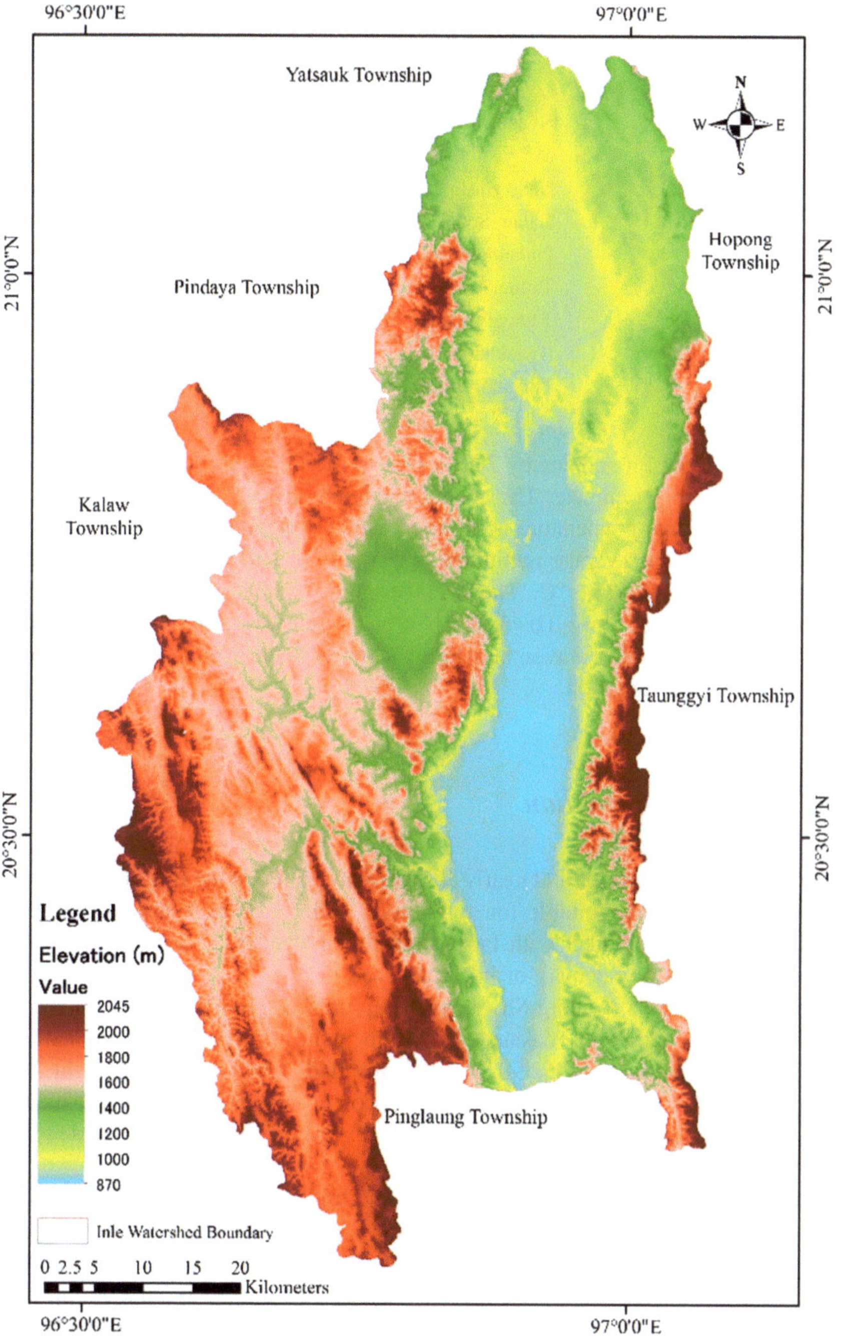

Fig. 13.2 Topography of the Inle Watershed

Table 13.1 Physical characteristics of Inle Watershed Area

Designation	Area (km^2)
Thamakham Plain	287.43
Lonpo Plain	186.68
Nyaung Shwe Valley	784.07
Mobye Valley	314.30
Heho Basin	98.81
Kalaw South Mountains	270.69
Pinlaung range	461.40
Taunglaylone range	393.81
Taunggyi range	393.81
Kyauktalone range	388.33
Thikaung ridge	90.45

and Catalan 2000; Su and Jassby 2000). Due to the location, the climate in the Inle region has a tropical climate. The annual maximum temperature is in the month of April and minimum temperature is in the month of January (May 2008). In Nyaung Shwe Township, the average minimum temperature is 2.8 °C and the average maximum temperature is 36.6 °C. The average annual rainfall for the last 4 years is 855 mm for Nyaung Shwe Township. Not only southwest monsoon brings rainfall into the region but also waves of whirlwind from South China Sea which brings rainfall in the late monsoon.

13.3.2 Social Condition

There are seven townships and nearly all of this area fall into the Inle lake watershed. All of these areas are suitable for agriculture, with favorable climatic conditions, easily accessible to market with fairly high living standard and moderately population. The population density of Inle watershed is shown in Fig. 13.3. The highest population density is Nyaung Shwe Township on the southeastern part of the Inle watershed and the lowest is Yaksauk Township on the north part. Total population of Inle watershed is about 805,590 in 1990 and had increased to 94,932 in 2000. Therefore, within one decade, the total population growth rate was 143,732. In 2010, total population was 1,011,084 and its growth was 61,762. The population density in this area was 52.12 per km^2 in 2010.

In 2010, agriculture is the occupation of the Inle watershed and it is extensively cultivated in Heho Basin and Nyaung Shwe Basin. Large parts around Inle Lake and Heho Basin are under irrigated agriculture. Rotation cultivation and shifting cultivation are practiced in some mountain slopes. In Inle area, paddy is the chief crop and cash crop such as tomato, potato, sugarcane, peas beans, maize, cabbage, cauliflower, vegetables, etc., are cultivated. Generally, land use in Inle watershed is shown in Fig. 13.4.

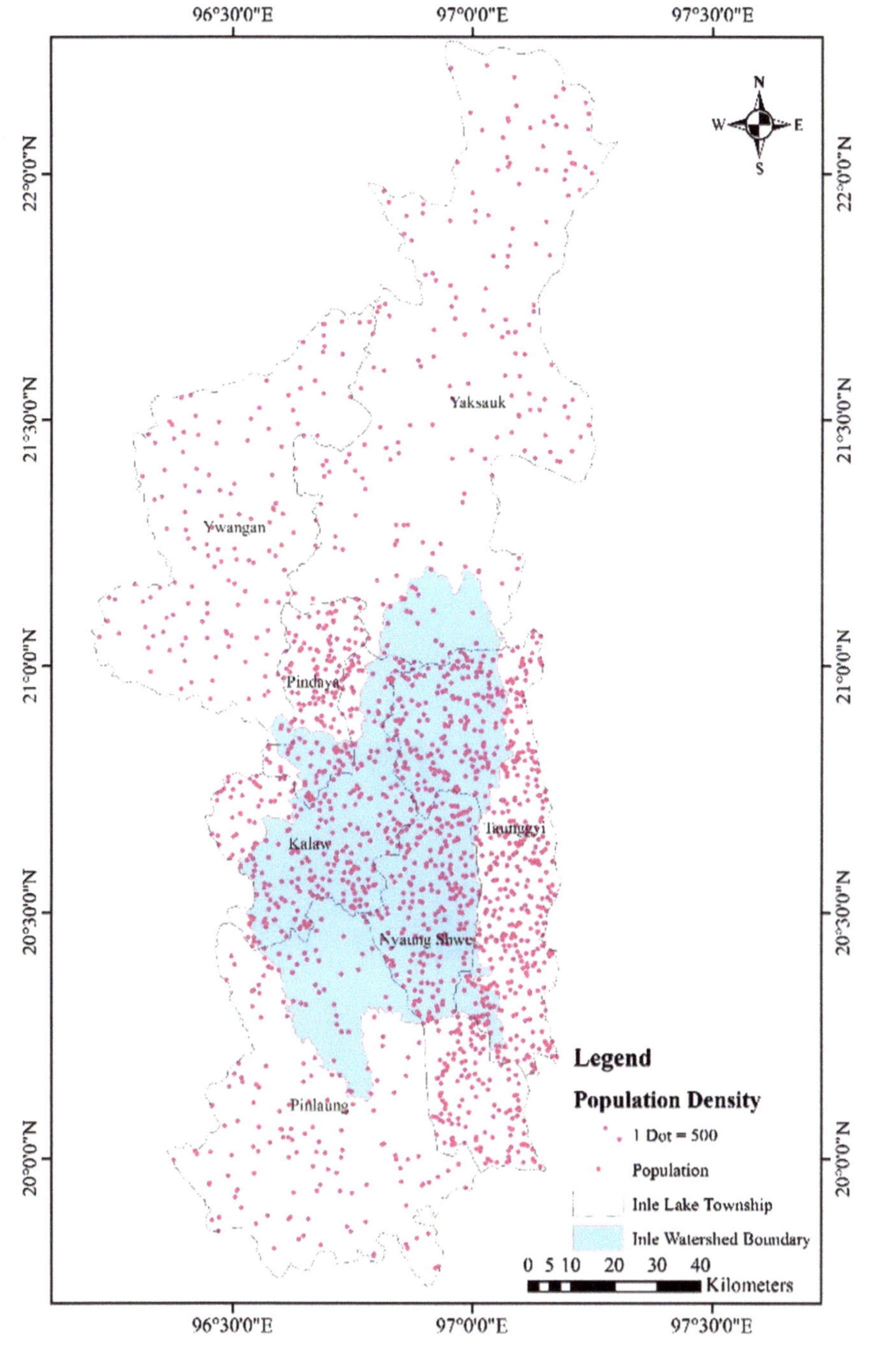

Fig. 13.3 Population density of the Inle Watershed

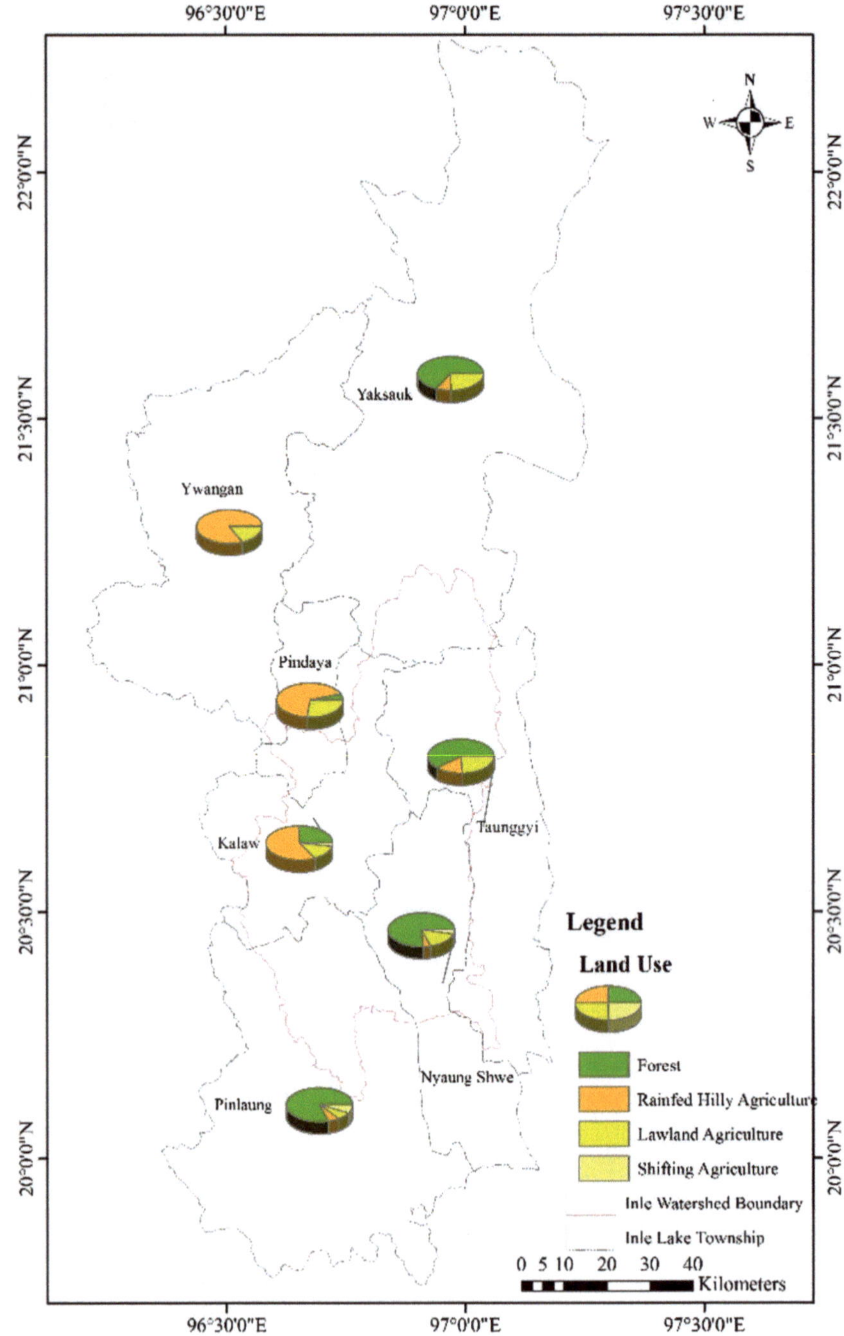

Fig. 13.4 Land use for Inle Watershed

13.4 Methodology

The present research relied on integration of geographical information systems (GIS) and remote sensing (RS) technologies. Three Landsat TM and Landsat ETM+ images (Path 132 Row 45 and Path 132 Row 46) of the Inle Watershed of Myanmar were used in the analysis.

13.4.1 Satellite Imagery and Image Processing

The satellite images (185 km × 185 km) are necessary in order to cover the whole study area to collect reliable data for assessing land cover changes affecting the Inle Watershed. Landsat 5 TM images acquired on February 10, 1990, Landsat 7 ETM + image acquired on January 24, 2000, and Landsat 7 ETM+ image acquired on March 5, 2010, were used for this study. The Landsat 5 TM and 7 ETM+ images were co-registered to each other using Arc Map 10 Software and a set of points selected on the 1990 image which matched the same locations on the 2000 and 2010 images. Experience showed that the near infrared (band 4), infrared (band 5), and red bands (band 3) were the best combination for the 1990 Landsat 5 TM image and Landsat 7 ETM+ image.

This band combination was used to produce image maps of both the Landsat 5 TM and the Landsat 7 ETM+ data. Geometric correction was undertaken to avoid geometric error from a distorted image. The Landsat 5 TM image and Landsat 7 ETM+ images were rectified using ground control points (GCPs), which must be measured accurately. The GCPs were taken from topographical maps and field surveys of the study area using GPS. The coordinates of these points and a transformation function were used to complete the co-registration of the 1990 image to within a half pixel of the root mean square error of the 2000 image. Landsat 5 TM and Landsat 7 ETM+ provide data with 30 m resolution in five spectral bands and eight spectral bands, and panchromatic data in 15 m resolution in one band. The thermal infrared sensor provides 60 m resolution in five specified bands. The topographic maps of Myanmar (scale 1:63,360) prepared by the Aerial Survey Department were used as the base maps to determine land use in the study area.

13.4.2 Ground Verification and Measurement of Ground Control Points

Field data collection was done through visits to the study area to measure ground control points in conjunction with the ground verification. With the aid of ground control points taken from the Global Positioning System (GPS) equipment, satellite images were geometrically corrected. The accuracy of the GPS is about 1–3 m for the differential method and 100 m for the averaging method. Photographs were taken during the field trips. The main objective of the ground verification was to validate the interpretation of the satellite images by assessing the accuracy of the classification. The seven layers containing field observation areas were traced on the corrected satellite images; homogeneous polygons with similar spectral reflectance by visual interpretation, when viewed in several band combinations, were digitized on screen based on terrain knowledge acquired during fieldwork; and eight polygons in each layer were distributed throughout the study areas. Each polygon layer created using this process was later used to check the accuracy of the land cover classified map.

13.5 Results for Land Cover Changes of Inle Lake Watershed Area

13.5.1 Land Use Categories and Land Use Change Matrix

In order to detect land use changes easily and conveniently, the predefined land use classes were grouped into six major land use categories: (1) closed forest, (2) opened forest, (3) scrub land, (4) marsh land, (5) agriculture land, (6) Floating garden, and (7) water body in Fig. 13.5 (photo taken during field observation).

The overall classification accuracy was about 86%. Supervised classification using all reflective bands of the three images acquired on 10 February, 1990, 24 January, 2000, and 5 March, 2010 was carried out using a maximum likelihood classifier. According to the results of satellite images processing, the watershed area of Inle Lake is approximately 4116.7 km^2. The results of land cover classification from Landsat TM and ETM+ are shown in Tables 13.2 and 13.3, which provide a measure of the reliability of the land cover map for 1990, 2000, and 2010. There is significant change both in spatial structure and in the distribution of vegetation types. Water surface of the lake is greatly influenced by its watershed and land cover condition. In order to consider the changes, water surface area and land cover changes should be assessed. Land cover or land use in the watershed area is greatly influenced by the quality and quantity of water. In this research, land cover classifications are basically divided into seven classifications in Inle Watershed.

These are closed forest, opened forest, scrub land, marsh land, agriculture land, floating garden, and water body. The most important differences are a decrease in the area of closed forest, and a large increase in the extent of opened forest, settlement, and water body. In 1990, 61.67% of the total area of the Inle Watershed was closed forest, 7.08% was opened forest, 8.53% was scrub lands, 3.86% was marsh land, 7.22% was agriculture land, 7.3% was floating garden land, and 5.19% was water body. In 2000, total areas under the above land use categories came to 13.85, 9.6, 41.84, 4.6, 15.73, 8.5, and 5.74%, respectively (Fig. 13.6). In 2010, total

Photo 1 Opened Forest

Photo 2 Scrub Land

Photo 3 Opened Forest

Photo 4 Scrub Land

Photo 5Agriculture Land

Photo 6 Floating Garden

Photo 7 Water Body

Photo 8 Settlement on the Inle Lake

Photo 9 Agriculture Land

Photo 10 Closed Forest

◂**Fig. 13.5** Land use/cover classification photos taken from field observation on 6 November, 2014 (1) Opened Forest: 40% canopy cover forest on the way to Pindaya Area, (2, 3) Opened Forest: 40–20% canopy cover forest in the western part of watershed area, (4, 5) Scrub Land: Scrubland affected by shifting cultivation grassland, (6, 7) Agriculture Land: rice cultivation, horticulture and gardens, (8) Floating Garden: Tomato plantation on the lake, (9) Water Body: Fishing on the lake, and (10) Settlement Area: Local people are living on the lake

Table 13.2 Status of land cover categories of Inle Watershed in 1990, 2000 and 2010

Land cover categories	Appraisal					
	1990 year		2000 year		2010 year	
	Area (km^2)	Percentage of total land cover	Area (km^2)	Percentage of total land cover	Area (km^2)	Percentage of total land cover
Close forest	2497.54	60.67	570.23	13.85	98.22	2.39
Open forest	291.59	7.08	395.09	9.60	1599.80	38.86
Scrub land	351.13	8.53	1722.40	41.84	1557.91	37.84
Marsh land	159.03	3.86	189.19	4.60	81.10	1.97
Agriculture land	297.13	7.22	647.76	15.73	406.90	9.88
Floating garden	300.72	7.30	349.87	8.50	249.66	6.06
Water body	213.56	5.19	236.15	5.74	117.11	2.84
Total	4116.70	100.00	4116.70	100.00	4116.70	100.00

Table 13.3 Changes of land cover condition in the Inle Watershed from 1990 to 2010 years

Land cover categories	Land cover changes					
	1990–2000		2000–2010		1990–2010	
	Area (km^2)	% of total land cover	Area (km^2)	% of total land cover	Area (km^2)	% of total land cover
Close forest	−1927.30	−46.82	−472.01	−11.47	−2399.32	−58.28
Open forest	103.50	2.51	1204.70	29.26	1308.21	31.78
Scrub land	1371.27	33.31	−164.49	−4.00	1206.78	29.31
Marsh land	30.16	0.73	−108.09	−2.63	−77.93	−1.89
Agriculture land	350.63	8.52	−240.86	−5.85	109.77	2.67
Floating garden	49.15	1.19	−100.21	−2.43	−51.07	−1.24
Water body	22.59	0.55	−119.04	−2.89	−96.44	−2.34

areas of the Inle Watershed under the above land use categories are only 2.39, 38.86, 37.84, 1.97, 9.88, 6.06, and 2.84%.

In 1990, there were 2497.54 km^2 of closed forest in watershed area of Inle Lake and it was changed to 376.6 km^2 of open forest, 79.91 km^2 of scrub and grassland, and 254.71 km^2 of agriculture land. Therefore, the closed forest was decreased

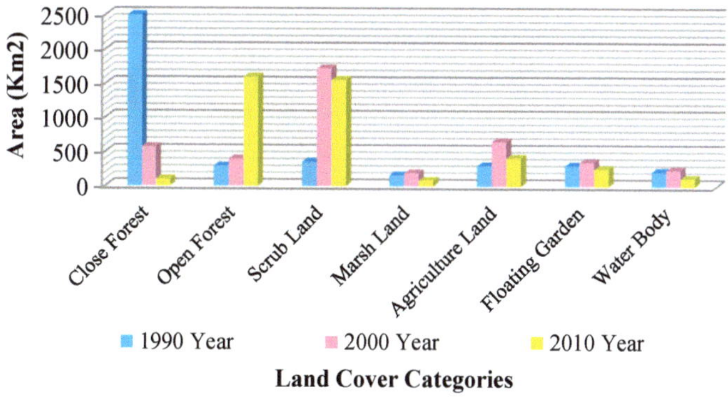

Fig. 13.6 Changes in extent from 1990 to 2010 by cover type. Areas (km^2) are shown over each cover type column and year. The percentage change is calculated using 1990 as reference

from 875.78 to 173.56 km^2 and, only 570.23 km^2 of the closed forest was left in the year 2000. In 2010, the closed forest was changed to 135.16 km^2 of opened forest and to 79.91 km^2 of scrub and grassland. But there was no change from closed forest to agriculture land in the years between 2000 and 2010. Based on the above source, closed forest was changed into open forest, scrub and grassland, and agriculture land. About 2399.32 km^2 of closed forest were reduced between 20 years. Among them, the most significant changes were from closed forest to agriculture land (Figs. 13.7, 13.8, 13.9; Tables 13.4, 13.5).

In Inle Watershed, about 1178.77 km^2 of opened forest was changed to 155.65 km^2 of scrub and grassland and 564.98 km^2 of agriculture land from 1990 to 2000. Therefore, opened forest of Inle Watershed changed from 291.59 km^2 in 1990 to 395.09 km^2 in 2000 and it increased to 1204.7 km^2 in these periods because about 376.6 km^2 of closed forest shifted to opened forest. In 2010, the open forest changed from 1196.11 to 451.18 km^2 of scrub land. But there was no change from open forest to agriculture land in this period. Therefore, the opened forest was reduced from 1196.11 to 1599.8 km^2 and only 98.22 km^2 was left in 2010. In this case, a noticeable point is that the opened forest was considerably increased between 1990 and 2000 with the consequences of changes from closed forest to open forest, but it was reduced again from 2000 to 2010. The opened forest increased to 1599.8 km^2 during 20 years (Figs. 13.7, 13.8, 13.9; Tables 13.4, 13.5).

Scrub and grassland were 351.13 km^2 in 1990 and it increased to 1722.4 km^2 in 2000 at the Inle Watershed. It was changed to 361.36 km^2 of open forest and 579.09 km^2 of agriculture land. Furthermore, the scrub land was continuously changed from 523.63 to 323.3 km^2 of agriculture land in 2010. These results illustrate that scrub land had shrunk to 1557.91 km^2 within 20 years (Figs. 13.6, 13.7, 13.8; Tables 13.4, 13.5).

Marsh land is one of the important factors of Inle Lake because it strongly affects the water surface of the Inle Lake. Between 1990 and 2000, mash land was changed

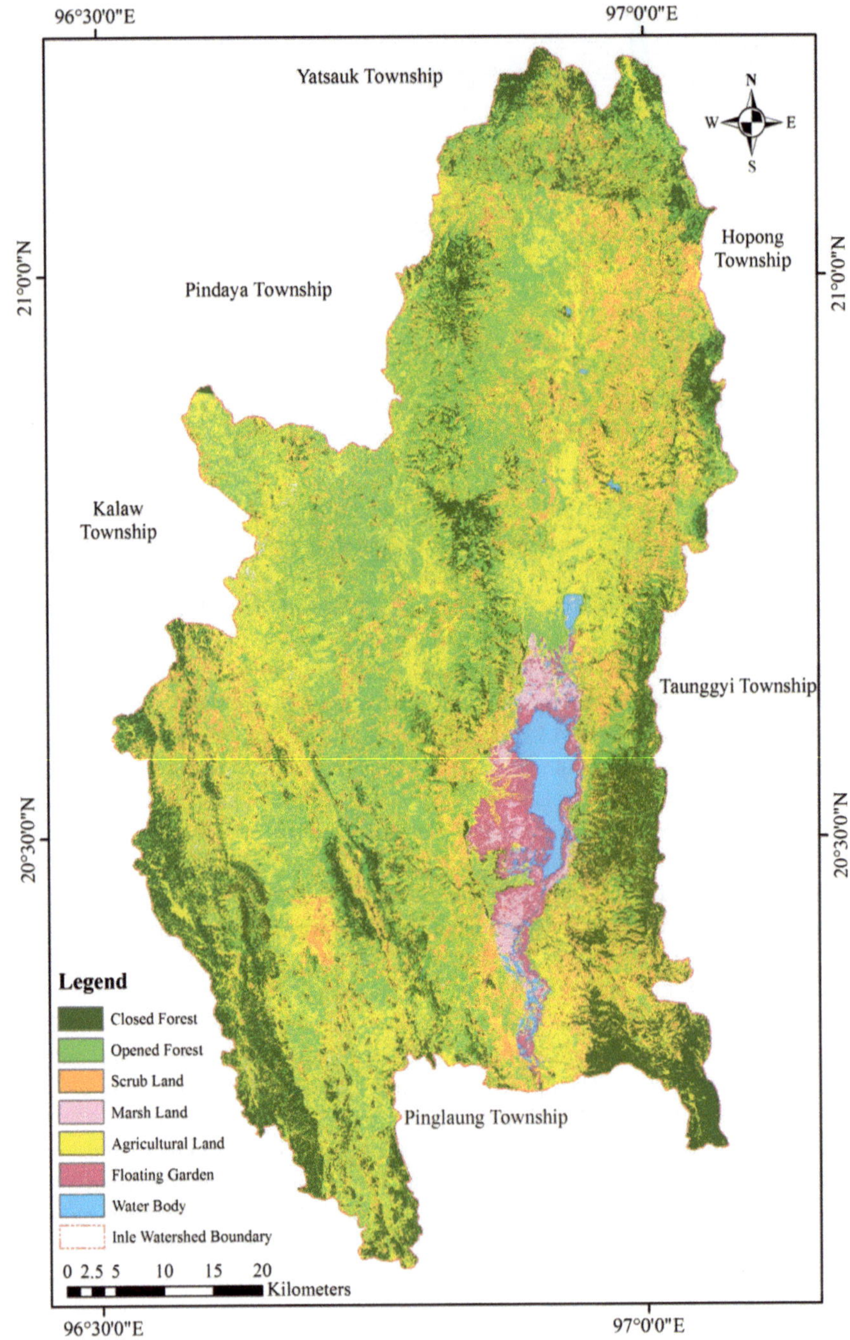

Fig. 13.7 Land use/cover classification map of the Inle Watershed Area in Myanmar using Landsat TM 1990

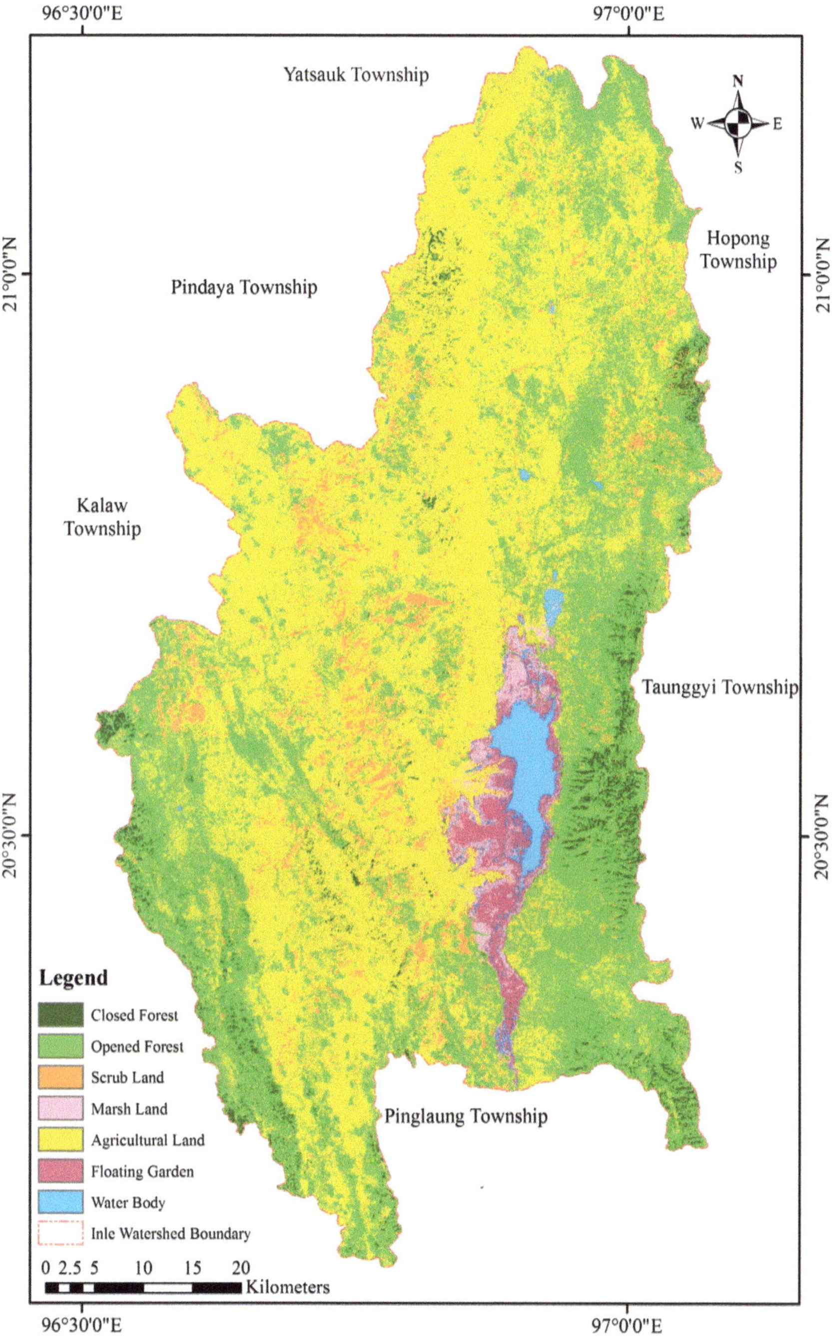

Fig. 13.8 Land use/cover classification map of the Inle Watershed Area in Myanmar using Landsat ETM + 2000

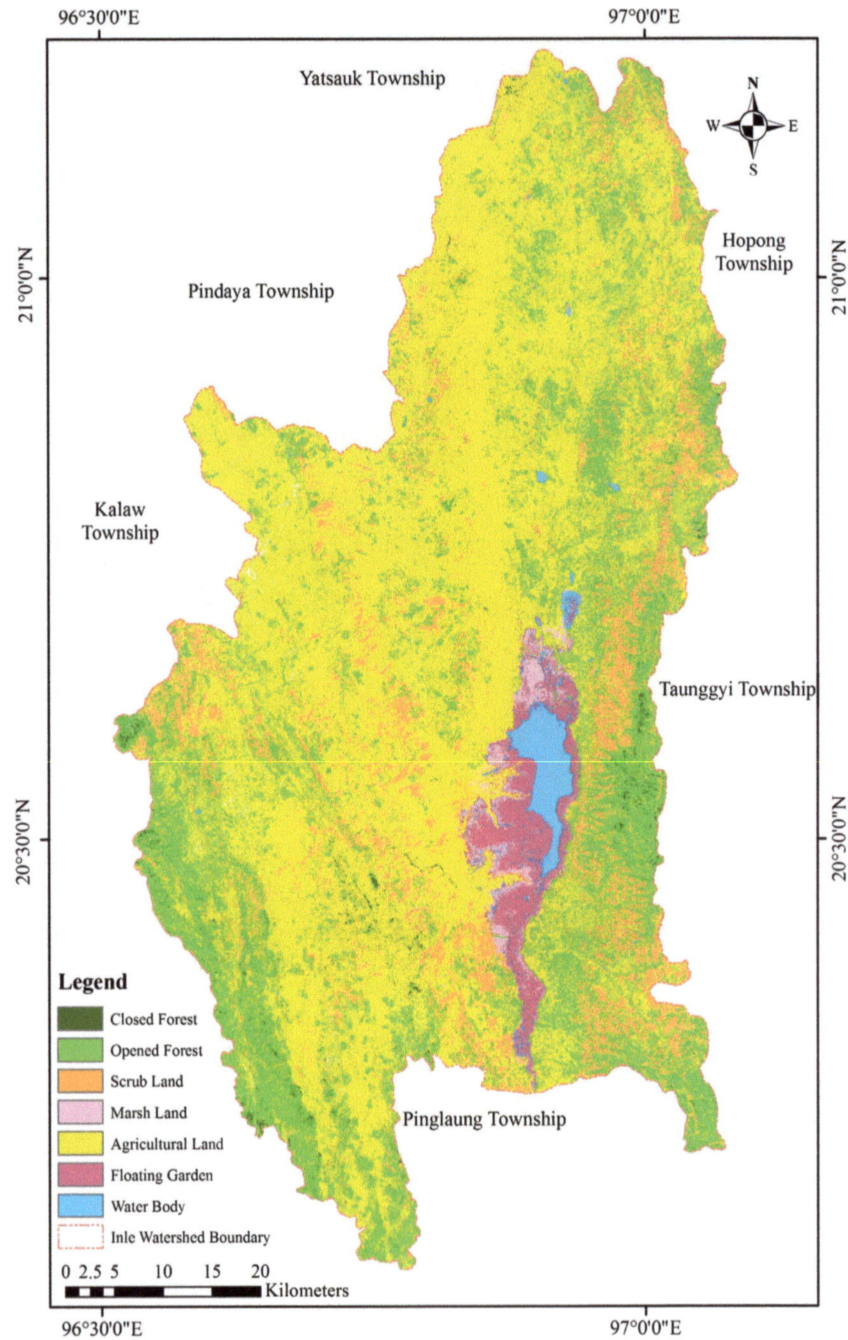

Fig. 13.9 Land use/cover classification map of the Inle Watershed Area in Myanmar using Landsat ETM + 2010

Table 13.4 Land Cover Change Matrix between 1990 and 2000

1990 Land cover categories	2000 Land cover categories							
	Closed forest	Open forest	Scrub grassland	Mash land	Agriculture land	Floating garden	Water body	Grand total
Closed forest	173.56	376.60	79.91	0.00	245.71	0.00	0.00	875.78
Open forest	0.00	458.15	155.65	0.00	564.98	0.00	0.00	1178.77
Scrub land	0.00	361.36	288.08	0.00	579.09	0.00	0.00	1228.53
Marsh land	0.00	0.00	0.00	47.17	2.38	23.88	0.00	73.43
Agriculture land	0.00	0.00	0.00	0.00	573.84	0.00	0.00	573.84
Floating garden	0.00	0.00	0.00	0.00	0.00	118.39	0.00	118.39
Water body	0.00	0.00	0.00	2.92	0.00	8.43	56.63	67.98
Grand total	173.56	1196.11	523.63	50.09	1965.99	150.70	56.63	4116.73

Table 13.5 Land Cover Change Matrix between 2000 and 2010

2000 Land cover categories	2010 Land cover categories							
	Closed forest	Open forest	Scrub grassland	Mash land	Agriculture land	Floating garden	Water body	Grand total
Closed forest	33.70	135.16	4.70	0.00	0.00	0.00	0.00	173.56
Open forest	0.00	744.94	451.18	0.00	0.00	0.00	0.00	1196.11
Scrub land	0.00	0.00	323.03	0.00	200.60	0.00	0.00	523.63
Marsh land	0.00	0.00	0.00	38.74	0.00	10.20	1.16	50.09
Agriculture land	0.00	0.00	0.00	0.00	1965.99	0.00	0.00	1965.99
Floating garden	0.00	0.00	0.00	0.00	0.00	150.70	0.00	150.70
Water body	0.00	0.00	0.00	0.00	0.00	4.34	52.30	56.63
Grand total	33.70	880.10	778.91	38.74	33.71	165.23	53.45	4116.73

from 153.03 to 189.19 km^2 because of 2.38 km^2 of agriculture land and approximately 23.88 km^2 of floating garden cultivation was reversed to marsh land. In 2010, marsh land was changed to 10.2 km^2 of floating garden cultivation and 1.16 km^2 of water body. Therefore, marsh land area was changing alternately with water body, floating garden cultivation, and agriculture land. Thus, marsh land was changed from 189.19 km^2 in 2000 to 81.1 km^2 in 2010. It was found that marsh land was reduced by 77.93 km^2 within 20 years (Figs. 13.7, 13.8, 13.9; Tables 13.4, 13.5).

In Inle Lake Watershed, there was 297.13 km^2 of agricultural land and it grew to 647.76 km^2 in 2000. The extension of agriculture in this region was 350.63 km^2 and it increased more than three times within 10 years. These agriculture lands were extended from 245.71 km^2 of closed forest, 564.98 km^2 of open forest, 579.09 km^2 of scrub land, and 2.38 km^2 of mash land. Agricultural land extension was steadily extended in 2010, in which 200.6 km^2 of scrub land was converted to agriculture land. Thus agricultural land was 406.9 km^2 in 2010. These results show that the agricultural land was dramatically increased during 1990–2000 but the increase rate was gradual to 2010. The extension of agriculture land was 109.77 km^2 during 20 years (Figs. 13.7, 13.8, 13.9; Tables 13.4, 13.5).

Floating garden cultivation (hydro phonic farm) is a distinguished characteristic of Inle Lake. In this classification, it includes all the cultivation types (paddy, vegetable, flower, etc.) that are influencing to the water surface area. In 1990, floating garden cultivation was 300.72 km^2 in Inle Lake and it gradually increased to 349.87 km^2 in 2000. This figure represented that floating garden cultivation was extended 49.15 km^2 from 1990 to 2000. This extension is caused by changes of mash land 23.88 km^2 and water body 8.43 km^2. In the year 2010, floating garden cultivation is decreased to 249.66 km^2. It was found that the floating garden cultivation was extended 300.72 km^2in 1990 to 249.66 km^2 in 2010. It represents that the extension was 51.07 km^2 within 20 years (Figs. 13.7, 13.8, 13.9; Tables 13.4, 13.5).

Water body or water surface area decreased by 96.44 km^2 within 20 years, from 67.98 km^2 in 1990 to 56.63 km^2 in 2010. In 1990, the water surface area was 213.56 km^2, it was increased to 236.15 km^2 in 2000. In 2010, the water surface area was decreased to 117.11 km^2 due to the extension of floating garden and marsh land propagation (Figs. 13.7, 13.8, 13.9; Table 13.4, 13.5).

13.6 Conclusion

Watershed management for Inle Lake is one of the import issues in Myanmar because there are very serious problem of water quality and water storage area degradation due to the effect of land cover change of its watershed. The land cover maps showed the changes very clearly. In 1990, closed forests were found in northern edge, eastern and southwestern part of the watershed area, near Pinlaung range, Kyauktalone range, and Pindaya-Ywangan area. But in 2000, these closed forests were changed to open forest, scrub grassland, and agriculture. In some part of the watershed area, near Taunggyi, scrub land was changed into open forest because of the reforestation program. Agriculture extension occurred in the lake watershed area especially in western and northwestern part of the lake, near Kalaw-Aung Ban, Heho Valley, and Thamakhan Plain.

One of the unique characteristics of Inle Lake is hydroponic cultivation (floating garden cultivation). Besides, tomato cultivation is an important economy of Inle Lake. This type of cultivation is practiced on the naturally floating island and it is effected the water surface area of the Inle Lake in the following ways such as after using the old floating garden islands, they are decomposed to the lake bottom and the lake will be shallower, and the extension of floating garden cultivation causes more shrinkage to the water surface area in the study area. The most intensive floating garden is found in Kela Village, Myaynigone Village, Kyaysar Village, Minchaung Village, Yemebin Village, Kyunegyi Village, etc. Most of villages are located in the western part of the lake. The floating garden area was increased from 1990 to 2010 years. On the other side, water surface area is decreased from 1990 to 2010 year because it is influenced by climatic condition, extension of floating garden, and population growth of the study area.

References

Asian Development Bank (2006) Myanmar case studies: (1) environmental performance in Mandalay City; (2) environmental performance assessment of Inlay Lake. ADB T.A. No. 6069-REG prepared by Myanmar National Community of Environment Affairs and Project Secretariat UNEP Regional Resource

BANCA (2006) Integrated multi-stakeholder ecosystem approach at Inle Lake Based on zoning principles and integration of eco-restoration and agro-farming practices. Biodiversity and Nature Conservation Association Publisher (Myanmar)

Butkus S, Myint S (2001) Pesticide use limits for protection of human health in Inle Lake watershed. Living Earth Institute Olympia. Washington, USA

May SY (2008) Changes of water quality and water surface area in inle lake: facts and perceptions. Unpublished Ph.D. dissertation. Department of Geography, University of Yangon, Myanmar

Ngwe Sint U, Catalan I (2000) Preliminary survey on potentiality of reforestation under clean development mechanism in Myanmar with particular reference to Inle region. Unpublished report by Karamosia International, Yangoon Final report. Myanmar Ministry of Forestry

Sidle RC, Ziegler AD, Negishi JN, Nik AR, Siew R, Turkelboom F (2006) Erosion processes in steep terrain-truths, myths, and uncertainties related to forest management in Southeast Asia. For Eco-Manag 224:199–225

Su M, Jassby AD (2000) Inle: a large Myanmar lake in transition. Lakes Reserv Resour Manag 5:49–54

Swe KL (2011) Development of clean water and sanitation in Inle Lake. In: Myanmar EE2 seminar: water and environment in Asia's developing communities. Singapore International Water Week 2011 Co-located Event Publisher

Than MM (2007) Community activities contribution to water environment conservation of Inle Lake. In: Proceeding of international forum on water environmental governance in Asia, December 2007 at Beppu, Oita, Japan, pp 215–221

Volk P, Heymann, J, Saradeth S, Bechstedt HD, Löffler E, Stuurman W, Aiblinger S, Carl S, Küpper A, Lamprecht S, Ringenberg H, Schönberg A (1996) Mapping and land use planning for watershed management

Win T (1996) Floating island agriculture of Inle Lake. Unpublished M.A. thesis, Department of Geography, University of Yangon

Chapter 14
The Ecological Footprint and Carrying Capacity in Northeast Asia

Zhang Bai and Liu Weijie

Abstract Under the goal of sustainable development, optimum population rests on the comprehensive carrying capacity of many factors, such as ecology, economy, and land, etc. Recently, the ecological environment of the Northeast Asia has been deteriorating seriously, because of the fall of its ecological carrying capacity resulted from human activities. The ecological carrying capacity of the Northeast Asia is directly related to its ecological environment and socioeconomic sustainability. The ecological carrying capacity is based on the net primary productivity (NPP) of natural vegetation which can reflect the productive and recovery capacity, and thus is the index of the ecological integrity of natural system. Based on the above purposes and the assessment method, this paper studies the distribution and the change of the ecological footprint (EF) and the ecological carrying capacity in the Northeast Asia. The change of per capita EF shows a trend of decline in the Far East of Russia, Japan, and Mongolia, but the original value is still higher in the front row. It is more than 3 hm^2 and showed an upward trend in the Northeast China and South Korea. North Korea is the most stable and the lowest EF is about 2 hm^2. As a whole situation of the Northeast Asia, we can see in addition to a small part where its ecological carrying capacity is near 0, in the northern areas and the most regions of Northeast Asia, the ecological carrying capacity is between 0 and 30 hm^2/km^2. In the most central region of the Northeast Asian the ecological carrying capacity is between 30 and 50 hm^2/km^2. In the southeastern and midwestern areas, the ecological carrying capacity is between 80 and 100 hm^2/km^2. The ecological carrying capacity even exceeded 150 hm^2/km^2 in southern areas. In the southwest region there is a large bareland area, the ecological carrying capacity is near 0.

Keywords Ecological footprint · Ecological carrying capacity · Northeast Asia

---

Z. Bai (✉) · L. Weijie
Northeast Institute of Geography and Agro-ecology, CAS, Changchun, China
e-mail: zhangbai@iga.ac.cn

Y. Himiyama (ed.), *Exploring Sustainable Land Use in Monsoon Asia*,
Springer Geography, https://doi.org/10.1007/978-981-10-5927-8_14

14.1 Basic Concept and Assessment Method

The ecological environment is the cradle of human survival and development, and the foundation of economic development, and the source for social progress (Costanza et al. 1992). In recent two decades, with the population proliferation and science progress, human beings have changed the environment at great scale and speed, and ecological environment has been affected and damaged unprecedentedly. The ecological environment problem is not just point source problem, but also has developed into regional, and national or even global problem. Northeast Asia is very rich in natural resources. However, with its rapid economic development, the regional environment problems have become more and more serious.

Northeast Asia is an important ecological protective screen in the eastern of Eurasia; its eco-environment quality has the direct impacts on the national and regional environment safety. This thesis is on the basis of the land use data and satellite remote sensing data, with the method of ecological carrying capacity, calculated the ecological carrying capacity in the Northeast Asia from 1990 to 2010.

14.1.1 The Model of Ecological Footprint

The ecological footprint (EF) is a measure of human demand on the Earth's ecosystems. It was developed in the early 1990s by M. Wackernagel and W. Rees. It is a standardized measure of demand for natural capital [(manufactured means of production) to goods and services relating to the natural environment] that may be contrasted with the planet's ecological capacity to regenerate. It represents the amount of biologically productive land and sea areas necessary to supply us with the resources for consumption, and to assimilate all the accompanied wastes (Rees 1992). Using this assessment model, it is possible to estimate how much of the Earth it would take to support our human beings if everybody followed a given lifestyle. For 2010, our human being's total ecological footprint was estimated at 1.5 planet Earths; that is, all of our human beings used ecological services 1.5 times as quickly as Earth can renew them. Every year, this number is recalculated to incorporate the 3-year lag due to the time it takes for the UN to collect and publish statistics and relevant researches (Yue et al. 2005).

The ecological footprint is a significant accounting tool for overall assessment of the status of sustainable development, and a comprehensive indicator for human resources consumption. The value of ecological footprint is based on the population, living condition, technical level, eco-production, and so on. According to the method of ecological footprint the ecological productive area is divided into five kinds: farmland, grassland, forest, water, and fossil energy land. The area's total EF formula is as follows:

$$\mathrm{EF} = N \times \mathrm{ef} = N \times \sum (a \times ai) = N \times \sum (ci/pi) \quad (1)$$

N population;
ef EF of per capita;
i different consumption item;
ai consumption item of per capita;
i ecological productive area;
Ci consumption of per capita in i consumption item;
Pi globe average productivity in i consumption item.

The ecological footprint includes two parts, the consumption of the living resources and the energy sources. The consumption of the living resources can be divided into agriculture production, animal husbandry production, forestry and aquatic production (FAO 1997). The consumption of the energy sources is divided into raw coal and oil (Chen et al. 2007).

14.1.2 The Model of Ecological Carrying Capacity

The concept of the ecological carrying capacity came from ecology. It was first applied in the field of human ecology in 1921 by Park and Burgess. It is the maximum value of individual existence under a particular environmental conditions (mainly refers to the combination of living space, nutrients, climate and other ecological factors).

In 1991, Hardin further clarified the definition of ecological capacity in the condition of the undamaged productivity and the function of ecosystem integrity, unlimited duration of maximum utilization of resources, and waste rate. Based on the scholars of ecological footprint, the ecological capacity of the region is defined as the land ecological productive area in a region for human beings, and as the indication of ecological capacity in a region.

According to the remote sensing data, the various biological productive land area of per capita and the ecological capacity of per capita are calculated (Seidl and Tisdell 1999). Based on the adjustment of equivalence factor and yield factor, the ecological capacity of per capita is adjusted. Then, based on the calculated data, the characteristics of ecological carrying capacity of per capita is presented and analyzed in each region and different years. In the Northeast Asia the ecological carrying capacity can be calculated by the following formula:

$$\mathrm{EC} = N \times \mathrm{ec} = N \times \sum aj \times rj \times yj \quad (j = 1, 2, 3 \ldots 6) \quad (2)$$

EC total capacity region;
ec regional ecological carrying capacity per capita;

aj a per capita bio-productive land area;
rj equilibrium factors of different land types;
yj the yield factor of some productive land.

GIS spatial analysis and visualization techniques provided the combination method for the remote sensing image interpretation and the net primary productivity (NPP) grid data. According to the type of land data in each grid, the ecological carrying capacity is calculated. Then, the different color is used to show the different ranges of ecological carrying capacity. The spatial differences and characteristics are displayed and analyzed on the more visual angle.

In this study, the spatial calculation of ecological carrying capacity is obtained by the spatial calculation of the equilibrium factors. The net primary productivity (NPP) reflects the production capacity of plant communities in a natural environment which directly refers to the amount of organic matter accumulation in green plants of unit time and unit area. The calculation results of the ecological carrying capacity can reflect the differences of the land cover types and production capacity.

The ecological carrying capacity of 1 km^2 land is calculated according to the following formula:

$$\mathrm{ec} = \sum aj \times rj \times yj = \sum aj \times \mathrm{NPP}j/\mathrm{NPP} \times yj \times 100 \times 0.88 \quad (j = 1, 2, 3\ldots 6) \tag{3}$$

ec regional ecological carrying capacity per capita;
aj a per capita bio-productive land area;
rj equilibrium factors of different land types;
yj the yield factor of some productive land;
0.88 the coefficients of the protective biodiversity.

14.2 Spatial Distribution and Temporal Change

14.2.1 The Basic Natural Eco-environment and the Land Use Change

Based on the characteristics of physical geography, the Northeast Asia including the Far East and East Siberia of Russia, Mongolia, Japan, D.P.R. Korea, R. Korea, and the Northeast of China. Its location is in the East longitude 87°45′–162°51′, North latitude 27°34′–74°37′. Its land area is about 10^7 km^2 and the population is about 450×10^6. There are the abundant forest, land, energy, mineral resources in this region. The forest area reaches 10.3×10^8 hm^2, accounts for 26.1% of the world in total. In 2007, the wood yield was 3.8×10^8 cubic meters, accounting for 17.2% of the world production in total. Oil, natural gas, and coal products, respectively, account for 17.4, 23, and 45.8% of the global output. Particularly, the timber production is more than

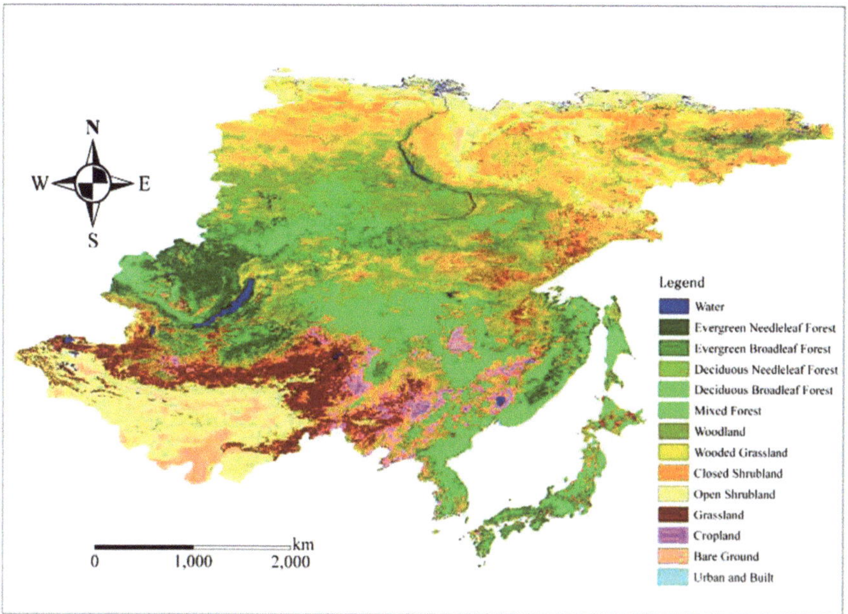

Fig. 14.1 Land use maps of Northeast Asia in 1990

one-fourth and one-fifth of the global timber volume and yield in the region. The forest resources have huge development potential in the area. Compared to the following situations, such as a widespread destruction and substantial reduction in the global forest resources, resource accumulation, the climate warming, and the international protection of forest resources and environment, the supply and demand of forest products will significantly increase in the Northeast Asia.

Using ArcGIS software to cut the global land use classification data in 1990 and generate the Shp. file. According to the classification standards of existing in 1990, the two classification land use data of Northeast Asia in 2010 was merged into one class category to ensure the comparability, as shown in Figs. 14.1 and 14.2.

14.2.2 The Ecological Footprint

According to the EF formula (1) and the statistical data of the yearbooks, the ecological footprint in Northeast Asia in 2010 is calculated. The results can be seen from Fig. 14.3, the figure shows the overall change situation of ecological footprint in the Northeast Asia from 1990 to 2010. The per capita EF is shown a trend of decline in the Far East of Russia, Japan, and Mongolia, but the original value is still higher in the front row. It is shown an upward trend in the Northeast China and

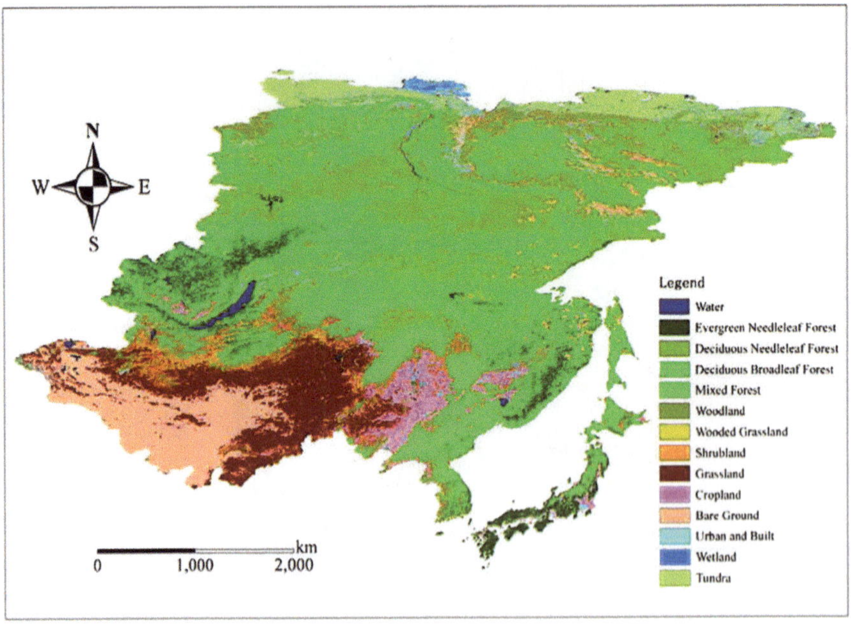

Fig. 14.2 Land use maps of Northeast Asia in 2010

Fig. 14.3 The change of ecological footprint in Northeast Asia from 1990 to 2010 (hm^2/per capita)

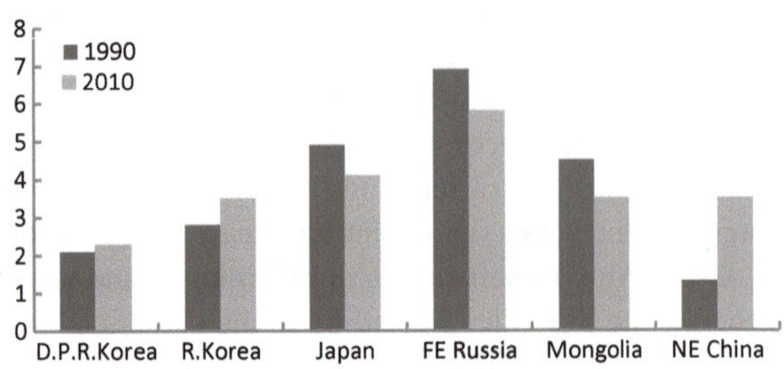

South Korea where is more than 3 hm^2 in 2010. The ecological footprint in North Korea is the most stable and the lowest, where it is about 2 hm^2.

14.2.3 The Ecological Carrying Capacity

From Figs. 14.4 and 14.5, we can see the whole situation of the Northeast Asia. In addition, a small part of the ecological carrying capacity is near 0, in the most of the regions of the northern area of Northeast Asia, the ecological carrying capacity is between 0 and 30 hm^2/km^2. In the most central region of the Northeast Asia, the ecological carrying capacity is between 30 and 50 hm^2/km^2. In the regions of

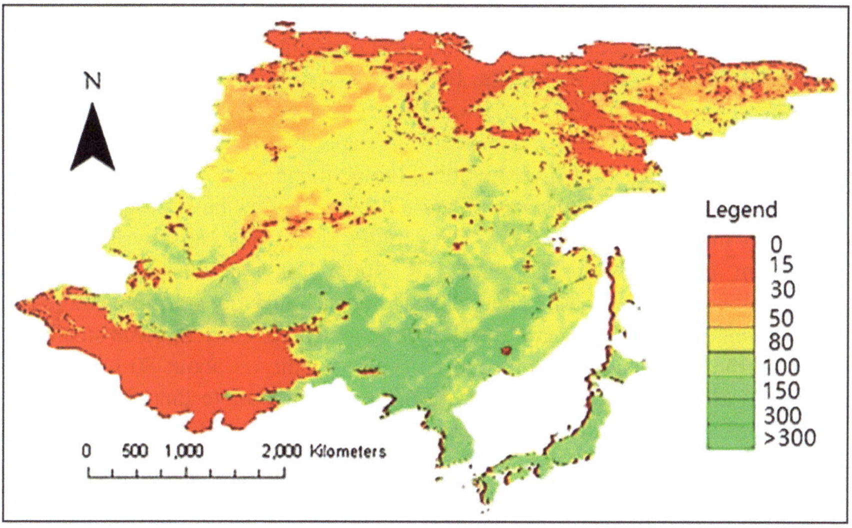

Fig. 14.4 The ecological carrying capacity of the Northeast Asia in 1990

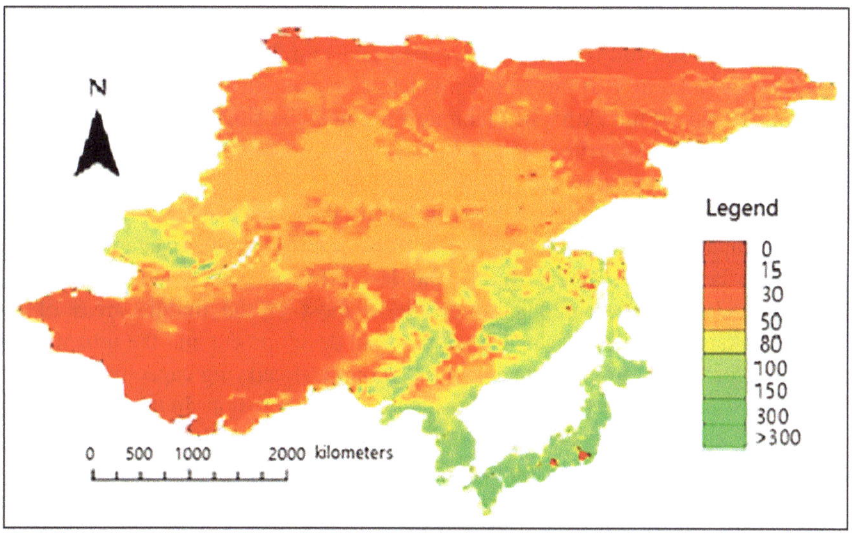

Fig. 14.5 The ecological carrying capacity of the Northeast Asia in 2010

southeastern and mid-western, the ecological carrying capacity is between 80 and 100 hm^2/km^2. The ecological carrying capacity even exceeded 150 hm^2/km^2 in the southern areas. In the southwest region, there is a large bareland area, the ecological carrying capacity is near 0.

Following is the situation of the ecological carrying capacity which is further observed in different countries. The ecological carrying capacity is between 15 and 80 hm^2/km^2 in the most part of the Far East of Russia. In the small parts of the northern region, the ecological carrying capacity is near 0. Also, in some small parts of the southern region, the ecological carrying capacity is between 80 and 100 hm^2/km^2. But in most areas of the Mongolia, the ecological carrying capacity is 15–50 hm^2/km^2. In some small parts of the northern region, the ecological carrying capacity is more than 80 hm^2/km^2. The ecological carrying capacity is more than 80 hm^2/km^2 in the most areas of the D.P.R. Korea. It is more than 100 hm^2/km^2 in some small parts of the coastal areas. The ecological carrying capacity is more than 100 hm^2/km^2 in the most area of the Republic of Korea. Only in some small parts of the northern region, the ecological carrying capacity is about 80 hm^2/km^2. In most parts of Japan, the ecological carrying capacity is more than 100 hm^2/km^2. The ecological carrying capacity is about 80 hm^2/km^2 in the Hokkaido of Japan. It is even more than 150 hm^2/km^2 in the Southern Island of Japan.

In the Inner Mongolia Autonomous Region of China, the ecological carrying capacity is similar to Mongolia due to the similarity of the climate situation. The ecological carrying capacity is about 15–50 hm^2/km^2 in the most parts and more than 80 hm^2/km^2 in some small parts. In the half of Heilongjiang Province, the ecological carrying capacity is about 15–50 hm^2/km^2, and in another half, the ecological carrying capacity is higher than 80 hm^2/km^2. The ecological carrying capacity is 80–100 hm^2/km^2 in the western and middle of Jilin Province, meanwhile in the eastern part of Jilin Province, the ecological carrying capacity is between 30 and 50 hm^2/km^2. In Liaoning Province, the situation is similar to Jilin Province, ecological carrying capacity in general is about 80 hm^2/km^2 which is higher than that in western and mid-part.

Comparing the 2 years data of 1990 and 2010, the general trend of the ecological carrying capacity change can be seen. Following the climate change and the large-scale land development during the last 20 years, the ecological carrying capacity had been decreased significantly in most of the middle areas of the Northeast Asia. In the Northeast China, the land development and the urbanization are the major factors to the decrease of the ecological carrying capacity. In the Far East of Russia, major factors are deforestation and soil erosion. In the west coast of Japan and the Sakhalin Peninsula of Russia, it may be the climate warming increases the ecological carrying capacity.

14.3 Conclusion

After 1990, the EF of per capita showed the approach motive trends in the different countries of Northeast Asia. Particularly, at the beginning of twenty-first century, the EF value is growing in the R. Korea and NE China, however, over the same period, it is dropping in the FE Russia, Japan, and Mongolia. The EF value is lower, which was only near 4 hm^2/per capita until 2010, but with the rapid economic development, the EF value will continue increase. Meanwhile, the huge population causes the total EF to be quite larger, even though the EF of per capita is at lower level, so that how to save the natural resources and protect eco-environment has become more and more important duty for the government and the public in the NE Asia.

The ecological carrying capacity holds a slightly different meaning when applied to human population growth. When discussing human populations, the ecological carrying capacity often refers to the number of individuals that the Earth could carry at different standards of living and levels of resource consumption (Zhang 2006). Thus, the Earth's ecological carrying capacity is smaller if the average living standard people achieved in the developed countries is closer to the average living standard people achieved in developing countries.

References

Chen B, Chen GQ, Yang ZF (2007) Ecological footprint accounting for energy and resource in China. Energy Policy 35:1599–1699 (in Chinese)

Costanza R, Norton BG, Haskell BD (1992) Ecosystem health: new goals for environmental management. Island Press, Washington DC, pp 223–238

FAO Proceedings (1997) Land quality indicators and their use in sustainable agriculture and rural development. In: Proceedings the workshop organized by the Land and Water Development Division FAO Agriculture Department, No. 2:5

FAOSTAT. http://faostat.fao.org/

Rees WE (1992) Ecological footprint and appropriated carrying capacity: what urban economics leaves out. Environ Urban 4(2):121–130

Seidl I, Tisdell CA (1999) Carrying capacity reconsidered: from Malthus' population theory to cultural carrying capacity. Ecol Econ 31:395–408

Yue D, Xu X, Li Z (2005) Spatiotemporal analysis of ecological footprint and biological capacity of Gansu, China 1991–2015: down from the environmental cliff. Ecol Econ 1:1–14 (in Chinese)

Zhang Y, Yang Z, Yu X (2006) Measurement and evaluation of interactions in complex urban ecosystem. Ecol Model 1:1–13

Chapter 15
Land Use and Spatial Policy Conflicts in a Rich-Biodiversity Rain Forest Region: The Case of Jambi Province, Indonesia

Ernan Rustiadi, Baba Barus, Laode Syamsul Iman, Setyardi Pratika Mulya, Andrea Emma Pravitasari and Dedy Antony

Abstract Jambi is one of the provinces in Indonesia with rainforests that support a rich biodiversity of flora and fauna. However, in the last two decades, Jambi Province has been experiencing rapid deforestation, expansion of monoculture plantation crops (especially palm oil and rubber), mining activities, and other types of natural resource exploitation. Various forms of anthropogenic disasters such as floods, peat subsidence, and forest fires have become more frequent in this region. Another serious problem is the conflict between different land use policies, especially governmental policies, land grabbing, and encroachment of forest and conservation areas. The objectives of this study are: (1) to overlay the existing land use maps over maps of the regional spatial plan, mining concession areas, and forest status, and (2) to analyze the land use and policy conflicts as well as their consequences. More than 2.2 million hectares (ha), or approximately 44.6% of land in Jambi province, which is located outside the forest area, is abandoned or is unproductive. Approximately 96% or more of the protected area (834,800 ha) is still maintained in accordance with its function. The space use conflicts mainly occur in the form of policy disagreements between the government institutions, namely between central government institutions of spatial planning, forestry, agriculture, as well as energy and mining. Furthermore, conflicts have also occurred because of the disagreement of local communities with the policies of the central government, local governments, and corporations (mostly mining and agricultural companies).

E. Rustiadi (✉) · B. Barus · L. S. Iman · A. E. Pravitasari
P4W/CRESTPENT, Bogor Agricultural University (IPB), Jl. Raya Pajajaran, Bogor, Indonesia
e-mail: ernan@indo.net.id

E. Rustiadi · B. Barus · S. P. Mulya · A. E. Pravitasari
Faculty of Agriculture, Department of Soil Science and Land Resources, Bogor Agricultural University (IPB), Jl. Meranti IPB Darmaga Campus, Bogor, Indonesia

D. Antony
Study Program of Soil Science, Faculty of Agriculture, University of Jambi, Jambi City, Indonesia

Y. Himiyama (ed.), *Exploring Sustainable Land Use in Monsoon Asia*, Springer Geography, https://doi.org/10.1007/978-981-10-5927-8_15

Keywords Land use · Spatial policy conflicts · Palm oil and rubber plantation
Mining concession · Deforestation

15.1 Introduction

15.1.1 The Richness of Jambi Province's Forest and Its Nature

Tropical rain forests are known for their rich ecosystems and abundant biodiversity in terms of both flora and fauna (FWI/GFW 2001). Located in the heart of Sumatra Island, the forests of Jambi Province have rich flora and fauna including some endangered species. The forests are home to threatened fauna species, such as the Sumatran Tiger (*Panthera Tigris Sumatra)*, Sumatran Elephant (*Elephas Maximus sumatranus)*, and Sumatran Rhinos (*Dicerorhinus sumatraensis)*. The forests are also the home of the world's largest flower, *Rafflesia arnoldii* (PHKA 2003).

Moreover, the Jambi forest area is important for indigenous people, such as the Orang Rimba Tribe, Kubu Tribe, and other tribes, who rely on the forest for their way of life. These people usually inhabit the buffer zone of the forest area and collect forest products, both for their own consumption and for selling it to the villages nearby (Li 2001; Colchester et al. 2011).

The forest area in Jambi encompasses approximately 2,179,440 hectares (ha) or 42.73% of the total area of the province (Ministry of Forestry 2011). Forest conservation in Jambi Province has strategic significance for both Indonesia and the world because of the extensive natural forests in the four national parks: the National Park of Kerinci Seblat (TNKS), which has been designated as a World Heritage Site; the National Park of Berbak, which is a Ramsar Convention wetland site with peat swamp forest landscapes that are relatively intact and the largest in Southeast Asia; the National Park of Bukit Duabelas; and the National Park of Bukit Tigapuluh. Thus, Jambi Province has a very important role in the carbon cycle and as a global biodiversity reservoir (Widayati and Suyanto 2013).

Economic development in Jambi still depends on the abundance of its natural resources, renewable and nonrenewable, sourced through farming, forestry, agriculture, and mining (Jambi 2011). In 2013, the Gross Regional Domestic Product (GRDP) at current market prices based on the agriculture, plantation, and forestry amounted to more than IDR 25,398,690 million (USD 1.95 billion), and mining to around IDR 14,662 trillion (USD 11.2 trillion) with growth rates (2011–2013) of 7.41 and 1.54%, respectively (BPS 2014). Forestry, agriculture, and plantations play a major role in shaping the structure of Jambi's economy.

15.1.2 The Trend of Natural Resources Exploitation: Deforestation, Expansion of Monoculture Plantation Crops (Palm Oil and Rubber), and Mining Concession

Jambi Province faces pressure from land use, land use change, and forestry (LULUCF) issues. The rates of deforestation and degradation in Jambi Province within and outside the forest area rose up to 76,522 ha and 9,431 ha per year, respectively, during the period from 2006 to 2009 (Ministry of Forestry 2011). Drivers of deforestation and forest degradation are usually defined as the overexploitation and the conversion of natural forests and peat lands into production forests (industrial timber, oil palm plantations), infrastructure development, mineral and coal mining, and other significant factors such as illegal logging, forest fires, and encroachment of protected forest area (Widayati and Suyanto 2013).

Jambi forests are by no means free from disturbances and exploitation for industrial crop plantation. Jambi is a province with a long tradition of smallholder rubber production (Potter and Lee 1998). Rubber or *Hevea brasiliensis* was introduced in Jambi province at the beginning of the twentieth century, progressively replacing rainfed cultivation and creating a farming system called agroforests (Feintrenie and Levang 2009). Rubber agroforest or Jungle rubber habitat has become more important for the conservation of biodiversity because the species richness and species accumulation curves for the seedling and sapling stages were similar to natural secondary forests (Beukema et al. 2007).

However, these agroforest systems are endangered due to the increasing demographic pressure, market integration, and household monetary needs (Feintrenie and Levang 2009; Villamor et al. 2014). Since the 1950s, rubber agroforests have been threatened by monoculture plantations (oil palm and rubber), which have higher profitability than agroforests (Feintrenie et al. 2010; Rist et al. 2010).

Monoculture oil palm plantation was first introduced in Jambi province in the early 1980s by the transmigration program (Feintrenie and Levang 2009). This program aimed at moving volunteers from the overpopulated islands of Java and Bali to the less populated islands of Sumatra, Kalimantan, and Sulawesi. Since the 1980s, the wealth of the Sumatran agriculture has attracted more migrants from Java, a move that was further encouraged by district and provincial authorities eager to increase the population density in their constituencies, especially since the passing of the regional autonomy laws in 2000 (Feintrenie and Levang 2009).

Oil palm expansion has major implications for rural Sumatrans. It implies a major reallocation of land and resources, dramatic changes to vegetation and local ecosystems, substantial investment and new infrastructure, movement of people and settlements, and major transformations of local and international trade. It requires the intervention of multiple government agencies. While palm oil can generate wealth and employment for local communities, oil palm estates can also lead to land alienation, loss of livelihoods, social conflicts, exploitative labor relations, and degraded ecosystems (Colchester et al. 2006).

Pressures on the forest areas and land conversions in Jambi Province have become increasingly higher in recent years compared to the time when mineral extraction has started (in early 2000 s). Forest Watch Indonesia (FWI 2015) stated that about 62,747 ha of forest areas in Jambi (2010–2013) have been handed to 31 companies for mining activities. Mining activities in Jambi Province are mostly performed by coal mining companies and to a lesser extent by gold, iron ore and quartz mining companies.

Since the 1970s, most primary forests and logged forests in Sumatra have been cleared and changed to oil palm, rubber, and timber plantations (World Bank 2001). During the period of 2000–2005, the rate of deforestation in Indonesia was estimated to be 1.1 million ha per annum. Although this annual forest loss is lower than that during the 1990s (c1.8 million ha p.a.), it is projected to increase again due to the growing demand in pulp, paper, oil palm, and rubber (DNPI 2011, Gunarso et al. 2013).

The implementation of the regional autonomy in 2000 has made provincial and district governments get a stronger control of the land and natural resources. The immediate impact was the increase of forest clearance for plantation expansion. Data from the BLHD (Environmental Agency) of Jambi Province (2014) stated that 366,964 ha of the forest area have already been turned into plantations. The expansion of large private oil palm companies has been encroaching on the state forestland or on agroforest smallholdings. The Provincial Government of Jambi announced plans to develop 1.3 million ha of oil palm plantation in the province by 2020 (Colchester et al. 2006).

Attractiveness of high return makes the farmers convert their agroforests into monoculture plantations using boom commodities such as rubber and oil palm. Various oil palm development schemes are ubiquitous in Jambi including the large-scale estates of 50,000 ha owned by international companies and 2 ha smallholdings owned by independent farmers (Feintrenie et al. 2010).

Mining concession has now become one of the sectors that contribute to the decrease of the forest areas. During 2008–2013, about 2.98 million ha of forestland was lost in Indonesia mostly by the mining activities that use the open pit method, which involves opening the vegetation cover and digging to a certain depth to extract the minerals. Mining companies have also been given an easy access to the protected forest areas that are subject to government regulations (FWI 2015).

15.1.3 Land Use and Policy Conflict: Conflict Between Governmental Policies, Land Grabbing, Encroachment of Forest, and Conservation Area

During the 1970s, virtually the entire forestland in Jambi was handed out to various companies in the form of large forestry concessions for the extraction of timber. In fact, there are many indigenous people who have been living in those lands for a long time, long before the declaration of the Indonesian National Independence in 1945. After the declaration of the independence, the government started handing out the lands of the indigenous people for logging, transmigration, and cocoa and palm oil projects without their consent. This seriously disrupted the peoples' connections with their ancestral territories, diminished the remaining forests, and deprived them of their land and livelihood. The land squeeze led to the out-migration of the indigenous people and the intrusion into their area by the settlers from Java (Colchester et al. 2011).

Many instances of social conflicts between the oil palm plantations and local communities or transmigrants, and between communities and district governments have been reported (FWI 2015). The reasons behind these conflicts are seldom linked to a rejection of the crop, but mostly to the promises not being kept or to unfair benefit sharing (Colchester and Chao 2013). Rist et al. (2010) argued that the conflicts between communities and companies resulted almost entirely from the lack of transparency, the absence of free, prior, and informed consent, and unequal benefit sharing, and have been exacerbated by the absence of clear land rights. Many oil palm plantations were expanded without clear land rights or protected by the strong law enforcement causing land use conflicts (Colchester and Chao 2013). In many cases, land conflicts and land clearance caused displacement of the original inhabitants from their traditional lands (Pichler 2010; Colchester et al. 2011).

Encroachment done by the companies is supported by the government, especially through the promotion of land development in official spatial plans. Moreover, many companies whose primary intent is to obtain timber harvesting concessions (logging activities) obtain land concession permits (Mongabay Indonesia 2014). People who do the encroachment typically live in the forest because their access to land ownership is low due to economic pressures. However, in general, the encroachment done by the people in the area is still limited compared to the company undertakings.

15.1.4 Anthropogenic Disaster of Jambi Province: Floods, Peat Land, and Forest Fires

Land conversions that dramatically decrease the vegetation cover have negative impacts on the hydrological cycle. For instance, peat soil clearance through draining the groundwater or the canalization of the peat ecosystem could result in fires and peat subsidence. In addition, the decrease or the loss of the swamp forest area will lower the quality of the environment and cause flooding during the rainy season as well as drought and fires in the dry season. BLHD (2014) showed that the flood disaster in Jambi in 2014 covered more than 1,300 ha and caused an economic loss of almost IDR 4 billion.

Over the last decade, there have been three major peat and forest fires in large areas of Jambi Province and these fires have become a serious problem both at the national and global level (Harrison et al. 2009; WHO 2006). Sometimes fires occur naturally or are used as a tool to clear land for agricultural purposes (e.g., forest to plantation) (Stolle et al. 2003). However, in this case, oil palm plantation estates as well as household farmers were mainly responsible for accelerating deforestation by the fires in the region since the 1990s. Large-scale land conversion was the largest single cause of the 1997–1998 fires, which burned nearly 5 million hactares of forest and caused USD 8 billion in economic losses on the part of the citizens and businesses of Indonesia (World Bank 2001). Policies on vegetation are also highlighted as another important cause of the fires (Stolle et al. 2003)

Forest fires tend to occur on an annual basis, particularly in the dry season (Harrison et al. 2009). The Ministry of Forestry (2006) recorded the number of hot spots as 1,678 points in Jambi Province in 2003 equivalent to a fire area of 3,025 ha. In 2006, the fires covered a total area of 2,408.10 ha of which 1,227.60 ha was in forest areas and the remaining 1,180.50 ha was in agricultural areas or outside the forest. In 2014, the forest fires covered 2,552 ha of the forest areas (Widodo 2014; BLHD 2014).

The massive carbon emissions from peat fires make them a significant contributor to the global increase in atmospheric CO_2 (Harrison et al. 2009; WHO 2006). The smoke resulting from the fires poses health risks and disrupts air and land transportation. Choking haze from forest and ground fires covering the southern parts of the Sumatra Island disrupted flight schedules in Jambi in the past. Furthermore, severe haze from peat land and forest fires has been affecting local populations as well as the neighboring countries such as Malaysia, Brunei, and Singapore.

15.1.5 Existing Land Use Versus Spatial Plan

Spatial planning is considered a key factor for sustainable development, but in reality, planning at various levels and sectors have inconsistencies in its implementation. The spatial planning process should be based on the environmental carrying capacity and the implementation must take into account the potential impacts on the environment.

The level of environmental damage of the spatial planning practices in Jambi Province can be seen from the decreased forest cover and increases in fires, floods, and droughts. These damages are not only due to the lack of public policies, but also related to the lack of administrative capacity of licensing (legal or illegal) in land utilization of industrial forests, plantations, and mines. This study focuses on the relationships between the policy/planning and the practice of licensing and land use. In the meantime, certain actors involved in these practices and policies will be observed at some of the sites, especially in local communities that are directly or indirectly affected by the spatial planning practices.

15.2 The Objectives and Study Area

The objectives of this study are: (1) to overlay the existing land use maps over the maps of regional spatial plans, mining concession areas, and forest areas in Jambi Province, (2) to analyze land use and policy conflicts as well as their consequences.

Jambi Province is located on the east coast of the central part of Sumatra Island between the coordinates of 0° 45’ and 2° 45’S latitude and 101° 10’ and 104° 55’E longitude. It is bounded by the province of Riau to the north, the Strait of Idols to the east, the province of South Sumatra to the south and the provinces of West Sumatra and Bengkulu to the west. Jambi Province consists of nine regencies and two municipalities, namely: Kerinci Regency, Merangin Regency, Sarolangun Regency, Batanghari Regency, Muaro Jambi Regency, Tanjung Jabung Timur Regency, Tanjung Jabung Barat Regency, Tebo Regency, Bungo Regency, Jambi Municipality, and Sungai Penuh Municipality (Fig. 15.1). The population of Jambi Province in 2012 and 2013 was 3,242,814 and 3,317,034, respectively, which corresponds to a population growth of 2.29% from 2012 to 2013. The annual mean temperature in the province ranges between 23°C and 31°C.

Its strategic location and abundant natural resources make this province an important one among other provinces. The needs of the industries and societies in the surrounding towns supported the supply of raw materials and material needs of the province. The total area of Jambi Province is 53,435 km^2 and approximately 60% of the land consists of plantations and forests. Jambi Province has one of the largest agricultural lands and forestry in Sumatra. Oil palm and rubber plantations cover 400,168 ha and 595,473 ha, respectively, producing 898.24 thousand tons of palm oil per year and 240.15 thousand tons of rubber per year. The production rates of the other crops are 119.34 thousand tons of coconut (virgin coconut) per year, 69.65 thousand tons of cassiavera per year, and 5.6 thousand tons of tea per year.

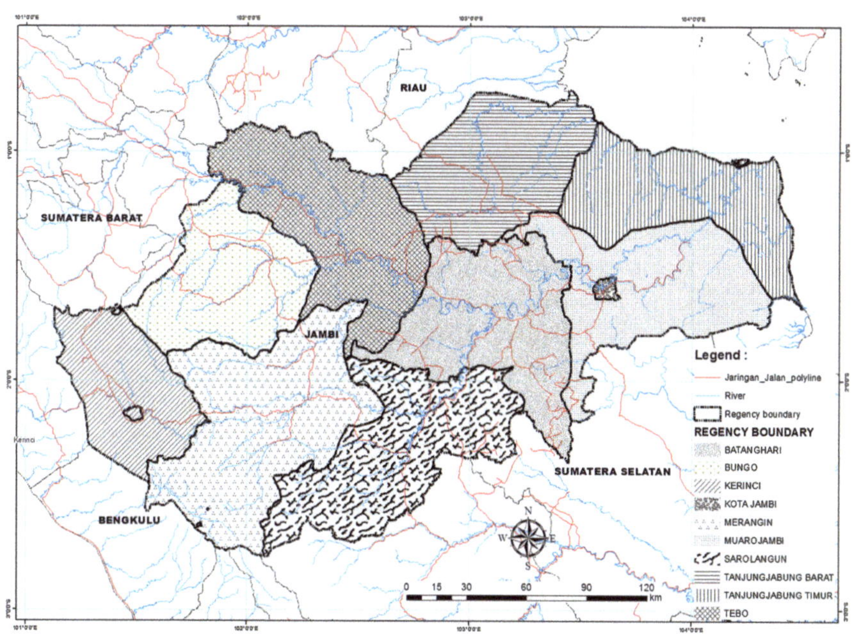

Fig. 15.1 The administrative map of Jambi Province

Agricultural products generated in the western part of Jambi Province include Kerinci rice, potatoes, cabbage, tomatoes, and soybeans.

According to the National Statistics Agency (BPS 2012), natural resources of Jambi province are oil, natural gas, coal, and tin. Total petroleum and gas potential of Jambi Province are 1,270.96 million m^3 and 3,572.44 billion m^3, respectively. Petroleum reserves in the Jambi Province are 408.99 million barrels and the natural gas reserves are 2,185.73 billion m^3.

Based on the data from the Forestry Service of Jambi Province, in 2012 the total forest area in Jambi was 2,118,298.15 ha, where the total area for protection forest and conservation forest were 181,425.00 ha and 677,232.50 ha, respectively. It means that the proportion of the protected area to the total forest area was around 31.97%. The distribution of the forest area in Jambi Province based on the forest status is given in Fig. 15.2. The main products of the forests of Jambi Province were pulp (717,101.87 m^3, 10.08% decrease from 2011) and small logs (wood) (440,788.21 tons of production, 81.97% increase from 2011).

Some thematic maps analyzed in this study include the land system map (Fig. 15.3), the peat map (Fig. 15.4), the map of the plantation area (Fig. 15.5), the map of the mining area (Fig. 15.6), the licensing map for companies (Fig. 15.7), and the land use/land cover map (Fig. 15.8).

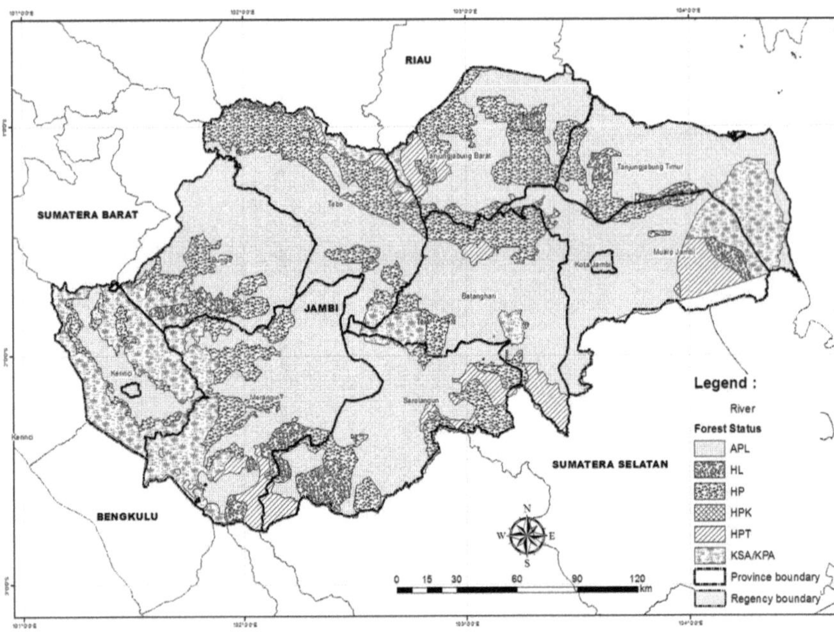

Fig. 15.2 The peat map of Jambi Province

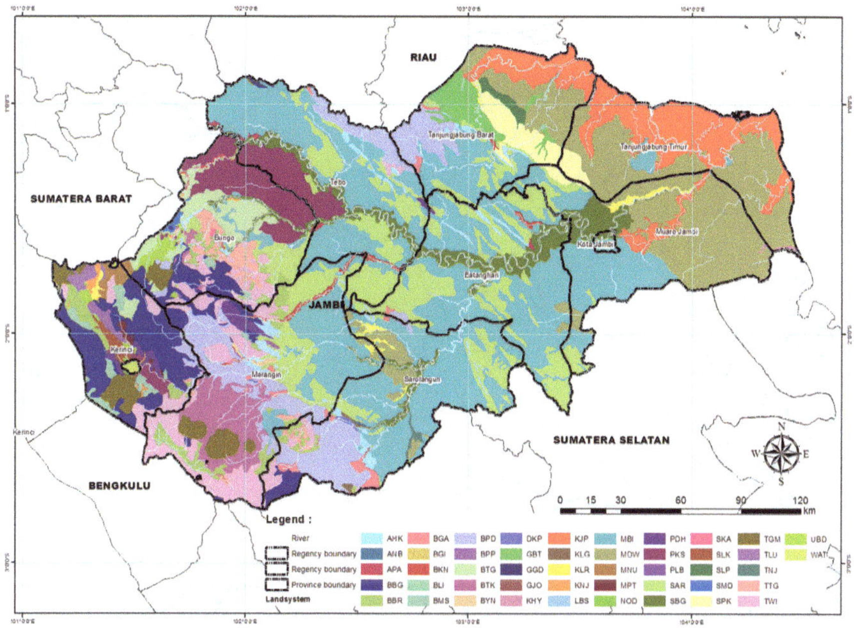

Fig. 15.3 The land system map of Jambi Province

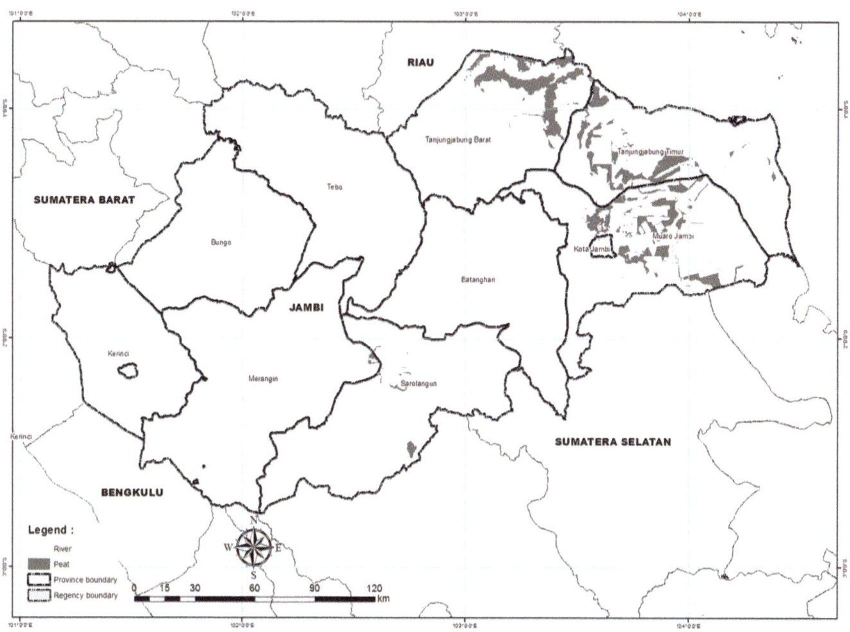

Fig. 15.4 Forest status map (APL = non forest, HL = protected forest, HPK = convertible production forest, HP = production forest, HPT = limited production forest, KSA/KPA = conservation forest)

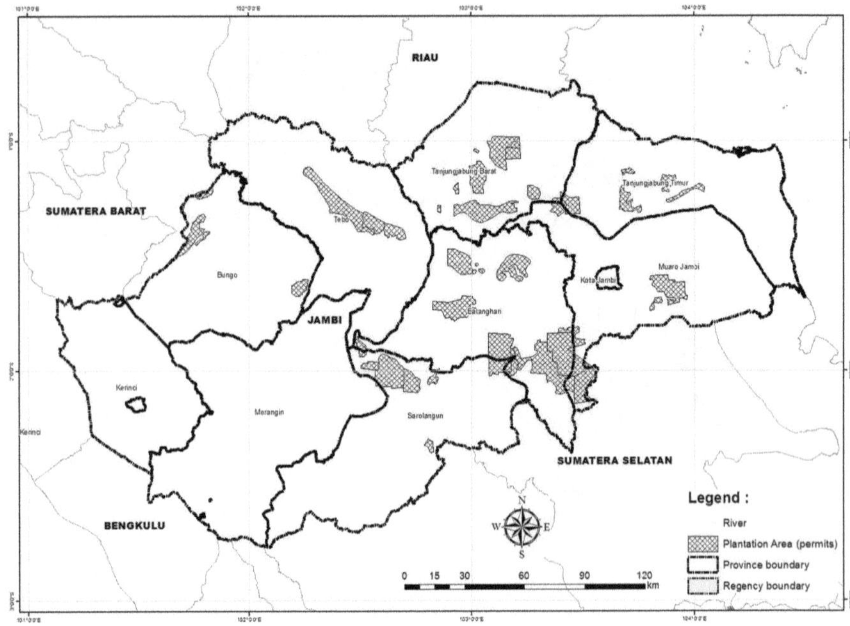

Fig. 15.5 Plantation area in Jambi Province

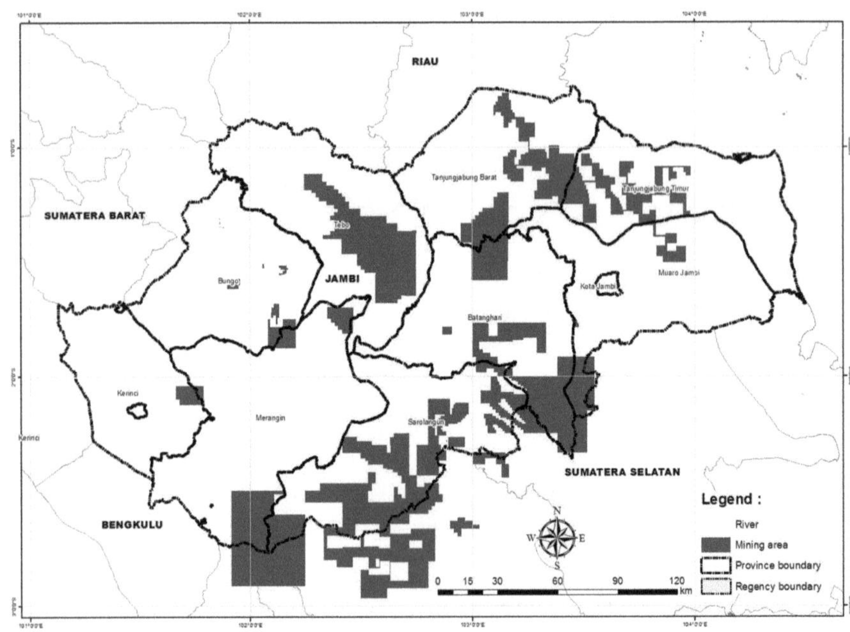

Fig. 15.6 Map of the mining area in Jambi Province

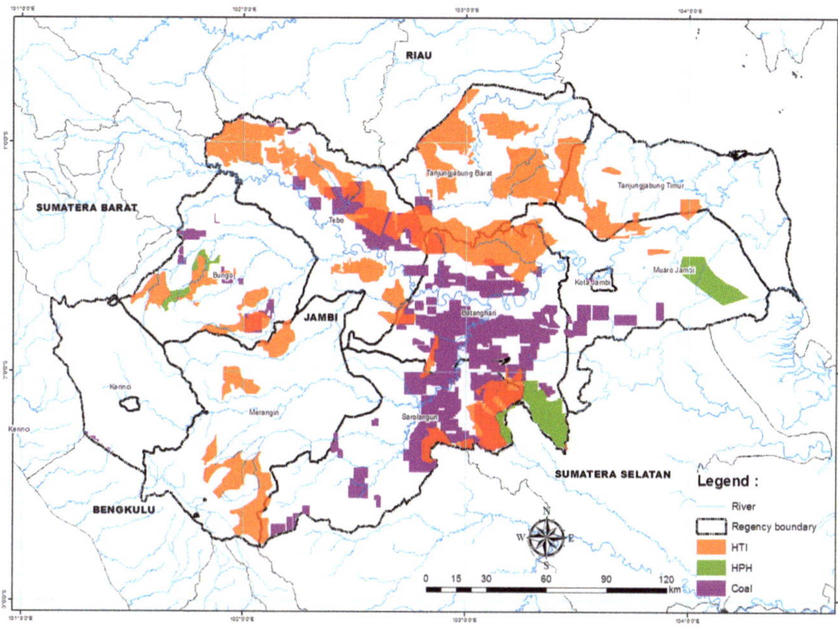

Fig. 15.7 Forest and mining licensing map of Jambi Province (HPH: forest concession license; HTI: industrial timber plantation)

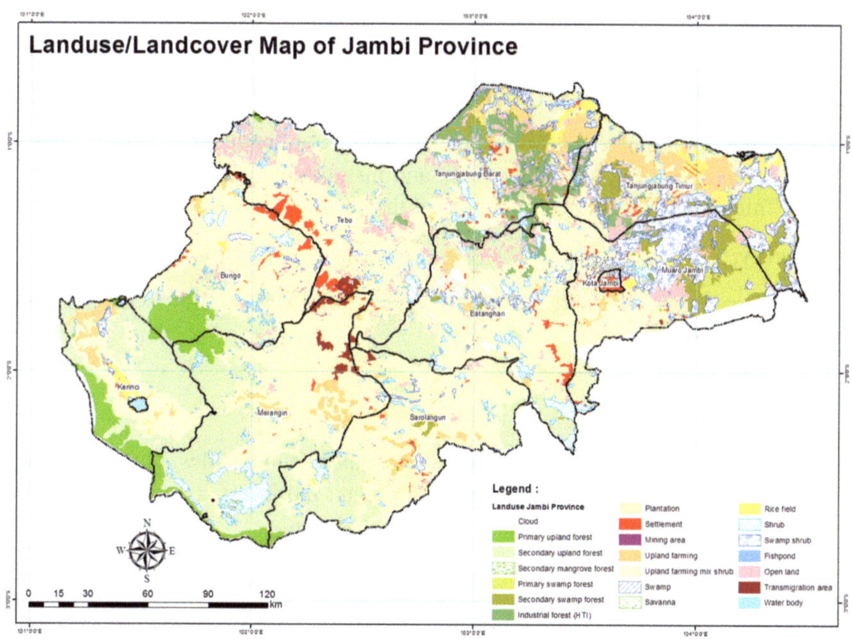

Fig. 15.8 Land use/land cover map of Jambi Province

15.3 Methods

Two types of data were collected in this study. Primary data included field observations in the form of field interviews, geotagged photos, and GPS monitoring results. Non-primary data were collected from government and nongovernment agencies and included policies and regulations on national and regional levels covering spatial planning provincial areas. Non-primary data also included the spatial plan of the island of Sumatra, the status of the forest management plan and the use of peat, eco-region environment, license utilization, medium-term development plan of the province, economic development master plan of Indonesia (MP3EI), and land policies.

Additional supporting data included environmental maps, peat soil map, maps of degraded land and forest/peat fire hotspots, climate and temperature maps, environmental carrying capacity of the Sumatra Island, land use map with information on economic development, customary land, tenurial conflicts on land, and information related to environmental conflicts. We analyzed remotely sensed satellite imagery data, namely Landsat ETM with +30 m spatial resolution and SPOT-6 imagery with 2.5 m resolution for detailed land use information.

The study started with the analysis of policy and regulatory documents using the information matrix, followed by GIS spatial analysis, conflict status analysis, and decision tree analysis. The entire process is structured with GIS-based spatial data and database processing. Analyses were conducted using Boolean spatial analysis, database categorization, overlain thematic maps, and calculating the value of deviation and changes in land use patterns.

The study stages are summarized as follows:

(1) **Preparation and data collection**:

   (a) Conducting an inventory of the secondary data and compiling the data for the biophysical, socioeconomic, and institutional potential of the region,
   (b) Initial analysis of the conditions of the region based on the secondary data,

(2) **Survey**:

   (a) Carrying out a field survey to verify the results of the secondary data analysis,
   (b) Primary data collection,

(3) **Analysis**:

   (a) Examining various official documents and planning files,
   (b) Spatial analysis of the official spatial plan, the forest status, land use, peat map, and license maps for plantation and mining.

15.4 Analysis and Discussion

15.4.1 Official Plans and Policies

The related policy documents consist of Midterm Regional Development Plan (RPJMD), the Master Plan for Acceleration of Indonesian Economic Development (MP3EI), Presidential Decree on Sumatera Island Spatial Plan (RTR), Provincial Level Spatial Plan (RTRWP) (Fig. 15.9), Land Resources Map of Sumatra, Carrying Capacity of Sumatra Island, and Capability and Land Policy (produced by Bappenas, National Development Planning Agency).

The content analysis was conducted to identify relevant government policies, especially land use, spatial planning, and economic development policies. Various implications of the potential use of space were the focus of the analysis. Global observation of the Sumatra region in terms of spatial planning and the carrying capacity of the island was also performed. Land policies were studied in terms of space utilization licensing.

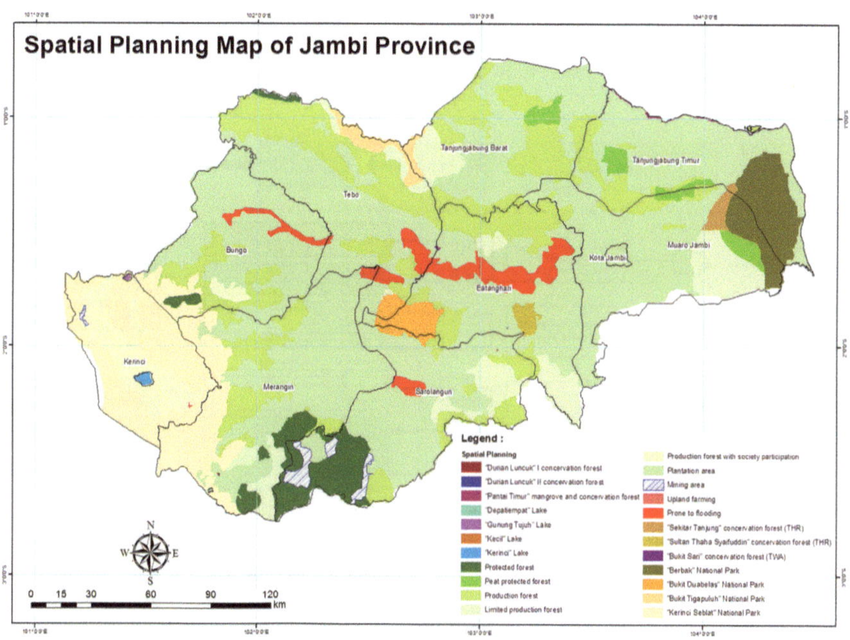

Fig. 15.9 Spatial planning map of Jambi Province

15.4.2 Analysis of Spatial Plan

The analysis is based on the hierarchical decision tree model. It consists of two stages. The first stage is to determine the process and priority decisions (Fig. 15.10). At this stage, a decision tree analysis is performed on the available data (Fig. 15.11). This is to assess the performance of the hierarchy based on the information at the decision level so that the process of establishing the planning space can be better structured. The assessment is done by combining the data and information on the spatial plan policies (RTWP), the forest status from the Ministry of Forestry, and other relevant official documents.

The concept of conflict (contradiction) is arisen from the inconsistencies between the spatial plan maps by the provincial government (draft), the forest status map by the Ministry of Forestry, and the permit granted (license) for investment of the companies. The three forms of data are reflected in the land use policies at the provincial level.

The contradictions between the allocated functions and the utilization of space are analyzed using the policy sources. In general, the nomenclature of the Ministry of Forestry includes Protected/Conservation area—K (National Parks, Conservation Forest, Sanctuary Forest, and protected forest), Production Forest—P1 (limited production forest, production forest, industrial crop forest), and non-state forest area (APL)—P2. According to the Province Spatial Plan (draft of RTRWP), the

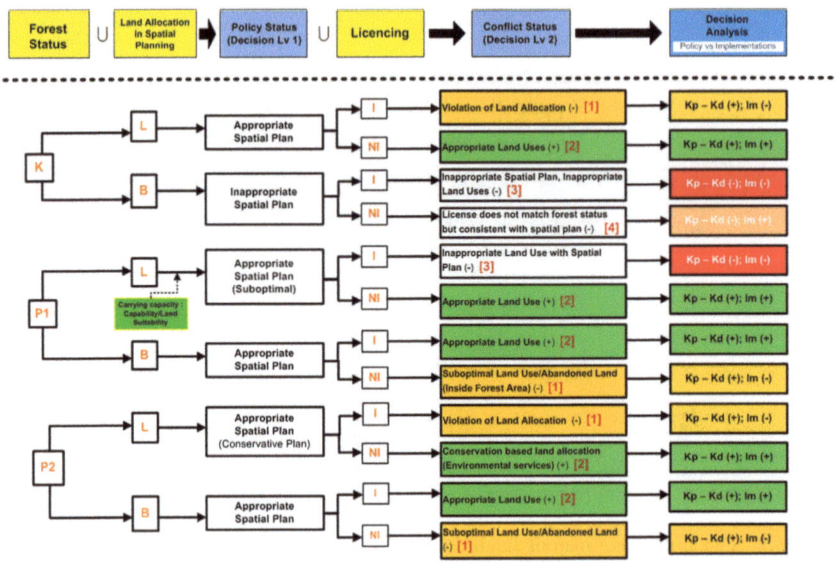

Fig. 15.10 System of decision-making for spatial planning policies (K: conservation forest; P1: limited production forest; P2: production forest; L; protection area; B: cultivation/culture area; I: licensed area for plantation/mining activities; NI: non-licensed area)

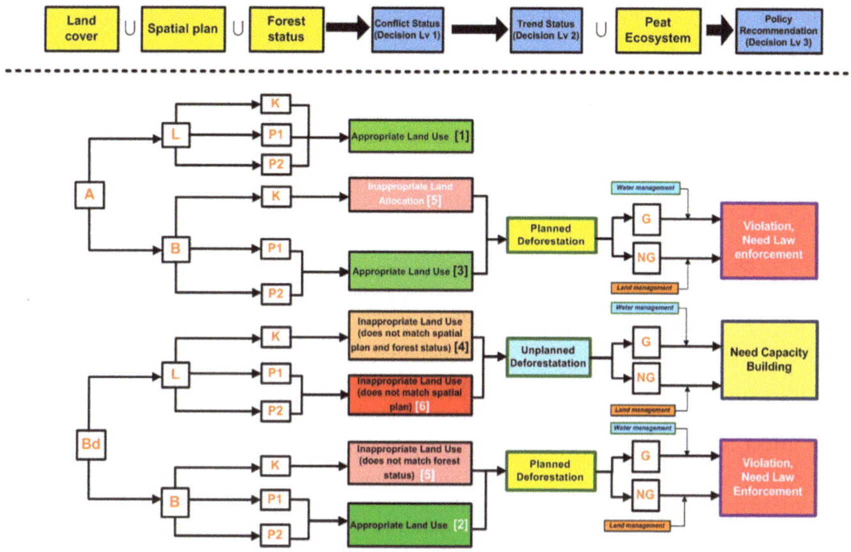

Fig. 15.11 Implementation stage of spatial utilization planning decision tree (Note: **A**: primary forest, secondary forest, mangroves, wetlands, riparian; **Bd**: Plantation, agriculture, settlements, transmigration, mining, ponds, water bodies; **L**: Protected areas; **B**: Cultivation area; **K**: Natural Reserve Area (KSA)/Nature Conservation Area (KPA)/Protected Forest (HL); **P1**: Production Forest (HP)/Limited Production Forest (HPT)/Convertible Production Forest (HPK); **P2**: Non Forest Area (APL); **G**: Peat)

Regional Planning Agency of Jambi Province divided the entire province into two allotments, the protected areas—L and cultivation area—B. The combination of two formal policies/plans, forest status, and spatial plans, is used to build the spatial decision system based on the legal regulations of spatial use. At the first decision level, the space allotment is set to have six spatial decisions consisting of one category that is not legally appropriate, and five suitable categories with four variations of suitability levels: allocated land use, carrying capacity or suboptimal use, conservative use, and productive use. At the second decision level, licensing data and information are used with two categories, namely, with a license—I and without a license—NI.

In this study, the status of land use and spatial policy conflicts were studied for two levels of decision-making. The first stage is the evaluation of compliance between the forest status and the spatial plan (RTRW, spatial plan). In the second stage, the policy at the first level of decision is overlain with the spatial licensing for the legality of land utilization for plantation or mining purposes. At this stage, 12 categories of decisions are determined, with five decisions of violation/inconsistency in inappropriateness and seven positive decisions (appropriate, conservatives, and others) (Table 15.1). Figure 15.10 shows the diagram of the decision-making system.

Table 15.1 Areas for the spatial-hierarchical decision model for Jambi Province

No	Code	Category, policy level 2	Area (Ha)	Area (%)
1	K->L->I	Violation of land allocation [1]	19,304.63	0.39
2	P1->B->NI	Suboptimal land use/ Abandoned land (Inside forest area); HKM, HTR [1]	342,494.78	6.89
3	P2->B->NI	Suboptimal land use/ Abandoned land; HR, KM [1]	2,217,767.01	44.60
4	K->L->NI	Appropriate land uses [2]	802,381.95	16.14
5	P1->B->I	Appropriate land uses [2]	1,222,602.09	24.59
	P2->B->I	Appropriate land uses [2]		
6	P1->L->NI	Appropriate land uses [2]	48,585.41	0.98
7	P2->L->NI	Conservation-based land allocation (Environmental services) [2]	133,295.60	2.68
8	K->B->I	Inappropriate spatial plan, Inappropriate land uses [3]	9,052.61	0.18
9	P1->L->I	Inappropriate land use with spatial plan [3]	31,663.57	0.64
	P2->L->I	Violation of land allocation [1]		
10	K->B->NI	License does not match forest status but consistent with spatial plan [4]	40,620.74	0.82
11	K->ND->I; K->ND->NI; ND->B->I; ND->B->NI; ND->L->I; ND->L->NI; P1->ND->I;	Not categorized	104,684.84	2.11
12	(blank)	No data	31.46	0.001
Grand total			**4,972,484.69**	**100,00**

15.4.3 Contradiction in the Implemented Spatial Policies and Utilization

The concept of contradiction is based on the alignment between land cover maps, spatial plans by the provincial government (draft), the forest status maps by the Ministry of Forestry, and the license and peat land ecosystems map. Contradictions or inconsistencies are examined for substance (content analysis), implementation, and policy.

The analysis was conducted by examining the content alignment between the actual use of space, the three policies, and the actual characteristics of the environment. In the first stage, the alignment was evaluated between land cover, forest status, and spatial plan with the result of seven decisions, consisting of four categories of suitability: spatial utilization, use, ownership and control, and three unsuitable categories in terms of control, ownership, existing land use, and/or utilization. All of these factors are grouped into two categories, "planned deforestation" and "unplanned deforestation".

In stage 2, all implementations were examined by whether they were in peat ecosystems or not, so that the planned recommendations could be related to the management of water and/or land, intended for the activities of law violation and enforcement, empowerment, and development (Fig. 15.11).

15.4.4 Spatial Conflicts

Plan and policies are made based on the information on all aspects of planning by a holistic approach. Spatial planning is a reflection of spatial policies (and the derivatives). The spatial conflicts are caused by the incompatible approaches to spatial planning/policies at the operational level because of the differences in defining and analyzing the space with different levels of information during planning. Strategic issues of spatial planning at the national level are separately translated into the directed policies and priority areas of development because they are not accompanied by suitable information on the spatial implementation as planned. Land use conflicts are caused by the disagreements between the government institutions on the decisions of development strategies. On the other hand, the spatial plan was formulated without considering the environmental carrying capacity.

Space utilization for mere development purposes without taking into account the sustainability of the ecosystem function has emerged due to the perspective that space is purely for economic function, without maintaining the elasticity of land with its limited carrying capacity. The general description of conflicts between the existing land use and planning/policies shows some biased information, particularly on the implementation on the field. The constraints and the definition of spatial function and utilization can be seen from the implementation stages of the decision tree in Fig. 15.11 as follows: (a) complying to the spatial function and utilization (current land cover), (b) utilization of nonproductive space (neglected land), (c) inconsistent licensing of spatial uses, (d) policies of spatial use/pattern of the conservative area (low emission), and (e) violation against the spatial function and utilization. The spatial distribution of the contextual information is given in Figs. 15.12 and 15.13.

Based on the spatial-hierarchical decision model on the policies of spatial ownership (property/tenurial rights) and control with the implemented space utilization, the decision is distinguished by the directive policy of the central and local governments towards its implementation. The decision follows the (a) spatial patterns of central and local policies that contradict in implementation, resulting in

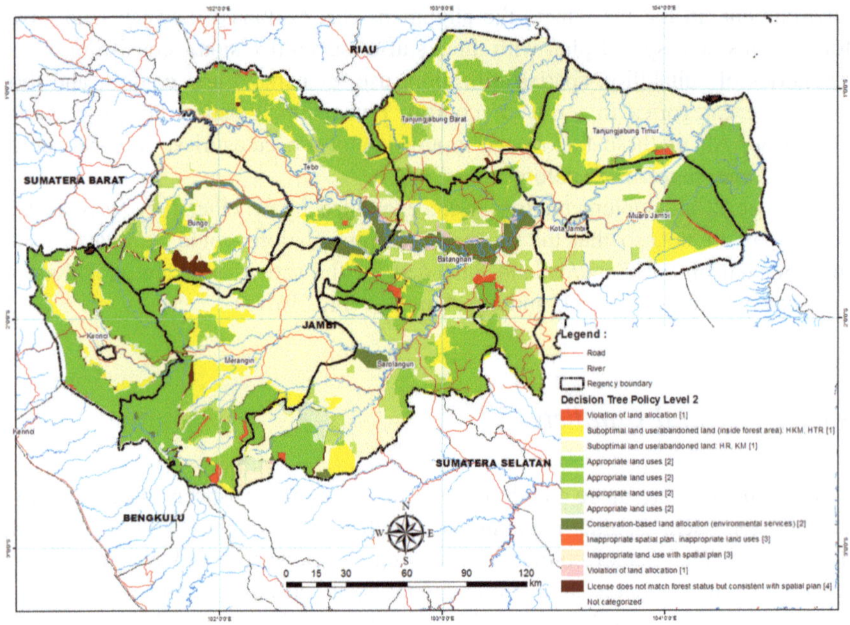

Fig. 15.12 Map of spatial-hierarchical decision model of Jambi Province

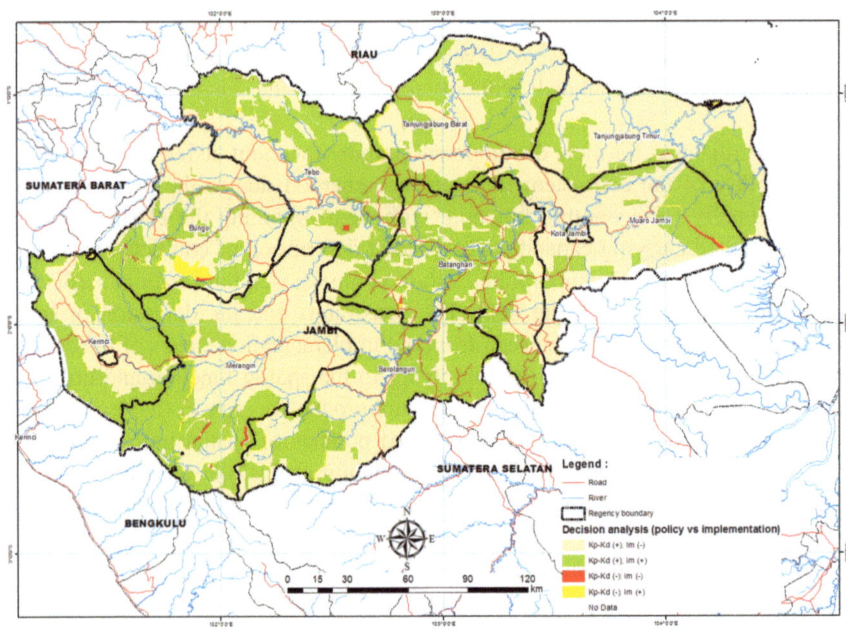

Fig. 15.13 Map of spatial-hierarchical decision model at the policy and implementation level in the province of Jambi

land use policies that degrade the natural environment, (b) spatial patterns with contradicting central and local policies but the implementation of the spatial function and utilization supports conservative environmental protection efforts (such as reducing carbon emissions and consideration of supporting capacity and specific ecosystem), and (c) spatial patterns of central and local policies that are not in line with the implementation of the different spatial designations and utilizations. The spatial distribution pattern of the hierarchical model of decision on policy and implementation can be seen in Fig. 15.13.

15.5 Concluding Remarks

This chapter has discussed one of the main issues causing environmental degradation in the form of land and forest fires and loss of biodiversity (deforestation and monoculture agricultural practices) in the world's biodiversity hotspot, the Jambi province, Indonesia. More than 2.2 million ha, or approximately 44.6% of the land in Jambi province located outside the forest area is abandoned or unproductive. Approximately more than 96% of the protected area (834,800 ha) is still maintained in accordance with its function. The land use conflicts in Jambi are the results of institutional disagreements among the governmental agencies of the spatial planning, forestry, agriculture, and mining. Besides, conflicts have occurred between the policies of the central government, local governments, corporate world (mostly mining and plantation companies), and local communities.

References

Beukema H, Danielsen F, Vincent G, Hardiwinoto S, van Andel J (2007) Plant and bird diversity in rubber agroforests in the lowlands of Sumatra. Indones Agrofor Syst 70:217–242

BLHD (2014) Buku Data Status Lingkungan Hidup Daerah Provinsi Jambi. 2014. Badan Lingkungan Hidup Daerah, Provinsi Jambi

BPS (2014) Biro Pusat Statistik. Jambi dalam Angka Tahun 2014. Jambi. Indonesia

Colchester M, Chao S (2013) Conflict or consent? The oil palm sector at a crossroads. FPP, Sawit Watch, TUK INDONESIA

Colchester M, Jiwan N, Andiko, Sirait M, Firdaus AY, Surambo A, Pane H (2006) Promised land: palm oil and land acquisition in Indonesia-implications for local communities and indigenous peoples. Forest Peoples Programme, Perkumpulan Sawit Watch, HuMA, World Agroforestry Centre, Bogor

Colchester M, Anderson P, Firdaus AY, Hasibuan F, Chao S (2011) Human rights abuses and land conflicts in the PT Asiatic Persada concession in Jambi: report of an independent investigation into land disputes and forced evictions in a palm oil estate November 2011. Forest Peoples Programme, Sawit Watch and HuMa, Bogor

DNPI (2011) DNPI green review on REDD+DNPI. Jakarta

Feintrenie L, Levang P (2009) Sumatra's rubber agroforests: advent, rise and fall of a sustainable cropping system. Small-scale For 8(3):325–335

Feintrenie L, Schwarze S, Levang P (2010) Are local people conservationists? Analysis of transition dynamics from agroforests to monoculture plantations in Indonesia. Ecol Soc 15(4):37. http://www.ecologyandsociety.org/vol15/iss4/art37/
FWI (2015) Intip hutan. Media Seputar Informasi Hutan Indonesia. Bogor
FWI/GFW (2001) Keadaan Hutan Indonesia. Bogor, Indonesia: Forest Watch Indonesia dan Washington D.C, Global Forest Watch, Bogor
Gunarso P, Hartoyo ME, Agus F, Killeen TJ (2013) Oil palm and land use change in Indonesia, Malaysia, and Papua New Guinea. Reports from the technical panels of the 2nd greenhouse gas working group of the roundtable on sustainable palm oil (RSPO). www.rspo.org. Accessed Nov 2013
Harrison ME, Page SE, Limin SH (2009) The global impact of Indonesian forest fires. Biologists 56(3):156–163
Jambi B (2011) Analisis Pertumbuhan Ekonomi Provinsi Jambi Tahun 2010. BAPPEDA Provinsi Jambi
Li TM (2001) Masyarakat adat, difference, and the limits of recognition in Indonesia's forest zone. Mod Asian Stud 35(3):645–676
Ministry of Forestry (2011) Forestry statistics of Indonesia, Jakarta
Mongabay Indonesia (2014) FOKUS LIPUTAN: Kelapa Sawit, antara kepentingan politik dan tata guna lahan (bagian II-selesai). http://www.mongabay.co.id/2014/04/08/fokus-liputan-kelapa-sawit-antara-kepentingan-politik-dan-tata-guna-lahan-bagian-ii-selesai/
PHKA (2003) Tropical rainforest heritage of Sumatera. Directorate General of Forest Protection and Nature Conservation, Ministry of Forestry
Pichler M (2010) Agrofuels in Indonesia: structures, conflicts, consequences, and the role of the EU. Austrian J South-East Asian Stud 3(2):175–193
Rist L, Feintrenie L, Levang P (2010) The livelihood impacts of oil palm: smallholders in Indonesia. Biodivers Conserv 19(4):1009–1024
Stolle F, Chomitz KM, Lambin EF, Tomich TP (2003) Land use and vegetation fires in Jambi Province, Sumatra, Indonesia. For Ecol Manag 179(2003):277–292
Villamor GB, Bao QBL, Djanibekov U, Vlek PLG, van Noordwijk M (2014) Biodiversity in rubber agroforests, carbon emissions, and rural livelihoods: an agent-based model of land-use dynamic in lowland Sumatra. Environ Model Softw 61:151–165. doi:10.1016/i.envsoft.2014.07.013
WHO (2006) Sumatera and Kalimantan fires and haze. Emergency situation report # 1, 29 Aug 2006
Widayati A, Suyanto S (eds) (2013) Emission reduction efforts in Jambi: towards nesting approach—substantive report of REALU II Indonesia. World Agroforestry Centre—ICRAF Southeast Asia Regional Program, Bogor
Widodo RB (2014) Pemodelan spasial resiko kebakaran hutan (studi kasus Provinsi Jambi, Sumatera). Jurnal Pembangunan Wilayah dan Kota 10(2):127–138
World Bank (2001) Indonesia Environment and natural resource management in a time of transition. World Bank, Washington

MIX
Papier aus verantwortungsvollen Quellen
Paper from responsible sources
FSC® C105338
FSC
www.fsc.org

If you have any concerns about our products,
you can contact us on
ProductSafety@springernature.com

In case Publisher is established outside the EU,
the EU authorized representative is:
Springer Nature Customer Service Center GmbH
Europaplatz 3, 69115 Heidelberg, Germany

Printed by Libri Plureos GmbH
in Hamburg, Germany